Mass Transfer

Anthony F. Mills
University of California at Los Angeles

Prentice
Hall

Prentice Hall, Upper Saddle River, NJ 07458

Library of Congress Cataloging-in-Publication Data

Mills, Anthony F.
  Mass transfer / Anthony F. Mills – [2nd ed.].
    p.   cm.
  Includes bibliographical references and index.
  ISBN 0–13–032829–4
  1. Mass transfer.   I. Mills, Anthony F. Heat and mass transfer.
    II. Title.

QC318.M3 M55 2001
530.5'15–dc21   00-066896

Vice President and Editorial Director, ECS: *Marcia J. Horton*
Acquisitions Editor: *Laura Curless*
Editorial Assistant: *Erin Katchmar*
Vice President and Director of Production and Manufacturing, ESM: *David W. Riccardi*
Executive Managing Editor: *Vince O'Brien*
Managing Editor: *David A. George*
Managing Editor: *Barbara A. Till*
Director of Creative Services: *Paul Belfanti*
Creative Director: *Carole Anson*
Art Director: *Jayne Conte*
Art Editor: *Adam Velthaus*
Cover Designer: *Bruce Kenselaar*
Manufacturing Manager: *Trudy Pisciotti*
Manufacturing Buyer: *Pat Brown*
Marketing Manager: *Holly Stark*
Marketing Assistant: *Karen Moon*

© 2001 by Prentice-Hall, Inc.
Upper Saddle River, New Jersey 07458

Printed in the United States of America
10 9 8 7 6 5 4 3 2 1

ISBN 0-13-032829-4

Prentice-Hall International (UK) Limited, *London*
Prentice-Hall of Australia Pty. Limited, *Sydney*
Prentice-Hall Canada Inc., *Toronto*
Prentice-Hall Hispanoamericana, S.A., *Mexico City*
Prentice-Hall of India Private Limited, *New Delhi*
Prentice-Hall of Japan, Inc., *Tokyo*
Pearson Education Asia Pte. Ltd. *Singapore*
Editora Prentice-Hall do Brasil, Ltda., *Rio de Janeiro*

To Brigid
For your patience and understanding

Contents

A   PROPERTY DATA   347

List of Tables   347

Preface

The first edition of the text *Heat and Mass Transfer* (Richard D. Irwin, 1995) was prepared by adding three chapters on mass transfer to the text *Heat Transfer* (Richard D. Irwin, 1992). In preparing second editions it became apparent that the added new material would result in *Heat and Mass Transfer* having over 1300 pages. Thus it was decided to publish the mass transfer chapters as a supplement to the second edition of *Heat Transfer* (Prentice Hall, 1999) and discontinue the single volume *Heat and Mass Transfer*. For this second edition, two new sections have been added to the former Chapter 10, renumbered in the supplement as Chapter 2. Section 2.7 deals with transport in multicomponent systems, and Section 2.8 develops conservation equations for a multicomponent gas mixture. With the addition of this more advanced material, the mass transfer supplement should prove useful as a text for a graduate-level mass transfer course suitable for mechanical, aerospace, and nuclear engineers. Other changes include some new exercises and numerous minor improvements.

Mass Transfer contains three chapters and additional appendix material:

      Chapter 1: Elementary Mass Transfer
      Chapter 2: High Mass Transfer Rate Theory
      Chapter 3: Mass Exchangers
      Appendix A: Property Data

Chapter 1 considers diffusion in a stationary medium, and low mass transfer rate convection. As was the case for heat convection in Chapter 4 of *Heat Transfer*, mass convection is introduced using dimensional analysis and the Buckingham pi theorem. The analogy between low mass transfer rate convection and heat transfer to an impermeable surface is thoroughly exploited. Simultaneous heat and mass transfer is considered, with an emphasis on problems involving evaporation of water, such as the wet- and dry-bulb psychrometer. Diffusion and chemical reaction in porous catalysts are analyzed in order to provide results needed for the study of automobile catalytic converters in Chapter 3. The chapter closes with methods for calculating transport properties, with special emphasis on gas mixtures. Chapter 2 is considerably more advanced than Chapter 1. Velocities and fluxes in a mixture or solution are rigorously defined and the general species conservation equation derived. Important problems such as diffusion with one component stationary and combustion of volatile hydrocarbon fuel droplets are carefully analyzed. The Couette-flow model is introduced to obtain blowing factors for high mass transfer rate convection. These factors are then improved to account for both flow geometry and variable properties to give a unique engineering problem-solving facility. Mass, momentum, and heat transfer in a constant laminar boundary layer on a flat plate are rigorously analyzed for a binary mixture, complementing the analysis of momentum and heat transfer in a pure fluid given in Chapter 5 of *Heat Transfer*. The chapter continues with a generalized formulation of steady convective heat and mass transfer, based on the well-known contributions of D. B. Spalding. The chapter closes with new advanced material on transport in multicomponent systems and conservation equations for multicomponent gas mixtures.

Chapter 3 deals with mass exchangers, such as catalytic converters and gas scrubbers; particle removal equipment, such as filters and electrostatic precipitators; and simultaneous heat and mass exchangers, such as humidifiers and cooling towers. All the analyses are based on low mass transfer rate theory, and the high mass transfer rate theory of Chapter 2 is not a prerequisite to the study of Chapter 3. Humidifiers and cooling towers are usually considered within the scope of mechanical engineering, but mass exchangers have been traditionally the province of chemical engineers. Particle removal equipment has not been emphasized in either discipline. However, environmental concerns are now having profound effects on the design of mechanical engineering systems. Tuning and operating of the modern automobile engine are now dictated by requirements of the catalytic converter used to reduce exhaust emissions. A significant portion of the investment in a modern coal-fired central power plant is for equipment used to clean the stack gas, including a scrubber to remove sulfur oxides, and cyclones, baghouses, or electrostatic precipitators to remove particulates. Incinerators for destroying toxic wastes have even more stringent requirements on stack gas cleanup. A major concern of nuclear engineers has become the design of safety systems to prevent escape of radioactive gases and particulates into the atmosphere in the event of an accident involving damage to the reactor core. Major aerospace engineering projects, such as the space station, require complex life-support systems that involve a variety of mass exchangers. My goal in Chapter 3 is to prepare mechanical, aerospace, and nuclear engineers for effective participation in the design of the systems described above, complementing rather than competing with chemical engineers on the design team.

As with the first edition, the fully integrated software package MT serves as a database and tool for the student, both at college and after graduation as practicing engineers. GASMIX, AIRSTE, and PSYCHRO facilitate the rapid and reliable calculation of gas mixture thermodynamic and transport properties. MCONV is a useful tool for calculating convective mass transfer and simultaneous convective heat and mass transfer using low mass transfer rate theory. FILTER, SCRUB, and CTOWER are design tools that enhance the students' interest in the design of practical equipment. With these interactive programs the student can easily perform parametric studies of the performance of filters, scrubbers, and cooling towers. At UCLA, CTOWER has proven indispensable when teaching the thermal-hydraulic design of forced and natural draft cooling towers. The MT software is available free for download from the Prentice Hall Website at www.prenhall.com.

Some of the material in *Mass Transfer*, mostly in the form of examples and exercises, has been taken from an earlier text co-authored by my former-colleagues at UCLA, D. K. Edwards and V. E. Denny (*Transfer Processes*, 1st ed., Holt, Rinehart

& Winston, 1973; 2d ed., Hemisphere-McGraw-Hill, 1979). D. B. Spalding introduced me to the subject of mass transfer as a student at the Imperial College of Science and Technology, London: his influence is surely in evidence in this text. I gratefully acknowledge the contributions of these gentlemen, both to this book and to my professional career. The computer software for *Mass Transfer* was expertly written by Hae-Jin Choi and Benjamin Tan. I also wish to acknowledge the contributions made by many others. The late D. N. Bennion provided a chemical engineering perspective to some of the material on mass exchangers. R. Greif of the University of California, Berkeley, and J. H. Lienhard, V. of the Massachusetts Institute of Technology provided detailed critiques of the manuscript. Reviewers commissioned by Richard D. Irwin, Inc., suggested numerous improvements, most of which I was able to incorporate in the final manuscript of the first edition. My students have been most helpful—in particular, S. W. Hiebert, R. Tsai, E. Myhre, B. H. Chang, A. Gopinath, P. Hwang, M. Tari, B. Tan, A. Na-Nakornpanom, J. Sigler, M. Demitrou, and M. Fabri. My special thanks to the secretarial staff at UCLA and the University of Auckland—in particular, Phyllis Gilbert, Joy Wallace, and Julie Austin, for their enthusiastic and expert typing of the manuscript.

ANTHONY F. MILLS
amills@ucla.edu

Elementary Mass Transfer

1.1 INTRODUCTION

Mass transfer may occur in a gas mixture, a liquid solution, or a solid solution. There are several physical mechanisms that can transport a chemical species through a phase and transfer it across phase boundaries. In this chapter we will consider only the two most important, which are **ordinary diffusion** and **convection.** Mass diffusion is analogous to heat conduction and occurs whenever there is a gradient in the concentration of a species. Mass convection is essentially identical to heat convection: a fluid flow that transports heat may also transport a chemical species. The similarity of the mechanisms of heat transfer and mass transfer results in the mathematics often being identical, a fact that can be exploited to advantage. For example, the solution to a problem of diffusion in a solid can often be written down immediately by referring to the analogous heat conduction problem. But, for the beginning student in particular, there are some significant differences between the subjects of heat and mass transfer. One difference is the much greater variety of physical and chemical processes that require mass transfer analysis, as illustrated by the following list.

1. Evaporation of water into air in a cooling tower
2. Drying of wood, paper, and textiles
3. Leakage of helium from the laser of a copying machine
4. Diffusion of carbon into iron during case-hardening of a gear wheel
5. Catalytic oxidation of carbon monoxide and unburnt hydrocarbons in an automobile catalytic converter
6. Measurement of humidity using wet and dry thermocouples
7. Combustion of pulverized coal in a power plant furnace
8. Combustion of kerosene droplets in a gas turbine combustion chamber
9. Combustion of iron in oxy-acetylene steel cutting
10. Absorption of sulfur dioxide into an alkaline solution in a power plant flue-gas scrubber
11. Doping of semiconductors for transistors
12. Discharge of a lead-acid battery
13. Power generation by a fuel cell
14. Aeration of sewage for biological treatment
15. Evaporation and condensation in gas-controlled heatpipes
16. Ablation of a heat shield on a reentry vehicle
17. Erosion of a graphite or rhenium nozzle in a solid-propellant rocket
18. Scrubbing of radioactive nuclides in a nuclear reactor pressure-suppression pool following an accident involving core degradation
19. Oxidation of Zircaloy fuel rod cans by steam following a loss-of-coolant accident in a boiling-water nuclear reactor
20. Separation of oxygen and nitrogen in the separation column of a gaseous oxygen plant
21. Blood oxygenation in a rotating-disk oxygenator during open-heart surgery
22. Desalination of brackish water by reverse osmosis

A second difference between mass and heat transfer is the extent to which essential features of a given process may depend on the particular chemical system involved, and on temperature and pressure. As an example of how a particular chemical system affects a given process, consider scrubbing of a chemical species from an air stream by water in a packed tower, as shown in Fig. 1.1. Inside the tower is a particle bed over which water is sprayed, while the air flows upward through the bed. If the species to be scrubbed is only slightly soluble, such as hydrogen sulfide, the process is controlled by diffusion into the water, and the precise details of the air flow are unimportant. However, if the species is ammonia, which is very soluble, the process is controlled by diffusion out of the air stream, and the precise details of the water flow are unimportant. As a second example, consider the combustion of a particle in air at a high temperature, as illustrated in Fig. 1.2. If it is a carbon particle, the reaction occurs on the carbon surface to form gaseous carbon monoxide, which diffuses away into the air stream. However, if it is an iron particle, the reaction can produce solid iron oxide, which remains on the particle surface to impede the

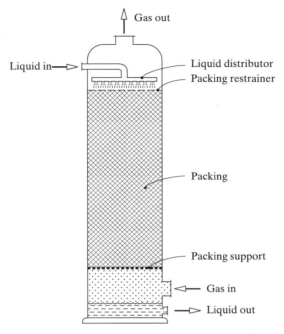

Figure 1.1   A packed-tower gas scrubber.

oxidation process. As an example of how temperature affects the process, consider the same carbon particle at a much lower temperature: the reaction product is then carbon dioxide, and not carbon monoxide. And at intermediate temperatures two reactions occur: on the surface carbon dioxide is reduced to carbon monoxide, while some distance away from the surface carbon monoxide reacts with oxygen to form carbon dioxide in a flame. Sometimes the essential features of the process will be well known and supported by experimental observation, but often physical and chemical thermodynamic data, and chemical kinetics data, must be employed to assist in the understanding of the process.

The foregoing discussion indicates that mass transfer might be both a challenging and interesting subject. Assuming that the students (and instructor!) take up the challenge, the remainder of Chapter 1 contains the following material. In Section 1.2 concentration is defined, and special attention is paid to phase interfaces where the concentration of a chemical species is almost always discontinuous. Then Fick's law of diffusion, which is the mass transfer analogy to Fourier's law, is introduced. In Section 1.3 diffusion in stationary media is analyzed, exploiting where possible the analogy between diffusion and conduction. In Section 1.4 mass convection is introduced. Attention is restricted to so-called **low mass transfer rate** theory, for which there is an exact analogy between heat and mass transfer. In Section 1.5 processes involving simultaneous consideration of heat and mass transfer are analyzed. Examples considered include the wet- and dry-bulb psychrometer and combustion of carbon particles. In Section 1.6 mass transfer in porous catalysts is analyzed, and examples are given for automobile catalytic converters. Methods for calculating

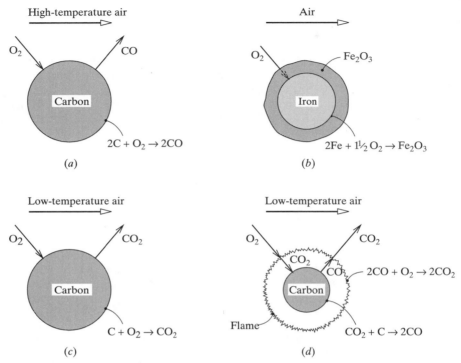

Figure 1.2   Some examples of particle oxidation. (a) High-temperature oxidation of carbon in air to form carbon monoxide. (b) Oxidation of iron in air to form solid ferrous oxide. (c) Oxidation of carbon at low temperatures to form carbon dioxide. (d) Oxidation of carbon at intermediate temperatures to form carbon dioxide.

transport properties of gas mixtures, liquid solutions, and small particles are given in Section 1.7.

1.2 CONCENTRATIONS AND FICK'S LAW OF DIFFUSION

Concentration is to mass transfer what temperature is to heat transfer. Concentration is a measure of composition. A number of such measures are in common use, and they will be defined carefully in Section 1.2.1. In general, temperature is continuous across an interface, whereas, in general, concentration is discontinuous across an interface. Clearly, the concentrations of air on the two sides of a water–air interface are very different. Concentration differences across interfaces are given by thermodynamic data and relations, and a number of different physical situations are considered in Section 1.2.2. Fick's law of diffusion, which is analogous to Fourier's law of heat conduction, is introduced in Section 1.2.3. Other diffusion phenomena are briefly discussed in Section 1.2.4.

1.2.1 Definitions of Concentration

In a gas mixture, or liquid or solid solution, the local **concentration** of a mass species can be expressed in a number of ways. Figure 1.3 shows an elemental volume ΔV surrounding the location of concern. The problem is to specify the composition of the material within ΔV. One method would be to determine, somehow, the number of molecules of each species present and divide by ΔV to obtain the number of molecules per unit volume; hence the **number density** is defined.

$$\text{Number density of species } i \;=\; \text{Number of molecules of } i \text{ per unit volume}$$

$$\equiv \mathcal{N}_i \text{ molecules/m}^3 \tag{1.1}$$

Alternatively, if the total number of molecules of all species per unit volume is denoted as $\mathcal{N}$, then we define the **number fraction** of species i as

$$n_i \equiv \frac{\mathcal{N}_i}{\mathcal{N}}; \quad \mathcal{N} = \sum \mathcal{N}_i \tag{1.2}$$

where the summation is over all species present, $i = 1, 2, \ldots, n$. Equations (1.1) and (1.2) describe *microscopic* concepts and are used, for example, when the kinetic theory of gases is used to describe transfer processes.

Whenever possible, it is more convenient to treat matter as a continuum. Then the smallest volume considered is sufficiently large for macroscopic properties such as pressure and temperature to have their usual meanings. For this purpose we also require *macroscopic* definitions of concentration. First, on a mass basis,

$$\textbf{Mass concentration} \text{ of species } i \;\equiv\; \text{Partial density of species } i$$

$$\equiv \rho_i \text{ kg/m}^3 \tag{1.3}$$

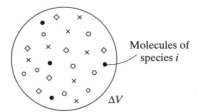

Figure 1.3 Elemental control volume used to define concentrations.

The total mass concentration is the total mass per unit volume, that is, the density $\rho = \sum \rho_i$. The **mass fraction** of species i is defined as

$$m_i = \frac{\rho_i}{\rho} \tag{1.4}$$

Second, on a molar basis,[1]

Molar concentration of species i $\equiv$ Number of moles of i per unit volume

$$\equiv c_i \ \text{kmol/m}^3 \tag{1.5}$$

If M_i [kg/kmol] is the molecular weight of species i, then

$$c_i = \frac{\rho_i}{M_i} \tag{1.6}$$

The total molar concentration is the molar density $c = \sum c_i$. The **mole fraction** of species i is defined as

$$x_i \equiv \frac{c_i}{c} \tag{1.7}$$

A number of important relations follow directly from these definitions. The mean molecular weight of the mixture or solution is denoted M and may be expressed as

$$M = \frac{\rho}{c} = \sum x_i M_i \tag{1.8a}$$

or

$$\frac{1}{M} = \sum \frac{m_i}{M_i} \tag{1.8b}$$

There are summation rules:

$$\sum m_i = 1 \tag{1.9a}$$

$$\sum x_i = 1 \tag{1.9b}$$

It is often necessary to have the mass fraction of species i expressed explicitly in terms of mole fractions and molecular weights; this relation is

$$m_i = \frac{x_i M_i}{\sum x_j M_j} = x_i \frac{M_i}{M} \tag{1.10a}$$

and the corresponding relation for the mole fraction is

$$x_i = \frac{m_i/M_i}{\sum m_j/M_j} = m_i \frac{M}{M_i} \tag{1.10b}$$

Equations (1.10a) and (1.10b) are to be derived as Exercise 1–1.

Although all the above definitions are not strictly necessary for the engineering analysis of mass transfer, it will be seen later that the use of an appropriate definition

[1] The SI unit for the amount of a substance is the mole (the gram mole of the cgs system). However, to be consistent with using the kilogram as the unit for mass, it is more convenient to use the kilogram mole (symbol kmol) rather than the mole (symbol mol). We will use kmol exclusively.

of concentration often can simplify a particular analysis. Indeed, these measures of concentration are not the only ones in common use. For example, partial pressure is often used in dealing with ideal gas mixtures. Dalton's law of partial pressures states that

$$P = \sum P_i, \quad \text{where } P_i = \rho_i R_i T \qquad \text{(1.11)}$$

Dividing partial pressure by total pressure and substituting $R_i = \mathcal{R}/M_i$ gives

$$\frac{P_i}{P} = \frac{\rho_i}{M_i}\frac{\mathcal{R}T}{P} = c_i\frac{\mathcal{R}T}{P} = x_i\frac{c\mathcal{R}T}{P} = x_i \qquad \text{(1.12)}$$

Thus, for an ideal gas mixture, the mole fraction and partial pressure are equivalent measures of concentration (as also is the number fraction).

In combustion problems, we often need to know the mass fraction of oxygen in atmospheric air. A commonly used specification of the composition of dry air is given on a volume basis as 78.1% N_2, 20.9% O_2, and 0.9% Ar. (The next largest component is CO_2, at 0.03%.) Since equal volumes of gases contain the same number of moles, specifying composition on a volume basis is equivalent to specifying mole fractions. Thus, the mass fraction of O_2 is

$$m_{O_2} = \frac{x_{O_2} M_{O_2}}{x_{N_2} M_{N_2} + x_{O_2} M_{O_2} + x_{Ar} M_{Ar}}$$

$$= \frac{(0.209)(32)}{(0.781)(28) + (0.209)(32) + (0.009)(46)} = 0.231$$

Similarly, $m_{N_2} = 0.755$, $m_{Ar} = 0.014$. The value of $m_{O_2} = 0.231$ for dry air should be remembered: it will be used often.

1.2.2 Concentrations at Interfaces

Although temperature is continuous across a phase interface, concentrations are usually discontinuous. In order to clearly define concentrations at interfaces, we introduce imaginary surfaces, denoted u and s, on both sides of the real interface, each indefinitely close to the interface, as shown on Fig. 1.4a for water evaporating into an air stream. Thus, the liquid-phase quantities at the interface are subscripted u, and gas-phase quantities are subscripted s. If we ignore the small amount of air dissolved in the water, $x_{H_2O,u} = 1$. Notice that the subscript preceding the comma denotes the chemical species, and the subscript following the comma denotes location. To determine $x_{H_2O,s}$ we make use of the fact that, except in extreme circumstances, the water vapor and air mixture at the s-surface must be in thermodynamic equilibrium with the water at the u-surface. Equilibrium data for this system are found in conventional steam tables: the saturation vapor pressure of steam at the water temperature, $T_s (T_s = T_u)$ is the required partial pressure $P_{H_2O,s}$. With the total pressure P known, $x_{H_2O,s}$ is calculated as $P_{H_2O,s}/P$. If $m_{H_2O,s}$ is required, Eq. (1.10a) is used. For example, at $T_s = 310$ K, the saturation vapor pressure is 0.06224×10^5 Pa

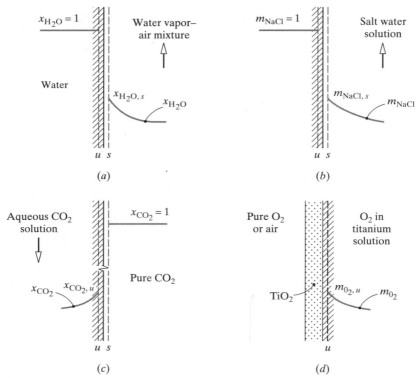

Figure 1.4 Concentrations at interfaces (a) water–air, (b) salt–water, (c) water–carbon dioxide, (d) the titanium–oxygen system.

(Table A.12a); if the total pressure is 1 bar ($= 10^5$ Pa),

$$x_{H_2O,s} = \frac{0.06224 \times 10^5}{10^5} = 0.06224 \quad \text{mole fraction}$$

$$m_{H_2O,s} = \frac{(0.06224)(18)}{(0.06224)(18) + (1 - 0.06224)(29)} = 0.0396 \quad \text{mass fraction}$$

where air has been taken to have a molecular weight of 29.

As another example, consider a slab of salt dissolving in water, as shown in Fig. 1.4b. If the salt is pure, $m_{NaCl,u} = 1$. Again, thermodynamic equilibrium must exist between the salt in water solution at the s-surface and solid salt at the interface temperature. Equilibrium data of this kind is often referred to simply as *solubility* data, found in chemistry handbooks (usually in cgs units) and in Table A.24. For example, at 30°C the solubility of salt in water is 36.3 g/100 g water. Thus at 30°C $= 303.15$ K,

$$m_{NaCl,s} = \frac{36.3}{100 + 36.3} = 0.266$$

Many processes, particularly those encountered in chemical engineering practice, involve absorption of a gas into a liquid. Many gases are only sparingly soluble, and for such dilute solutions solubility data are conveniently represented by **Henry's law**, which states that the mole fraction of the gas at the s-surface is proportional to its mole fraction in solution at the u-surface, the constant of proportionality being the **Henry number**, He. For species i,

$$x_{i,s} = He_i x_{i,u} \qquad (1.13)$$

The Henry number is inversely proportional to total pressure and is also a function of temperature. The product of Henry number and total pressure is the **Henry constant**, C_{He}, and for a given species is a function of temperature only:

$$He_i P = C_{He_i}(T) \qquad (1.14)$$

Selected data for the Henry constant are given in Table A.21. Figure 1.4c shows absorption of carbon dioxide from a stream of pure CO_2 at 3 bar pressure into water at 300 K. From Table A.21, $C_{He} = 1710$ bar; thus

$$He_{CO_2} = \frac{1710}{3} = 570; \qquad x_{CO_2,u} = \frac{1}{570} = 0.00175$$

where the small partial pressure of water vapor at the s-surface has been ignored. Advantage is taken of the increase in solubility with pressure in the carbonation of soft drinks. For highly soluble gases, such as sulfur dioxide and ammonia in water, a simple linear solubility relation does not hold, and solubility data are usually available in tables relating gas-phase partial pressure to liquid-phase mole fraction, in the form $P_{i,s} = P_{i,s}(x_{i,u}, T)$. Some such data are given in Table A.22.

Dissolution of gases into metals is characterized by varied and rather complex interface conditions. Provided temperatures are sufficiently high, hydrogen dissolution is reversible (similar to CO_2 absorption into water); hence, for example, titanium–hydrogen solutions can exist only in contact with a gaseous hydrogen atmosphere. As a result of hydrogen going into solution in atomic form, there is a characteristic square-root relation,

$$m_{H_2,u} \propto P_{H_2,s}^{1/2}$$

The constant of proportionality is strongly dependent on temperature, as well as on the particular titanium alloy: for Ti–6Al–4V alloy it is twice that for pure titanium. Table A.23 gives selected data. In contrast to hydrogen, oxygen dissolution in titanium is irreversible and is complicated by the simultaneous formation of a rutile (TiO_2) scale on the surface, as shown in Fig. 1.4d. Provided some oxygen is present in the gas phase, the titanium–oxygen *phase diagram* (found in a metallurgy handbook) shows that $m_{O_2,u}$ in alpha-titanium is 0.143, a value essentially independent of temperature and O_2 partial pressure. Dissolution of oxygen in zirconium alloys has similar characteristics to those discussed above for titanium.

All the preceding examples of interface concentrations are situations where thermodynamic equilibrium can be assumed to exist at the interface. Sometimes thermodynamic equilibrium does not exist at an interface: a very common example is when a chemical reaction occurs at the interface, and temperatures are not high enough for equilibrium to be attained. Then the concentrations of the reactants and products at the s-surface are dependent both on the rate at which the reaction proceeds—that is, the *chemical kinetics*—as well as on mass transfer considerations. An example of such a situation is discussed in Section 1.3.3.

1.2.3 Fick's Law of Diffusion

It is convenient to introduce **Fick's law** as a phenomenological relation and later examine the physical basis of the law in greater detail. In 1855 Fick proposed what has become known as his **first law**, a linear relation between the rate of diffusion of a chemical species and the local concentration gradient of that species. The concept of a linear relation between a flux and a corresponding driving force, or potential, had already been introduced by Newton in his law of viscosity, by Fourier in his law of heat conduction, and by Ohm in his law of electrical conduction. We will first restrict our attention to a stationary medium—for example, a solid or stagnant gas consisting of only two species, 1 and 2—and suggest the following form of Fick's law:

$$\mathbf{j}_1 = -\rho \mathcal{D}_{12} \nabla m_1 \qquad (1.15)$$

where $\mathbf{j}_1$ [kg/m^2 s] is the diffusive mass flux of species 1, ρ is the local mixture or solution density [kg/m^3], m_1 is the mass fraction of species 1, and $\mathcal{D}_{12}$ [m^2/s] is the constant of proportionality, called the **binary diffusion coefficient** or sometimes the **mass diffusivity.** (Notice that no comma separates the subscripts 1 and 2 on $\mathcal{D}_{12}$ since both 1 and 2 refer to chemical species. We use a comma only when the subsequent subscript refers to location.)

The corresponding law for species 2 is

$$\mathbf{j}_2 = -\rho \mathcal{D}_{21} \nabla m_2 \qquad (1.16)$$

where it will be shown in Section 2.2.4 that $\mathcal{D}_{21} = \mathcal{D}_{12}$. As is the case with Fourier's law of heat conduction, Eq. (3.2) of *Heat Transfer*, Fick's law is a vector equation, and the vector $\mathbf{j}_1$ is in the same direction as the negative gradient of concentration, $-\nabla m_1$. For a one-dimensional situation with a gradient of concentration of the z-direction only, Eq. (1.15) simplifies to

$$j_1 = -\rho \mathcal{D}_{12} \frac{dm_1}{dz} \qquad (1.17)$$

Ordinary diffusion of a chemical species down its concentration gradient results from random motion of molecules of the species. Figure 1.5 illustrates this point for one-dimensional diffusion in a model binary system. Looking at the plane located at $z = Z$, random motion of the molecules results in more molecules of species 1 crossing the plane from left to right than from right to left. Hence there is a net flux of species 1 from left to right, that is, in the direction of decreasing concentration. A similar argument holds for species 2.

A valid question at this point is the suitability of Eq. (1.15) as a statement of Fick's law. After all, we have introduced six measures of the composition of a mixture (number density, number fraction, partial density, mass fraction, molar concentration, and mole fraction), so why should mass fraction be the appropriate driving potential for diffusion? And why should diffusion be expressed as a mass flux; would not a molar flux be more suitable? We can, simply by introducing the pertinent relations, derive

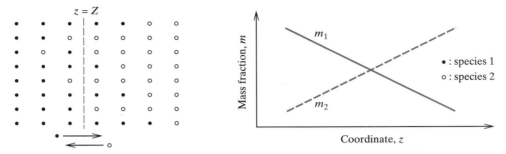

Figure 1.5 Schematic of one-dimensional diffusion in a model binary system.

alternative forms of Eq. (1.15). After we have carefully examined what we mean by a stationary medium, we may, for example, derive

$$\mathbf{J}_1 = -c\mathcal{D}_{12}\nabla x_1 \tag{1.18}$$

where $\mathbf{J}_1$ [kmol/m^2 s] is the diffusive molar flux of species 1. On the other hand, if we were to simply write down the equation

$$\mathbf{j}_1 = -\mathcal{D}_{12}\nabla \rho_1 \tag{1.19}$$

we would find that, although the diffusion coefficient defined by Eq. (1.19) has the same dimensions as the one appearing in Eq. (1.15), it will in general have a different value. The problem is that $\rho\nabla m_1 \neq \nabla\rho_1$, unless ρ is constant. It is usually possible to assume a constant density for solids and dilute liquid solutions, and then Eq. (1.19) is indeed often used. On the other hand, it is seldom possible to assume a constant density for gas mixtures and concentrated liquid solutions, so either Eq. (1.15) or Eq. (1.18) should be used. It cannot be shown in a simple manner that Eq. (1.15) is indeed the most appropriate mathematical statement of Fick's law. There is no single physical mechanism of diffusion; in particular, there are radical differences between the mechanisms in solids, liquids, and gases. The problem must be examined in the light of physical theory, the results of irreversible thermodynamics, and experimental data. Here it will be stated, without demonstration, that the kinetic theory of gases shows that Eq. (1.15) is appropriate for binary gas mixtures at normal pressures; furthermore, experimental data show that it is appropriate for dilute liquid and solid solutions. Since the majority of engineering problems fall within these categories, it is convenient to accept Eq. (1.15) as a fundamental statement of Fick's law of diffusion.

Diffusion coefficients of gases at normal pressures are almost composition independent, vary inversely with pressure, and increase with temperature. Liquid and solid diffusion coefficients are markedly concentration dependent and generally increase with temperature. Table 1.1 shows some typical values, and Appendix A contains further data. Methods for calculating diffusion coefficients are given in Section 1.7. Notice that Table A.17b, for selected gas species in dilute mixture in air, and Table A.18, for diffusion in dilute aqueous solutions, give the **Schmidt number**, rather than the diffusion coefficient directly. The Schmidt number, Sc, is defined as $Sc = \mu/\rho\mathcal{D}$ or $\nu/\mathcal{D}$, and plays a role in convective mass transfer analogous to the role played by the Prandtl number in convective heat transfer. Since the mixture or

Table 1.1 Selected data for binary diffusion coefficients.

System	Temperature K	Pressure atm	$\mathcal{D}_{12}$ m^2/s
Gas Mixtures			
H_2-air	300	1	7.77×10^{-5}
H_2-air	300	0.01	7.77×10^{-3}
CO_2-air	300	1	1.55×10^{-5}
CO_2-air	1000	1	1.24×10^{-4}
H_2O-air	373	1	3.71×10^{-5}
H_2O-air	300	0.01	2.57×10^{-3}
Dilute Liquid Solutions			
NH_3-water	300	—	2.18×10^{-9}
SO_2-water	300	—	1.67×10^{-9}
SO_2-water	400	—	8.09×10^{-9}
Sucrose-water	300	—	5.20×10^{-10}
CO_2-water	300	—	1.95×10^{-9}
CO_2-methanol	300	—	8.37×10^{-9}
CO_2-ethanol	300	—	3.88×10^{-9}
CO_2-propanol	300	—	2.73×10^{-9}
Solid Solutions			
He-Pyrex glass	300	—	8.70×10^{-13}
He-Pyrex glass	700	—	4.42×10^{-10}
H_2-steel	300	—	3.27×10^{-13}
H_2-steel	700	—	2.21×10^{-9}
O_2–alpha-phase Ti–6A1–4V alloy	1500	—	5.75×10^{-11}

solution is dilute, μ, ρ, and ν can be evaluated for pure air when using Table A.17b, and for pure water when using Table A.18. Notice also that gas-phase diffusion coefficients tend to be a few orders of magnitude larger than liquid- or solid-phase diffusion coefficients.

1.2.4 Other Diffusion Phenomena

Molecules in a gas mixture, and in a liquid or solid solution, can diffuse by mechanisms other than ordinary diffusion governed by Fick's law. *Thermal diffusion* is diffusion due to a temperature gradient and is often called the *Soret effect*. Thermal diffusion is usually negligible compared with ordinary diffusion, unless the temperature gradient is very large. However, some important processes depend on thermal diffusion, the best known being the large-scale separation of uranium isotopes. *Pressure diffusion* is diffusion due to a pressure gradient and is also usually negligible unless the pressure gradient is very large. Pressure diffusion is the principle underlying the operation of a centrifuge. Centrifuges are used to separate liquid solutions and are increasingly being used to separate gaseous isotopes as well. *Forced diffusion* results from an external force field acting on a molecule. Gravitational force fields do not cause separation since the force per unit mass of molecule is constant. Forced diffusion occurs when an electrical field is imposed on an electrolyte (for example, in charging an automobile battery), on a semiconductor, or on an ionized gas (for example, in a neon tube or metal-ion laser). Depending on the strength of the electric field, rates of forced diffusion can be very large.

Some interesting diffusion phenomena occur in porous solids. When a gas mixture is in a porous solid, such as a catalyst pellet or silica-gel particle, the pores can be smaller than the mean free path of the molecules. Then the molecules collide with the wall more often than with other molecules. In the limit of negligible molecule-molecule collisions we have *Knudsen diffusion*, also called *free molecule flow* in the fluid mechanics literature. If the pore size approaches the size of a molecule, then Knudsen diffusion becomes negligible, and *surface diffusion*, in which adsorbed molecules move along the pore walls, becomes the dominant diffusion mechanism. Knudsen diffusion will be considered in Section 1.6.

Very small particles of 10^{-3}–10^{-1} μm size—for example, smoke, soot, and mist—behave much like large molecules. Ordinary diffusion of such particles is called *Brownian motion* and is described in most elementary physics texts. Diffusion of particles due to a temperature gradient is called *thermophoresis* and plays an important role for larger particles, typically in the size range 10^{-1}–1 μm. Diffusion of particles in a gas mixture due to concentration gradients of molecular species is called *diffusiophoresis*. *Forced diffusion* of a charged particle in an electrical field is similar to that for an ionized molecular species. Thermal and electrostatic precipitators are used to remove particles from power plant and incinerator stack gases, and depend on thermophoresis and forced diffusion, respectively, for their operation. Diffusion phenomena are unimportant for particles of size greater than about 1 μm in air at 1 atm; the motion of such particles is governed by the laws of Newtonian mechanics. Brownian motion is dealt with in Section 1.7, and in Chapter 3 electrostatic precipitators and particle filters are analyzed as examples of mass exchangers.

1.3 MASS DIFFUSION

Mass diffusion is analogous to heat conduction. The simplest mass diffusion problem is steady diffusion through a plane wall, which is analyzed from first principles in Section 1.3.1. Often the solution of a mass diffusion problem can be obtained directly from the solution of the analogous heat conduction problem. The solution for transient diffusion in a semi-infinite solid is obtained this way in Section 1.3.2. As mentioned in the introduction to this chapter, mass transfer problems usually differ from heat transfer problems in their variety and complexity of boundary conditions. Heterogeneous catalysis, analyzed in Section 1.3.3, is an example of a more complicated boundary condition.

1.3.1 Steady Diffusion through a Plane Wall

Figure 1.6 shows a plane solid wall of surface area A and thickness L. Imaginary u- and v-surfaces are located within the solid adjacent to the interfaces at $z = 0$ and $z = L$, respectively, and the corresponding mass fractions of species 1 are $m_{1,u}$ and $m_{1,v}$, which are steady in time. We wish to determine the rate of transfer of species 1 through the wall. For example, the wall may be that of a hydrogen tank, with hydrogen leaking through the wall into the atmosphere. Note that our approach is similar to that used in Chapter 1 of *Heat Transfer* in analyzing elementary heat

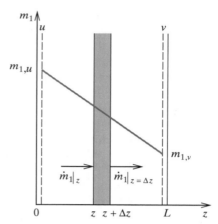

Figure 1.6 Schematic of steady
diffusion through a plane wall.

conduction: we initially take the mass fractions $m_{1,u}$ and $m_{1,v}$ as known, and later concern ourselves with how these are related to conditions on either side of the wall. An elemental control volume $\Delta V = A\Delta z$ is located between z and $z + \Delta z$; if the composition is unchanging in time, and if there are no chemical reactions producing or consuming species 1, then the **principle of conservation of mass species** requires that the flow of species 1, denoted $\dot{m}_1$ [kg/s], across the face at z must equal that across the face at $z + \Delta z$,

$$\dot{m}_1|_z = \dot{m}_1|_{z+\Delta z}$$

or

$$\dot{m}_1 = j_1 A = \text{Constant} \tag{1.20}$$

We shall see that the mass flow of species 1, $\dot{m}_1$ [kg/s], is the mass transfer analog to the heat flow $\dot{Q}$ [J/s] of heat transfer. Substituting for j_1 from Fick's law, Eq. (1.17),

$$\frac{\dot{m}_1}{A} = -\rho\mathcal{D}_{12}\frac{dm_1}{dz} = \text{Constant}$$

Then integrating across the wall,

$$\frac{\dot{m}_1}{A}\int_0^L dz = -\int_{m_{1,u}}^{m_{1,v}} \rho\mathcal{D}_{12}\, dm_1$$

where $\dot{m}_1$ and A are outside the integral sign since both are constants. If any variations of ρ and $\mathcal{D}_{12}$ are ignored for the present,

$$\dot{m}_1 = \frac{\rho\mathcal{D}_{12}A}{L}(m_{1,u} - m_{1,v}) = \frac{m_{1,u} - m_{1,v}}{L/\rho\mathcal{D}_{12}A} \tag{1.21a}$$

Comparison of Eq. (1.21a) with Ohm's law, $I = E/R$, suggests that $\Delta m_1 = m_{1,u} - m_{1,v}$ can be viewed as a driving potential for flow of mass species 1, analogous to

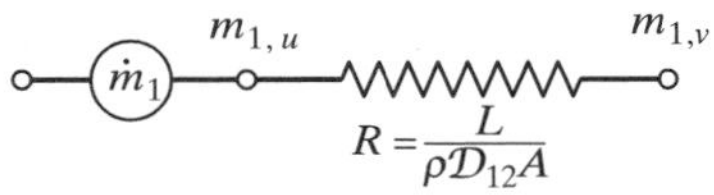

Figure 1.7 Mass flow circuit resistance for diffusion across a plane wall.

voltage difference being a driving potential for current (and temperature difference being the driving potential for heat flow); then $R = L/\rho \mathcal{D}_{12} A$ can be viewed as a *diffusion resistance* analogous to an electrical or conduction resistance. The mass flow through a slab can therefore be represented by the circuit shown in Fig. 1.7. Examination of the analysis in Section 1.3.1 of *Heat Transfer* leading to Eq. (1.9) shows the close analogy between heat conduction and mass diffusion; however, the fact that concentrations are usually discontinuous at phase interfaces, whereas temperature is not, means that circuits for species flow through diffusion resistances in series cannot be constructed in the same way as for heat or current flow (however, see Exercise 1–14).

If the preceding analysis is repeated on a molar basis, the result is

$$\dot{M}_1 = \frac{c\mathcal{D}_{12}A}{L}(x_{1,u} - x_{1,v}) \tag{1.21b}$$

where $\dot{M}_1$ [kmol/s] is the molar flow of species 1. Analysis of steady diffusion through a plane wall yields a simple result, as did the analogous problem of steady heat conduction through a plane wall. However, application of Eqs. (1.21) to mass transfer problems is complicated by the need to use solubility data to obtain the concentrations at the u- and v-surfaces. Furthermore, as was seen in Section 1.2.2, solubility data obtained from different sources are expressed in a variety of forms.

Permeability

Consider gas-solid systems for which Henry's law applies. It is common practice to define the *solubility S* as the volume of solute gas (at STP of 0°C and 1 atm) dissolved in unit volume of solid, when the gas is at a partial pressure of 1 atm. Since 1 kmol of gas at STP occupies 22.414 m^3, it follows that the molar concentration of gas dissolved in a solid is related to the partial pressure of the adjacent gas as $c_{1,u} = S(P_{1,s}\text{ atm})/22.414$ [kmol/m^3]. Also, the volume flow rate $\dot{V}_1$ is related to the molar flow rate $\dot{M}_1$ as $\dot{V}_1 = 22.414\dot{M}_1$. Then substituting in Eq. (1.21b) with c assumed constant gives the volume rate of gas flow across the wall as

$$\dot{V}_1 = \frac{\mathcal{P}_{12}A}{L}(P_{1,s} - P_{1,s'})\text{ m}^3\text{ (STP)/s} \tag{1.21c}$$

where $P_{1,s}$ and $P_{1,s'}$ are the partial pressures of species 1 in atmospheres on each side of the wall, and $\mathcal{P}_{12} = \mathcal{D}_{12}S$ is the *permeability* of the gas in the solid. The units of permeability are [m^3 (STP)/m^2 s (atm/m)]. Physicists often prefer to tabulate data for permeability, rather than for solubility and diffusion coefficient separately, and use Eq. (1.21c) to calculate steady transfer across shells and membranes. Table A.23 contains selected solubility and permeability data for gases in solids. Notice that other units and definitions of solubility are used for gas-solid systems. Often, the solubility S is given in units of kmol/m^3 bar. Also, a dimensionless solubility coefficient is defined as $S' = c_{1,u}/c_{1,s}$.

Two examples demonstrating the application of Eqs. (1.21) follow. The first is worked on a mass basis and the second on a molar basis, the choice being determined by the form of the solubility data. Both examples concern leakage of a gas through the walls of a container. Such problems are encountered when gases with small molecules, for example, H$_2$ and He, are stored at high temperatures, owing to the relatively high diffusion coefficients (for solids). It is of interest to note that, although steady heat conduction through a plane wall is encountered in a great range of engineering problems, relatively few mechanical engineering mass transfer problems involve steady ordinary diffusion across a plane wall. In chemical engineering and the life sciences there are many examples of mass transfer across thin membranes (see Exercises 1–13 and 1–14). However, often the mass transfer mechanism is not simple ordinary diffusion governed by Fick's law—for example, reverse osmosis for desalination of saline water, or blood oxygenation in the human lung.

Example 1.1   Loss of Hydrogen from a Storage Tank

A spherical steel tank of 1 liter capacity and 2 mm wall thickness is used to store hydrogen at 400° C. The initial pressure is 9 bar, and there is a vacuum outside the tank. Calculate the time required for the pressure to drop to 5 bar. Data for the hydrogen–steel (1–2) system include the following items.

$$\text{Diffusion coefficient: } \mathcal{D}_{12} \simeq 1.65 \times 10^{-6} e^{-4630/T} \, \text{m}^2/\text{s}$$

$$\text{Solubility relation: } m_{1,u} \simeq 2.09 \times 10^{-4} e^{-3950/T} P_{1,s}^{1/2}$$

$$\text{for } T \text{ in kelvins and } P \text{ in bars.}$$

Solution

Given: High-temperature hydrogen gas stored in a 1-liter steel tank.
Required: Time required for pressure to drop from 9 bar to 5 bar.
Assumptions:

1. Quasi-steady diffusion through the steel.
2. Wall curvature effects are negligible.
3. The ideal gas law applies.

To obtain a differential equation governing the rate of change of pressure with time, we equate the rate of change of mass of H$_2$ stored in the tank to the rate at which H$_2$ diffuses across the tank wall. For an ideal gas the mass of gas in the tank, w, is

$$w = \frac{PV}{(\mathcal{R}/M)T}$$

Differentiating with respect to time and recognizing that V and T are constant gives

$$\frac{dw}{dt} = \frac{V}{(\mathcal{R}/M)T}\frac{dP}{dt}$$

or

$$\frac{dP}{dt} = \frac{\mathcal{R}T}{MV}\frac{dw}{dt} \tag{1}$$

Since the wall thickness is small compared with the tank diameter, Eq. (1.21a) for diffusion across a plane wall can be used.

$$\frac{dw}{dt} = -\dot{m}_1 = -\frac{\rho\mathcal{D}_{12}A}{L}(m_{1,u} - m_{1,v}) \tag{2}$$

where $m_{1,u}$ and $m_{1,v}$ are H_2 mass fractions in the steel at the inner and outer surfaces of the tank, $A = 4\pi R^2$ is the surface area, and L is the wall thickness. Combining Eqs. (1) and (2),

$$\frac{dP}{dt} = -\frac{\mathcal{R}T}{MV}\frac{\rho\mathcal{D}_{12}A}{L}(m_{1,u} - m_{1,v})$$

Since there is pure hydrogen in the tank, $P_{1,s} = P$, and for a constant temperature the solubility relation can be used to write $m_{1,u} = KP^{1/2}$; also, since there is a vacuum outside the tank, $P_{1,s'} = 0$, so that $m_{1,v} = 0$. Thus we can write

$$\frac{dP}{dt} = -BP^{1/2}; \quad B = \frac{\mathcal{R}T}{MV}\frac{\rho\mathcal{D}_{12}A}{L}K$$

Integrating with $P = P_0$ at $t = 0$ gives

$$P = \left[P_0^{1/2} - \frac{Bt}{2}\right]^2 \quad \text{or} \quad t = \frac{2}{B}\left(P_0^{1/2} - P^{1/2}\right)$$

Next we evaluate the constant B. The tank volume is $(4/3)\pi R^3 = 10^{-3}$ m^3; hence $R = 6.20$ cm and $A/V = 3/R$.

$$K = 2.09 \times 10^{-4} e^{-3950/673} = 5.90 \times 10^{-7} \text{ bar}^{-1/2} = 1.87 \times 10^{-9} \text{ Pa}^{-1/2}$$

$$\mathcal{D}_{12} = 1.65 \times 10^{-6} e^{-4630/673} = 1.70 \times 10^{-9} \text{ m}^2/\text{s}$$

From Table A.1 in *Heat Transfer*, the density of steel is $\rho \simeq 7800$ kg/m^3; thus

$$B = \frac{(8314)(673)(7800)(1.70 \times 10^{-9})(3)(1.87 \times 10^{-9})}{(2)(0.0620)(0.002)} = 1.68 \times 10^{-3} \text{ Pa}^{1/2}/\text{s}$$

The time taken for the pressure to drop to 5 bar $= 5 \times 10^5$ Pa is, then,

$$t = \frac{2[(9 \times 10^5)^{1/2} - (5 \times 10^5)^{1/2}]}{(1.68 \times 10^{-3})} = 2.88 \times 10^5 \text{ s} \simeq 80 \text{ h}$$

Comments

1. Since atmospheric air contains very little hydrogen (0.5 parts per million by volume), the above result will be unchanged if the tank is located in a ventilated enclosure.
2. Strictly speaking, ρ in Eq. (2) is the density of the solution. However, since the amount of H_2 dissolved in the steel is very small, we simply use the steel density.
3. In assuming quasi-steady diffusion we are neglecting the change in the amount of H_2 stored in solution in the tank wall. Check this assumption by comparing the amount stored to the leakage rate.
4. Note the analogy to the lumped thermal capacity heat transfer analysis of Section 1.5 in *Heat Transfer*. In Section 1.5 we used the energy conservation principle to derive the governing differential equation; here we used the mass conservation principle.
5. Since Henry's law does not apply to dissolution of H_2 in steel, the permeability method for calculating diffusion across a wall, Eq. (1.21c), is inappropriate.
6. Check the stresses in the tank wall at 9 bar: is the tank safe?   ▲

Example 1.2   Loss of Helium from a He-Cd Laser

A helium–cadmium (blue) laser used in a copying machine printer contains He at a nominal pressure of 460 Pa. The glass outer shell has a volume of 150 cm^3, an area of 550 cm^2, and a wall thickness of 1.52 mm. At normal operating conditions the average temperature of the gas in the tube is 225°C and the shell temperature is 115°C. Estimate (i) the helium leak rate and (ii) the time required to lose 1% of the helium inventory. For the helium–shell glass (1–2) system the solubility coefficient and diffusion coefficient are

$$S' = \frac{c_{1,u}}{c_{1,s}} \simeq 0.007 + 3.0 \times 10^{-5}(T°C);$$

$$\mathcal{D}_{12} \simeq 1.40 \times 10^{-8} e^{-3280/(T \text{ K})} \text{ m}^2/\text{s}$$

Solution

Given: Glass shell containing helium.
Required: (i) Helium leak rate. (ii) Time to lose 1% of helium inventory.
Assumptions:

1. Molar density c and diffusion coefficient $\mathcal{D}_{12}$ constant across glass shell.
2. Curvature effects are negligible.

   i. Equation (1.21b) applies to quasi-steady diffusion of helium across the glass shell. It was derived assuming c and $\mathcal{D}_{12}$ constant. Since the helium concentration in the glass is very small, the molar density c can certainly be assumed constant. There is some variation of $\mathcal{D}_{12}$ due to the temperature gradient through the wall, but this variation will be ignored. The solubility data are given in terms of molar concentrations; thus it is convenient to rewrite Eq. (1.21b) as

$$\dot{M}_1 = \frac{\mathcal{D}_{12} A}{L}(c_{1,u} - c_{1,v}) \tag{1}$$

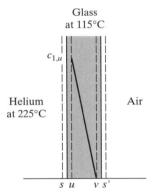

The molar concentrations $c_{1,u}$ and $c_{1,v}$ are calculated first. At the s-surface there is pure helium at $115°C = 388$ K, and 460 Pa. Using the ideal gas law,

$$c_{1,s} = \frac{P_{1,s}}{\mathcal{R} T_s} = \frac{(460 \text{ N/m}^2)}{(8314 \text{ J/kmol K})(388 \text{ K})} = 1.426 \times 10^{-4} \text{ kmol/m}^3$$

Substituting in the given solubility relation,

$$\frac{c_{1,u}}{c_{1,s}} = 0.007 + (3.0 \times 10^{-5})(115) = 1.05 \times 10^{-2}$$

Thus $c_{1,u} = (1.426 \times 10^{-4})(1.05 \times 10^{-2}) = 1.50 \times 10^{-6} \text{ kmol/m}^3$.
   Since the outer shell is cooled by an air flow, the helium concentration at the s'-surface is kept at a very low value. Thus

$$c_{1,s'} \simeq 0 \quad \text{and} \quad c_{1,v} \simeq 0$$

Next the diffusion coefficient for helium in the glass at $115°C = 388$ K is calculated.

$$\mathcal{D}_{12} = 1.40 \times 10^{-8}e^{-3280/388} = 2.98 \times 10^{-12} \text{ m}^2/\text{s}$$

Substituting in Eq. (1), the molar flow rate of helium through the shell is

$$\dot{M}_1 = \frac{(2.98 \times 10^{-12} \text{ m}^2/\text{s})(550 \times 10^{-4} \text{ m}^2)(1.50 \times 10^{-6} \text{ kmol/m}^3)}{(1.52 \times 10^{-3} \text{ m})}$$

$$= 1.62 \times 10^{-16} \text{ kmol/s}$$

ii. The molar inventory in the laser is $W = cV$. The bulk helium is at approximately $225°C = 498$ K and 460Pa; hence the molar concentration is

$$c \simeq \frac{P}{\mathcal{R}T} = \frac{460}{(8314)(498)} = 1.111 \times 10^{-4} \text{kmol/m}^3$$

$$W = (1.111 \times 10^{-4})(1.5 \times 10^{-4} \text{m}^3) = 1.67 \times 10^{-8} \text{ kmol}$$

Thus the time for 1% of the inventory to be lost is approximately

$$t \simeq \frac{(0.01)W}{\dot{M}_1} = \frac{(0.01)(1.67 \times 10^{-8})}{(1.62 \times 10^{-16})} = 1.03 \times 10^6 \text{ s } (\simeq 286 \text{ h})$$

Comments

1. In practice, such lasers are fitted with a high-pressure helium reservoir separated from the laser by a glass diaphragm, as shown. When the pressure falls below a set value, a control system switches on a heater, which raises the diaphragm temperature to allow helium to leak from the reservoir to the laser. When the pressure in the laser rises above a second set value, the heater is switched off.

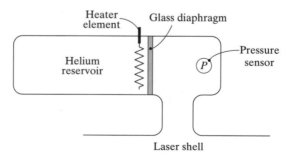

2. The composition of glass used in practice varies considerably. Typically, B_2O_3, Al_2O_3, TiO_2, Na_2O, and so on are added to the SiO_2 in varying proportions. For example, Pyrex glass contains 81% SiO_2, 13% B_2O_3, 4% Na_2O, and 2% Al_2O_3. Helium solubility and diffusion coefficients depend on the glass composition: in particular, as the amount of B_2O_3 increases, $\mathcal{D}$

decreases markedly (about tenfold for an increase of 13% to 30%). Thus special care must be taken to ensure that appropriate data are used when calculating helium diffusion through glass. Conversely, a particular glass can be chosen to give a higher or lower diffusion rate.

3. See Exercise 1–12 for use of Eq. (1.21c) and permeability to solve this problem. ▲

1.3.2 Transient Diffusion

The Mass Diffusion Equation
Transient diffusion in a stationary medium is governed by an equation that is analogous to the heat conduction equation derived in Section 3.2 of *Heat Transfer*. For simplicity we will consider one-dimensional diffusion in Cartesian coordinates with no chemical reactions. Referring to Fig. 1.8, the principle of conservation of mass species applied to the elemental control volume ΔV located between z and $z + \Delta z$ requires that in a time interval Δt,

$$\text{Storage of species 1} \; = \; \text{Net inflow of species 1}$$

$$\rho_1 \Delta V|_{t+\Delta t} - \rho_1 \Delta V|_t = (\dot{m}_1|_z - \dot{m}_1|_{z+\Delta z}) \, \Delta t$$

Substituting $\Delta V = A\Delta z$ and $\dot{m}_1 = j_1 A$, dividing through by $A\Delta z\Delta t$, and letting $\Delta z \Delta t \to 0$,

$$\frac{\partial \rho_1}{\partial t} = -\frac{\partial j_1}{\partial z}$$

Introducing Fick's law of diffusion, Eq. (1.17), gives

$$\frac{\partial \rho_1}{\partial t} = \frac{\partial}{\partial z}\left(\rho \mathcal{D}_{12}\frac{\partial m_1}{\partial z}\right) \tag{1.22}$$

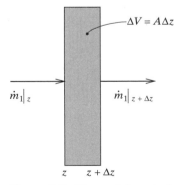

Figure 1.8 Elemental control volume used to derive the mass diffusion equation.

where the partial derivatives signify that ρ_1 and m_1 are functions of the two independent variables t and z. In the case of dilute solutions in solids, it is usually possible
to assume that the density is constant; also, if the solid is isothermal, the diffusion
coefficient $\mathcal{D}_{12}$ can also be taken to be a constant: then substituting $\rho_1 = \rho m_1$ and
canceling ρ,

$$\frac{\partial m_1}{\partial t} = \mathcal{D}_{12}\frac{\partial^2 m_1}{\partial z^2} \tag{1.23}$$

which in some texts is called *Fick's second law of diffusion.* However, Eq. (1.23) is
simply a statement of the principle of conservation of mass species rather than a new
phenomenological law, and thus current practice is to drop this name and to simply
call the phenomenological law of diffusion, Eq. (1.15), Fick's law. Equation (1.23) is
of similar form to the one-dimensional transient heat conduction equation

$$\frac{\partial T}{\partial t} = \alpha\frac{\partial^2 T}{\partial z^2} \tag{1.24}$$

which is valid for constant thermal properties and no internal heat generation. We
see also that both the mass diffusivity $\mathcal{D}_{12}$ and thermal diffusivity α have the same
units $[\text{m}^2/\text{s}]$. It follows that, when the various assumptions made hold true, the
solution of a diffusion problem may be obtained directly from the corresponding
heat conduction problem. In the mathematics literature both the heat conduction
equation and Fick's second law of diffusion are usually referred to as the **diffusion
equation**,

$$\frac{\partial \phi}{\partial t} = \alpha\frac{\partial^2 \phi}{\partial z^2} \tag{1.25}$$

where ϕ is any scalar potential function.

Boundary conditions for mass diffusion are more complicated than those for
heat conduction. In general concentrations are discontinuous at phase interfaces.
Dirichlet-type boundary conditions for Eq. (1.23) must be specified at u-surfaces,
not at s-surfaces. In the event that the s-surface concentration is known, solubility data are required to relate s- and u-surface concentrations (see Exercises 1–17
and 1–18). Third-kind boundary conditions involve mass convection at a surface
described by the mass transfer analog to Newton's law of cooling, but are also made
more complicated by a discontinuous concentration at the surface. Mass convection
is introduced in Section 1.4, and the formulation of third-kind boundary conditions
is explained in Section 1.4.5. Notice that there is no mass transfer analog to the
fourth-kind (radiation) heat transfer boundary condition specified by Eq. (3.16) of
Heat Transfer. On the other hand, diffusion into a solid is often accompanied by a
chemical reaction at the surface to give a more complicated boundary condition (see,
for example, Exercises 1–20 and 1–61).

Transient Diffusion in a Semi-Infinite Solid
In Section 3.4.2 of *Heat Transfer* we encountered heat conduction problems where
the conduction process is confined to a thin region near the surface into which temperature changes have penetrated. The typically low diffusion coefficients character-
izing diffusion in solids ($\sim 10^{-9}$–10^{-12} m^2/s) result in analogous diffusion problems

being quite common in mass transfer practice. Recall that the *penetration depth* for heat conduction was of the order $(\alpha t)^{1/2}$; analogously, it is of the order $(\mathcal{D}_{12}t)^{1/2}$ for mass diffusion. A well-known process is case-hardening of mild steel by packing the steel component in a carbonaceous material in a furnace (to increase the diffusion coefficient). The gem industry uses diffusion treatment to color clear stones. Figure 1.9*a* shows the cross section of a clear sapphire that was packed in titanium and iron oxide powders and heated for 600 hours near 2000°C: penetration of titanium and iron atoms is about 0.4 mm, which is sufficient to give the stone a brilliant blue color.

Figure 1.9*b* depicts a semi-infinite solid with species 1 in dilute solution in species 2. For example, the process might be case-hardening of steel with species 1 as carbon and species 2 as iron. The initial concentration of species 1 is $m_{1,0}$. At time $t = 0$ the concentration of species 1 at the u-surface is suddenly raised to $m_{1,u}$. The problem is to describe the concentration profiles of species 1 as a function of time and position, and to determine the rate of dissolution of species 1 as a function of time. Equation (1.23) is to be solved,

$$\frac{\partial m_1}{\partial t} = \mathcal{D}_{12}\frac{\partial^2 m_1}{\partial z^2} \tag{1.23}$$

with initial condition

$$m_1 = m_{1,0} \quad \text{at } t = 0$$

and boundary conditions

$$m_1 = m_{1,u} \quad \text{at } z = 0$$
$$m_1 \to m_{1,0} \quad \text{as } z \to \infty$$

The analogous heat conduction problem was solved in Section 3.4.2 of *Heat Transfer*, and hence the solution can be written down immediately. From Eq. (3.58) of *Heat Transfer*,

$$\frac{m_1 - m_{1,0}}{m_{1,u} - m_{1,0}} = \text{erfc}\frac{z}{(4\mathcal{D}_{12}t)^{1/2}} \tag{1.26}$$

Usually solid-phase diffusion coefficients are very low, and thus Eq. (1.26) finds wide utility in situations such as case-hardening, where the carbon penetrates only a short distance into the solid. A useful definition of the penetration depth δ_c of species 1 is the location where a tangent to the concentration profile at $z = 0$ intercepts the line $m_1 = m_{1,0}$, as shown in Fig. 1.9*c*. The concentration gradient at $z = 0$ is found by differentiation of Eq. (1.26),

$$-\frac{\partial m_1}{\partial z}\bigg|_{z=0} = \frac{m_{1,u} - m_{1,0}}{(\pi \mathcal{D}_{12}t)^{1/2}} \equiv \frac{m_{1,u} - m_{1,0}}{\delta_c}$$

and hence

$$\delta_c = 1.772(\mathcal{D}_{12}t)^{1/2} \sim (\mathcal{D}_{12}t)^{1/2} \tag{1.27}$$

That is, the penetration depth is proportional to the square root of both the diffusion coefficient and time. Note that other definitions of penetration depth are used—for example, the depth at which $(m_1 - m_{1,0})/(m_{1,u} - m_{1,0}) = 0.02$. However, the dependence on diffusion coefficient and time remains the same. The rate of dissolution

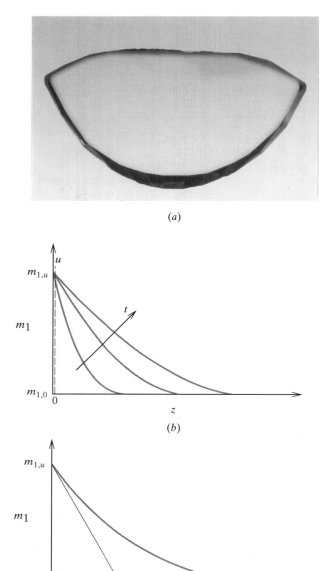

(a)

(b)

(c)

Figure 1.9 Transient diffusion in a semi-infinite solid.
(a) A clear sapphire heated for 600 h near $2050°$C
packed in titanium and iron oxide powders.
(b) Expected concentration profiles. (c) Definition of
penetration depth δ_c. (Photograph courtesy Gem Book
Publishers, Bethesda, Maryland.)

per unit surface area is the diffusion flux $j_{1,u}$:

$$j_{1,u} = -\rho \mathcal{D}_{12} \left. \frac{\partial m_1}{\partial z} \right|_{z=0} = \frac{\rho \mathcal{D}_{12}(m_{1,u} - m_{1,0})}{(\pi \mathcal{D}_{12} t)^{1/2}}$$

$$= \rho \left(\frac{\mathcal{D}_{12}}{\pi t} \right)^{1/2} (m_{1,u} - m_{1,0}) \tag{1.28}$$

and the rate of dissolution for a surface area A is $\dot{m}_1 = j_{1,u} A$. When the penetration depth is very small compared with the size of the solid, the shape of the solid has no effect on the diffusion process. Notice that $j_{1,u} \propto \mathcal{D}_{12}^{1/2}$: thus, the permeability approach used in Eq. (1.21c) for steady diffusion cannot be used for transient diffusion.

Transient Diffusion in Slabs, Cylinders, and Spheres

Transient heat conduction in slabs, cylinders, and spheres received considerable attention in Chapter 3 of *Heat Transfer*, owing to the practical utility of the results obtained. As was the case for the semi-infinite solid, the solutions for analogous mass diffusion problems can be obtained from solutions presented in that chapter, and no further mathematical analysis is required. An example involving a liquid droplet is given as Exercise 1–18. However, the practical utility of these solutions for mass diffusion is rather limited. For most problems of engineering relevance involving mass transport in solids, the transport mechanism is not simple ordinary diffusion described by Fick's law with a constant diffusion coefficient. If Fick's law does apply, the diffusion coefficient usually varies significantly due to its dependence either on concentration or on temperature. But, more often, Fick's law does not apply, and there are additional complications. In particular, Fick's law is seldom applicable to the movement of moisture in porous solids or to diffusion of oxygen through oxide layers.

Perhaps the most important engineering problems involving transient mass transport in solids are the great variety of drying processes encountered in practice. Drying of coal, timber, paper, textiles, food, and pharmaceutical products involves large-scale engineering operations that consume enormous amounts of energy. Drying of brush is a critical precursor to ignition in a brush fire and affects the propagation of such fires. Dehumidification of gases using desiccants, such as silica gel, depends on moisture transport in a porous solid. The student has surely had practical experience of drying processes, whether it was the baking of a cake or waiting for clothes to dry after falling into a river!

Figure 1.10 shows a typical *drying-rate curve* of a porous solid for constant drying conditions. When a large amount of moisture is present, water moves through the solid easily by capillary action, and the drying rate is controlled by evaporation from the wet surface. In turn, the evaporation is controlled by convective mass transfer of water vapor from the surface into an air flow, as well as by heat transfer to the surface to supply the enthalpy of vaporization. Such problems involve simultaneous convective heat and mass transfer and are dealt with in Section 1.5. In the terminology of drying technology, this phase of a drying process is called the *constant-rate period*, shown as *B–C* in Fig. 1.10. Actually, when the solid is placed in a drying oven,

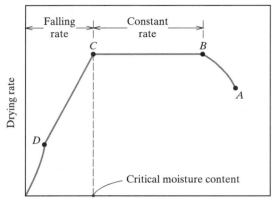

Figure 1.10   A characteristic drying-rate curve for a porous solid with constant drying conditions. (The free moisture content is the moisture in excess of the equilibrium content corresponding to the air condition.)

there is an initial short transient A–B in which the drying rate increases as the solid temperature increases: at point B the solid has reached an equilibrium temperature, and thereafter the drying rate is constant.

Past point C, the moisture content of the solid is below a critical value: dry regions can form on the surface, and the location of evaporation recedes into the solid. This phase of drying is called the *falling-rate period*, since the drying rate steadily decreases. In this period, the drying rate tends to be controlled by moisture transport through the porous structure. The mechanisms of this transport are very complex: they are not well understood and are the subject of current research. Diffusion mechanisms such as ordinary, Knudsen, and surface diffusion may be important, depending on pore size. Liquid water may move by capillary action. There may be bulk motion of both water and gas due to pressure gradients induced by temperature gradients in the solid, and simultaneous heat conduction and mass transport must be considered. In some cases, these mechanisms act in opposite directions. Recent progress for wood is reported by Plumb et al. [1] and Perre et al. [2], and for silica gel by Pesaran and Mills [3]. Often the falling-rate period is divided into two portions by point D, at which the surface becomes completely dry. Depending on the nature of the porous solid, there may or may not be a discontinuity in slope at D; also, one of the two portions may not even exist.

Many texts present falling-rate drying analyses using a Fick's-type law with a constant effective diffusion coefficient to describe moisture transport: Exercise 1–19 is an example of such an approach. However, it is now generally accepted that such analyses usually lead to inconsistent results, and that the model is too simplified to be of much value. The bibliographies for this chapter and Chapter 3 contain a variety of source material for the analysis and design of drying processes; most aspects are too specialized or advanced to be included in this introductory text.

Example 1.3   Case-Hardening of Steel

A mild steel rod with an initial carbon concentration 0.2% by weight is to be case-hardened. It is packed in a carbonaceous material and maintained at a high temperature. Thermodynamic data for the chemical system indicate an equilibrium concentration of carbon in iron at the phase interface of 1.5%. Calculate the time required for the location 1 mm below the surface to have a carbon mass concentration of 0.8%. The diffusion coefficient of carbon in steel at the process temperature is approximately 5.6×10^{-10} m^2/s.

Solution

Given: Mild steel exposed to carbon at high temperature.

Required: Time required to have a carbon mass fraction of 0.8% 1 mm below surface.

Assumptions: Since the penetration is small, the rod can be modeled as a semi-infinite solid.

Equation (1.26) for the instantaneous concentration profiles may be used with $m_{1,0} = 0.002$, $m_{1,u} = 0.015$, and a value of $m_1 = 0.008$ required at $z = 0.001$ m:

$$\frac{0.008 - 0.002}{0.015 - 0.002} = 0.462 = \text{erfc}\left(\frac{z}{(4\mathcal{D}_{12}t)^{1/2}}\right)$$

Using a table for the error function,

$$\frac{z}{(4\mathcal{D}_{12}t)^{1/2}} = 0.52$$

Solving,

$$t = \frac{z^2}{4\mathcal{D}_{12}(0.52)^2} = \frac{(0.001)^2}{(4)(5.6 \times 10^{-10})(0.52)^2} = 1650 \text{ s } (0.46 \text{ h})$$

Comments

1. The diffusion coefficient of carbon in iron increases exponentially with temperature. Hence, the process is carried out at high temperature to reduce the time required.

2. Notice that the u-surface concentration of carbon was specified in order to simplify the problem. Determination of $m_{1,u}$ requires solubility data, and possibly consideration of diffusion in the carbonaceous material.

3. The diffusion coefficient of carbon in steel depends on both the steel microstructure and the carbon concentration.   ▲

1.3.3 Heterogeneous Catalysis

The phenomenon of *heterogeneous catalysis* is often encountered in engineering problems. It is the process whereby a chemical reaction is promoted by contact of the reactants with a suitable surface, termed the *catalyst*. (The adjective *heterogeneous* means the reaction takes place on the surface, whereas the adjective *homogeneous* means that the reaction takes place within the medium.) For example, carbon monoxide and unburnt hydrocarbons can be oxidized in oxygen-rich exhaust gases of an automobile by passing the gases through a catalytic reactor in which the gases contact platinum metal. Also, the relatively cool metal of the nose of a hypersonic flight vehicle can act as a catalyst for the recombination of oxygen and nitrogen, the atomic species having resulted from dissociation in the high-temperature region behind the bow shock. The essential features of such processes can be demonstrated by analyzing a simple model experiment involving one-dimensional diffusion, as shown in Fig. 1.11. In Section 1.4, we shall see how the results obtained here can also be applied to the convection situations characteristic of typical applications.

The rate at which the catalytic reaction proceeds can often be expressed in a simple form as

$$\begin{array}{c}\text{Rate of consumption}\\ \text{of species 1}\end{array} = \begin{array}{c}\text{Rate}\\ \text{constant}\end{array} \times \left(\begin{array}{c}\text{Concentration of species 1}\\ \text{adjacent to the surface}\end{array}\right)^n$$

The rate at which species 1 is consumed is equal to the rate at which it diffuses to the catalyst surface. Referring to Fig. 1.11, we can write on a mass basis

$$-j_{1,s} = k''(\rho_{1,s})^n \tag{1.29}$$

where the rate constant k'' (the double prime signifies a surface reaction per unit area) depends on properties of the catalyst surface and on temperature, often as an *Arrhenius relation* $k'' = A\exp(-E_a/\mathcal{R}T)$, where A is the preexponential constant and E_a is the *activation energy*. The exponent n is the *order* of the reaction. Data for k'' and n are found in the chemical kinetics literature (though engineers often have to measure such data when developing a new process). The first-order reaction is particularly simple, inasmuch as the rate of reaction is directly proportional to the rate at which molecules of reactant collide with the surface. Appropriate units for k'' depend on the order of the reaction: for a first-order reaction, examination of Eq. (1.29) shows that k'' has units [m/s]. The reaction

$$2CO + O_2 \rightarrow 2CO_2 \tag{1.30a}$$

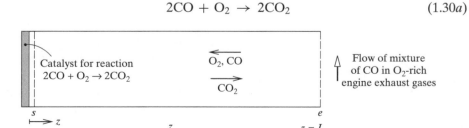

Figure 1.11 A model experiment for investigating heterogeneous catalysis with one-dimensional diffusion.

which is promoted in oxygen-rich automobile exhaust gases by platinum or cupric oxide in an *oxidation converter*, can be modeled as a first-order reaction. On the other hand, the reaction

$$2NO + 2CO \rightarrow N_2 + 2CO_2 \tag{1.30b}$$

which is promoted in oxygen-free exhaust gases by platinum–rhodium alloy in a *three-way converter*, has a more complex behavior: it cannot be adequately modeled as a single reaction of first order. Since analysis of this problem is relatively simple for a single first-order reaction, we will focus our attention on CO removal in an oxidation converter.

Figure 1.11 shows a tube with length L, and over its mouth flows a mixture of oxygen-rich gases including CO. The surface at the base of the tube is a catalyst for the reaction given by Eq. (1.30a). We label CO as species 1. Now Fick's law, as we have defined it, is strictly valid only for binary mixtures—hence the use of a binary diffusion coefficient $\mathcal{D}_{12}$. However, when species 1 is in small concentration, we may safely define an **effective binary diffusion coefficient** for species 1 diffusing through the mixture, which we denote $\mathcal{D}_{1m}$. This diffusion coefficient is an appropriate average value for the various species in the mixture [see also Eq. (1.119)]. The flow across the mouth of the tube can be assumed to maintain the concentration of CO there equal to $m_{1,e}$, its value in the bulk flow, but is not rapid enough to induce spurious convection currents inside the tube. Thus transport of CO down the tube is by diffusion only.

The Governing Equation and Boundary Conditions

For a timewise steady state and no reactions in the gas phase (no homogeneous reactions), Eq. (1.20) applies,

$$\dot{m}_1 = j_1 A = \text{Constant}$$

or

$$\frac{dj_1}{dz} = 0$$

Substituting from Fick's law, Eq. (1.17), with $\mathcal{D}_{1m}$ replacing $\mathcal{D}_{12}$,

$$\frac{d}{dz}\left(-\rho\mathcal{D}_{1m}\frac{dm_1}{dz}\right) = 0 \tag{1.31}$$

We now assume that the system is approximately isothermal so that $\mathcal{D}_{1m}$ is approximately constant, as is the mixture density ρ, since the range of species molecular weights is not large. Thus Eq. (1.31) becomes

$$\frac{d^2 m_1}{dz^2} = 0 \tag{1.32}$$

This differential equation is to be solved subject to the boundary conditions

$$z = 0: \; j_1 = -\rho\mathcal{D}_{1m}\frac{dm_1}{dz}\bigg|_{z=0} = -k''\rho m_1|_{z=0} \tag{1.33a}$$

$$z = L: \; m_1 = m_{1,e} \tag{1.33b}$$

The boundary condition at $z = 0$ states that the rate at which CO diffuses to the surface must equal the rate at which it is consumed in the chemical reaction. It is a *third-kind* boundary condition.

Solution for the Concentration Profile and Reaction Rate

Integrating Eq. (1.32) yields a linear concentration profile,

$$m_1 = C_1 z + C_2$$

Application of the boundary conditions gives

$$-\rho \mathcal{D}_{1m} C_1 = -k'' \rho C_2$$

$$m_{1,e} = C_1 L + C_2$$

Solving for C_1 and C_2 and substituting back gives the concentration profile as

$$\frac{m_1}{m_{1,e}} = \frac{1 + k'' z / \mathcal{D}_{1m}}{1 + k'' L / \mathcal{D}_{1m}} \tag{1.34}$$

The rate of reaction may be calculated most easily from Eq. (1.33a):

$$j_1 = -k'' \rho m_1 |_{z=0} = -k'' \rho m_{1,s} = \frac{-k'' \rho m_{1,e}}{1 + k'' L / \mathcal{D}_{1m}}$$

Dividing by $\rho k''$ and dropping the minus sign from here on, we obtain the *removal rate* for species 1, $\dot{m}_1 = -j_1 A$, where A is the tube cross-sectional area:

$$\dot{m}_1 = \frac{m_{1,e}}{1/\rho k'' A + L/\rho \mathcal{D}_{1m} A} \tag{1.35}$$

A circuit representation is given in Fig. 1.12a and shows the kinetics and diffusion resistances in series. Notice that the driving potential is $m_{1,e} = (m_{1,e} - 0)$. The zero potential appears because we assumed that the backward reaction ($2CO_2 \rightarrow 2CO + O_2$) is negligibly slow; that is, we wrote the reaction rate as $\rho k'' m_{1,s} - 0 = \rho k'' m_{1,s}$.

Two limiting cases can be distinguished, one in which the total resistance is dominated by the kinetics resistance, and the other in which the total resistance is dominated by the diffusion resistance.

1. *Rate-controlled limit.* If the reaction rate is slow, that is, $k'' \ll \mathcal{D}_{1m}/L$, the reaction is said to be *rate controlled.* From Eq. (1.35),

$$\dot{m}_1 = k'' \rho m_{1,e} A$$

and $m_{1,s} \rightarrow m_{1,e}$ in this limit.

2. *Diffusion-controlled limit.* If the reaction rate is fast, that is, $k'' \gg \mathcal{D}_{1m}/L$, the reaction is said to be *diffusion controlled.* From Eq. (1.35),

$$\dot{m}_1 = \frac{\rho \mathcal{D}_{1m} A}{L} m_{1,e}$$

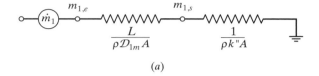

(a)

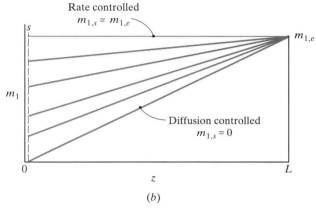

(b)

Figure 1.12 Heterogeneous catalysis with one-dimensional diffusion. (a) Mass flow circuit. (b) Concentration profiles.

and it is seen that the rate at which the reaction proceeds is governed by the diffusion of species 1 to the catalyst. Also, $m_{1,s} \to 0$ in this limit. Concentration profiles are shown in Fig. 1.12b.

If the reaction is rate controlled, it can be sped up by increasing k'', perhaps by increasing the temperature: decreasing the diffusion resistance by decreasing L will have no effect. Conversely, if the reaction is diffusion controlled, it can be sped up by decreasing L; increasing k'' will have no effect. The concept of a reaction being rate controlled or diffusion controlled is relevant to all types of chemical reactions. For example, the metal worker of ancient times used bellows to blow air over his bed of burning charcoal, because the reaction was close to being diffusion controlled, and the reaction rate and temperature could be increased by reducing the diffusion resistance for oxygen transport to the carbon surface. On the other hand, reactions in automobile catalytic converters are close to being rate controlled, and it is important to operate these converters at high temperatures. A recent development is for automobiles to have two converters. One is located very close to the exhaust manifold, where it can heat up rapidly after the motor is started. A second converter is in a more convenient location, underneath the passenger compartment. This design is very effective in reducing pollutant emissions in the first minutes of operation of the automobile.

Example 1.4   Catalytic Conversion of Carbon Monoxide

An experimental rig is set up to verify the theory described in Section 1.3.3. In a particular test, the tube is of length 10 cm and cross-sectional area 0.785 cm^2, the density of the gas mixture is 0.442 kg/m^3, and the diffusion coefficient of CO in the mixture is 97.8×10^{-6} m^2/s. When the mass fraction of CO in the gas mixture flowing over the tube is 0.001, the rate of removal of CO is measured to be 0.01855 μg/s. Determine the rate constant of the reaction on the particular catalyst used.

Solution

Given: Removal rate of CO due to catalytic reaction.
Required: Rate constant for the reaction.
Assumptions:

1. The flow of gas over the tube does not induce convection inside the tube.

2. The reaction is first-order.

Equation (1.35) gives the rate of removal of CO, $\dot{m}_1$. It can be solved for k'' to give

$$\frac{1}{k''} = \frac{\rho A m_{1,e}}{\dot{m}_1} - \frac{L}{\mathcal{D}_{1m}}$$

$$= \frac{(0.442)(0.785 \times 10^{-4})(0.001)}{(1.855 \times 10^{-11})} - \frac{0.1}{97.8 \times 10^{-6}}$$

$$= (1.870 - 1.022)10^3 = 0.848 \times 10^3$$

Hence $k'' = 1.179 \times 10^{-3}$ m/s.

Notice that this should be a fairly reliable determination of k'', since the kinetics and diffusion resistances are nearly of the same magnitude, namely,

$$\frac{1}{\rho k'' A} = \frac{1}{(0.442)(1.179 \times 10^{-3})(0.785 \times 10^{-4})} = 2.44 \times 10^7 \text{ (kg/s)}^{-1}$$

$$\frac{L}{\rho \mathcal{D}_{1m} A} = \frac{0.1}{(0.442)(97.8 \times 10^{-6})(0.785 \times 10^{-4})} = 2.95 \times 10^7 \text{ (kg/s)}^{-1}$$

Comments

1. If the diffusion resistance is much larger than the kinetics resistance, it is impossible to determine k'' accurately; in fact, if the objective is to determine rate constants accurately, then the diffusion resistance is usually made so small as to be negligible.

2. Suggest a possible technique for measuring the rate of removal of CO in this experiment.

3. A good textbook on chemical kinetics will give descriptions of methods commonly used to measure chemical kinetics data.   ▲

1.4 MASS CONVECTION

The terms **mass convection** or **convective mass transfer** are generally used to describe the process of mass transfer between a surface and a moving fluid, as shown in Fig. 1.13. The surface may be that of a falling water film in an air humidifier, of a coke particle in a gasifier, or of a silica-phenolic heat shield protecting a reentry vehicle. As was the case for heat convection, the flow can be *forced* or *natural, internal* or *external*, and *laminar* or *turbulent*. In addition, the concept of whether the mass transfer rate is *low* or *high* plays an important role: when mass transfer rates are low, there is a simple analogy between heat transfer and mass transfer that can be efficiently exploited in the solution of engineering problems.

1.4.1 The Mass Transfer Conductance

Analogous to convective heat transfer, the rate of mass transfer by convection is usually a complicated function of surface geometry and s-surface composition, the fluid composition and velocity, and fluid physical properties. For simplicity, we will restrict our attention to fluids that are either binary mixtures or solutions, or situations in which, although more than two species are present, diffusion can be adequately described using effective binary diffusion coefficients, as was discussed in Section 1.3.3. Referring to Fig. 1.13, we define the **mass transfer conductance** of species 1, g_{m1}, by the relation

$$j_{1,s} = g_{m1}\Delta m_1; \quad \Delta m_1 = m_{1,s} - m_{1,e} \tag{1.36}$$

and the units of g_{m1} are seen to be the same as for mass flux [kg/m^2 s]. Equation (1.36) is of a similar form to Newton's law of cooling, Eq. (1.20) of *Heat Transfer*, which defined the heat transfer coefficient h_c. Why we do not use a similar name and notation (e.g., mass transfer coefficient and h_m) will become clear later. On a molar basis, we define the **mole transfer conductance** of species 1, G_{m1}, by a corresponding relation,

$$J_{1,s} = G_{m1}\Delta x_1; \quad \Delta x_1 = x_{1,s} - x_{1,e} \tag{1.37}$$

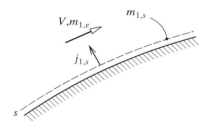

Figure 1.13 Notation for convective mass transfer into an external flow.

However, it usually proves advantageous to analyze convective transfer on a mass basis, rather than on a molar basis, so that Eq. (1.37) will be used infrequently.

1.4.2 Low Mass Transfer Rate Theory

To introduce the concept of a low mass transfer rate, we consider, as an example, the evaporation of water into air, as shown in Fig. 1.14. The water–air interface might be the surface of a water reservoir, or the surface of a falling water film in a cooling tower or humidifier. In such situations the mass fraction of water vapor in the air is relatively small; the highest value is at the s-surface, but even if the water temperature is as high as $50°C$, the corresponding value of $m_{H_2O,s}$ at 1 atm total pressure is only 0.077. From Eq. (1.36) the driving potential for diffusion of water vapor away from the interface is $\Delta m_1 = m_{1,s} - m_{1,e}$ and is small compared to unity, even if the free-stream air is very dry such that $m_{1,e} \simeq 0$. We then say that the mass transfer rate is *low* and the rate of evaporation of the water can be approximated as $j_{1,s}$; for a surface area A,

$$\dot{m}_1 \simeq j_{1,s} A \tag{1.38}$$

In contrast, if the water temperature approaches its boiling point, $m_{1,s}$ is no longer small, and of course, in the limit of $T_s = T_{BP}$, $m_{1,s} = 1$. The resulting driving potential for diffusion, Δm_1, is then large, and we say that the mass transfer rate is *high*. Then the evaporation rate cannot be calculated from Eq. (1.38), as will be explained in Section 2.2. For water evaporation into air, the error incurred in using low mass transfer rate theory is approximately $(1/2)\Delta m_1$, and a suitable criterion for application of the theory to engineering problems is $\Delta m_1 < 0.1$ or 0.2. A more general criterion is developed in Section 2.4.

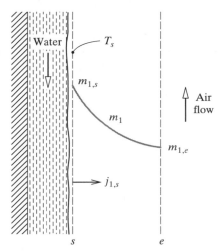

Figure 1.14 Schematic of evaporation from a falling water film into an air stream.

A large range of engineering problems can be adequately analyzed assuming low mass transfer rates. These problems include cooling towers and humidifiers, as mentioned above, gas absorbers for sparingly soluble gases, and catalysis. In the case of catalysis, the *net* mass transfer rate is actually zero. Reactants diffuse toward the catalyst surface and the products diffuse away, but the catalyst only promotes the reaction and is not consumed. On the other hand, problems that are characterized by high mass transfer rates include condensation of steam containing a small amount of noncondensable gas, as occurs in most power plant condensers; combustion of volatile liquid hydrocarbon fuel droplets in diesel engines and oil-fired power plants; and ablation of phenolic-based heat shields on reentry vehicles. In this chapter we will restrict our attention to low mass transfer rate problems.

1.4.3 Dimensional Analysis

As an example of dimensional analysis of a mass convection problem, we consider flow through a particle bed, as shown in Fig. 1.15. Packed particle beds for heat transfer were discussed in Section 4.5.2 of *Heat Transfer*. Such beds are used widely for mass transfer operations; for example, some types of automobile catalytic converters are in the form of packed beds of catalyst pellets. The flow field in the bed is expected to depend on the upstream superficial velocity, V, the bed volume void fraction, ε_v, and the effective particle diameter, d_p (defined in Section 4.5.2 of *Heat Transfer*), as well as fluid density, ρ, and dynamic viscosity, μ. The locally averaged mass transfer conductance g_{m1} is expected to depend on the aforementioned variables and, in addition, the diffusion coefficient $\mathcal{D}_{12}$ for a binary mixture. The effect of bed diameter D will be negligible for $D \gg d_p$ and is ignored. Thus, the functional dependence of g_{m1} is

$$g_{m1} = f(V, d_p, \rho, \mu, \mathcal{D}_{12}, \varepsilon_v)$$

From Appendix B of *Heat Transfer* the SI units of these variables are

g_{m1}	V	d_p	ρ	μ	$\mathcal{D}_{12}$	ε_v
$\dfrac{\text{kg}}{\text{m}^2\,\text{s}}$	$\dfrac{\text{m}}{\text{s}}$	m	$\dfrac{\text{kg}}{\text{m}^3}$	$\dfrac{\text{kg}}{\text{m s}}$	$\dfrac{\text{m}^2}{\text{s}}$	—

There are three primary dimensions or units—kg, m, and s—and seven variables. Using the Buckingham pi theorem, we expect $7 - 3 = 4$ independent dimensionless groups. We write

$$\Pi = g_{m1}^a V^b d_p^c \rho^d \mu^e \mathcal{D}_{12}^f \varepsilon_v^g$$

and the sums of exponents of each primary dimension are

$$\text{kg: } a + d + e = 0$$

$$\text{m: } -2a + b + c - 3d - e + 2f = 0$$

$$\text{s: } -a - b - e - f = 0$$

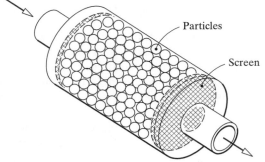

Figure 1.15 Flow through a packed particle bed.

The **Reynolds number** $\rho V d_p / \mu = \Pi_1$ can be found as for forced-convection heat transfer. We are free to choose values for four exponents; to obtain g_{m1} in a dimensionless group we set $a = 1, b = 0, e = 0, g = 0$:

$$1 + d = 0 \qquad \text{hence } d = -1, f = -1, c = 1$$

$$-2 + c - 3d + 2f = 0$$

$$-1 - f = 0 \qquad \Pi_2 = \frac{g_{m1} d_p}{\rho \mathcal{D}_{12}}$$

which is the **Sherwood number**, Sh (also called the mass transfer Nusselt number). The third group is found by seeking a fluid properties group that affects mass transfer in the fluid. Setting $e = 1, a = 0, b = 0, g = 0$,

$$d + 1 = 0 \qquad \text{hence } d = -1, f = -1, c = 0$$

$$c - 3d - 1 + 2f = 0$$

$$-1 - f = 0 \qquad \Pi_3 = \frac{\mu}{\rho \mathcal{D}_{12}}$$

which is the **Schmidt number**, Sc. The Schmidt number can also be written as $\nu / \mathcal{D}_{12}$, where $\nu = \mu / \rho$ is the kinematic viscosity. The last independent group is simply the dimensionless volume void fraction, ε_v. The result of the dimensional analysis can be written as

$$\text{Sh} = f(\text{Re}, \text{Sc}, \varepsilon_v) \qquad (1.39)$$

The Sherwood number can be viewed as a dimensionless mass transfer conductance. In general,

$$\text{Sh} = \frac{g_{m1} L}{\rho \mathcal{D}_{12}} \qquad (1.40)$$

where L is an appropriate characteristic length. As was the case for heat transfer, it is sometimes convenient to use a Stanton number. The **mass transfer Stanton number** is related to the Sherwood number as

$$\text{St}_m = \frac{\text{Sh}}{\text{ReSc}}$$

Hence

$$\text{St}_m = \frac{g_{m1}}{\rho V} \tag{1.41}$$

Molar equivalents of Eqs. (1.40) and (1.41) are

$$\text{Sh} = \frac{G_{m1}L}{c\mathcal{D}_{12}} \tag{1.42}$$

$$\text{St}_m = \frac{G_{m1}}{cV} \tag{1.43}$$

Dimensional analysis of natural-convection mass transfer leads to the result

$$\text{Sh} = f(\text{Gr}, \text{Sc}) \tag{1.44}$$

where the Grashof number is expressed in terms of density difference across the boundary layer,

$$\text{Gr} = \frac{(\Delta\rho/\rho)gL^3}{\nu^2}$$

The term $(\Delta\rho/\rho)$ cannot be replaced by $\beta\Delta T$, as was done for heat transfer, since the density difference $\Delta\rho$ is due to a concentration gradient.[2]

Table 1.2 gives selected data for the Schmidt number for binary gas mixtures and dilute liquid solutions. Additional data may be found in Appendix A. Notice that the Schmidt number is about unity for gas mixtures, whereas for liquids it is large. Strictly speaking, the Schmidt number should be subscripted the same as the diffusion coefficient—that is, Sc_{12} or Sc_{1m}, as the case might be. Also, the Sherwood and Stanton numbers should be subscripted the same as the mass transfer conductance— Sh_1, for example. Such practices will be followed only when it is necessary to avoid ambiguity.

1.4.4 The Analogy between Convective Heat and Mass Transfer

A close analogy exists between convective heat and convective mass transfer owing to the fact that conduction and diffusion in a fluid are governed by physical laws of identical mathematical form—that is, Fourier's and Fick's laws, respectively. As a result, in many circumstances the Sherwood or mass transfer Stanton number can be obtained in a simple manner from the Nusselt or heat transfer Stanton number for the same flow conditions. Indeed, in most gas mixtures Sh and St_m are nearly equal to their heat transfer counterparts. For dilute mixtures and solutions and low mass transfer rates, the rule for exploiting the analogy is simple: *The Sherwood or Stanton number is obtained by replacing the Prandtl number by the Schmidt number in the*

---

[2] If there is simultaneous natural-convection heat and mass transfer—for example, cold water evaporating into warm air—there is a density difference $\Delta\rho$ due to both concentration and temperature gradients. Dimensional analysis then yields $\text{Sh} = f(\text{Gr}, \text{Sc}, \text{Pr})$. See Example 1.8.

Table 1.2 Selected Schmidt-number data (the first species of each pair is in dilute concentration).

System	Temperature K	Schmidt Number
Dilute Gas Mixtures		
H_2-air	300	0.20
H_2O-air	300	0.61
H_2O-air	400	0.55
O_2-air	300	0.83
O_2-air	1000	0.77
CO_2-air	300	1.00
Heptane-air	300	2.1
Air-H_2O	300	0.48
Naphthalene-air	300	2.35
Dilute Liquid Solutions		
He-water	300	120
H_2O-water	300	340
SO_2-water	300	520
Isopropanol-water	300	730
Sucrose-water	300	1670
CO_2-water	287	813
CO_2-water	298	458
CO_2-water	313	236
CO_2-methanol	300	83
CO_2-ethanol	300	360
CO_2-propanol	300	890
CO_2–aqueous ethylene glycol solution, 0.20 mole fraction glycol	300	2700

appropriate heat transfer correlation. For example, in the case of fully developed turbulent flow in a smooth pipe, Eq. (4.44) of *Heat Transfer* is

$$\text{Nu}_D = 0.023 \text{Re}_D^{0.8} \text{Pr}^{0.4}; \quad \text{Pr} > 0.5$$

which for mass transfer becomes

$$\text{Sh}_D = 0.023 \text{Re}_D^{0.8} \text{Sc}^{0.4}; \quad \text{Sc} > 0.5 \tag{1.45}$$

Also, for natural convection from a heated horizontal surface facing upward, Eqs. (4.95) and (4.96) of *Heat Transfer* give

$$\overline{\text{Nu}} = 0.54(\text{Gr}_L \text{Pr})^{1/4}; \quad 10^5 < \text{Gr}_L \text{Pr} < 2 \times 10^7 \quad \text{(laminar)}$$

$$\overline{\text{Nu}} = 0.14(\text{Gr}_L \text{Pr})^{1/3}; \quad 2 \times 10^7 < \text{Gr}_L \text{Pr} < 3 \times 10^{10} \quad \text{(turbulent)}$$

which for isothermal mass transfer with $\rho_s < \rho_e$ become

$$\overline{\text{Sh}} = 0.54(\text{Gr}_L \text{Sc})^{1/4}; \quad 10^5 < \text{Gr}_L \text{Sc} < 2 \times 10^7 \quad \text{(laminar)} \tag{1.46a}$$

$$\overline{\text{Sh}} = 0.14(\text{Gr}_L \text{Sc})^{1/3}; \quad 2 \times 10^7 < \text{Gr}_L \text{Sc} < 3 \times 10^{10} \quad \text{(turbulent)} \tag{1.46b}$$

With evaporation, the condition $\rho_s < \rho_e$ will be met when the evaporating species has a smaller molecular weight than the ambient species—for example, water evaporating into air. Mass transfer correlations can be written down in a similar manner for almost all the heat transfer correlations given in Chapter 4 of *Heat Transfer*.[3] There are some exceptions: for example, there are no fluids with Schmidt numbers much less than unity, and thus there are no mass transfer correlations corresponding to those given for heat transfer to liquid metals with $\text{Pr} \ll 1$. In most cases it is important for the wall boundary conditions to be of analogous form—for example, laminar flow in ducts. A uniform wall temperature corresponds to a uniform concentration $m_{1,s}$ along the s-surface, whereas a uniform heat flux corresponds to a uniform diffusive flux $j_{1,s}$.

A situation often encountered in mass transfer equipment that has no analogy in Chapter 4 of *Heat Transfer* is gas absorption in a falling film column. If the film falls down the inside of a vertical tube, or down parallel plates, and if there is an upward counterflow of gas, the following correlation of experimental data is recommended for the gas-side Sherwood number [4]:

$$\text{Sh}_{D_h} = 0.00814\text{Re}_{D_h}^{0.83}\text{Sc}^{0.44}(4\Gamma/\mu_l)^{0.15} \tag{1.47}$$

where Γ [kg/m s] is the film flow rate per unit width (see Section 7.2.2 of *Heat Transfer*). Equation (1.47) is valid for $2000 < \text{Re}_{D_h} < 20{,}000$, and $4\Gamma/\mu_l < 1200$. Equation (1.47) gives higher values of the Sherwood number than those given by Eq. (1.45) for turbulent flow in a smooth tube. The effect of waves on the film surface is to increase the convective transfer, much in the same way as does roughness of a solid surface.

Further insight into this analogy between convective heat and mass transfer can be seen by writing out Eqs. (4.44) of *Heat Transfer* and (1.45) above as, respectively,

$$\frac{(h_c/c_p)D}{k/c_p} = 0.023\text{Re}_D^{0.8}\left(\frac{\mu}{k/c_p}\right)^{0.4} \tag{1.48a}$$

$$\frac{g_m D}{\rho\mathcal{D}_{12}} = 0.023\text{Re}_D^{0.8}\left(\frac{\mu}{\rho\mathcal{D}_{12}}\right)^{0.4} \tag{1.48b}$$

When cast in this form, the correlations show that the property combinations k/c_p and $\rho\mathcal{D}_{12}$ play analogous roles; these are **exchange coefficients** for heat and mass, respectively, both having units [kg/m s], which are the same as those for dynamic viscosity μ. Also, it is seen that the ratio of heat transfer coefficient to specific heat plays an analogous role to the mass transfer conductance and has the same units [kg/m² s]. Thus it is appropriate to refer to the ratio h_c/c_p as the **heat transfer conductance**, g_h, and for this reason we chose not to refer to g_m as the mass transfer *coefficient*.

---

[3] In chemical engineering practice, the analogy between convective heat and mass transfer is widely used in a form recommended by Chilton and Colburn in 1934, namely, $\text{St}_m/\text{St} = (\text{Sc}/\text{Pr})^{-2/3}$. The Chilton–Colburn form is of adequate accuracy for most external forced flows but is inappropriate for fully developed laminar duct flows.

The Mass Transfer Biot Number

The heat transfer Biot number Bi was first introduced in Section 1.5.1 of *Heat Transfer*. Equation (1.40) of that section defined the Biot number as the ratio of the internal conduction resistance to the external convection resistance. When the Biot number is small (<0.1), the lumped thermal capacity model can be used to calculate the transient temperature response of a solid. Subsequently, the Biot number was introduced in Section 3.4.3 of *Heat Transfer*: in the analysis of convective cooling of slabs, cylinders, and spheres, the Biot number appears naturally when the convection boundary condition is put in dimensionless form, for example, Eq. (3.66) of *Heat Transfer*. For a slab, $\text{Bi} = h_c L / k$, where L is the slab half width.

Analogously, the mass transfer Biot number Bi_m is defined as the ratio of internal diffusion resistance to external convection resistance. However, the mathematical form of Bi_m is not the same as Bi because, whereas temperatures are continuous across phase interfaces, concentrations are generally discontinuous. For example, consider a spray of small water droplets injected into an air flow containing a small amount of a gas, species 1. If the droplets are too small for any liquid circulation to occur, the gas absorption process can be modeled as transient diffusion in a sphere with a convective boundary condition. We will denote water as species 2, and work on a molar basis. Following chemical engineering practice, gas- and liquid-phase mole fractions are denoted y and x, respectively: the advantage of making this distinction will become apparent in the analysis that follows. The initial concentration of species 1 dissolved in the droplet is $x_{1,0}$, and the bulk concentration of species 1 in the air is $y_{1,e}$. Referring to Fig. 1.16, the interface boundary condition is

$$-c\mathcal{D}_{12} \left. \frac{\partial x_1}{\partial r} \right|_{r=R} = \mathcal{G}_m (y_{1,s} - y_{1,e}) \tag{1.49}$$

To obtain a dimensionless form of Eq. (1.49) that is analogous to the corresponding heat transfer boundary condition, we introduce $\eta = r/R$ as a dimensionless radial coordinate; also, we must choose a dimensionless concentration that varies from an

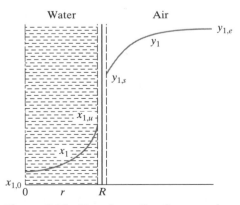

Figure 1.16 Gas absorption from an air mixture into a water droplet: concentration profiles for species 1.

initial value of 1 when $x_1 = x_{1,0}$, to 0 when the droplet is in equilibrium with the gas phase. At equilibrium the concentration of the dissolved gas in the droplet will be uniform and, from Henry's law, Eq. (1.13), will equal $y_{1,e}/\text{He}$. Thus, we define the dimensionless concentration as $\phi = (x_1 - y_{1,e}/\text{He})/(x_{1,0} - y_{1,e}/\text{He})$. Substituting in Eq. (1.49) gives

$$-\left.\frac{\partial \phi}{\partial \eta}\right|_{\eta=1} = \frac{\mathcal{G}_m R}{c \mathcal{D}_{12}} \frac{y_{1,s} - y_{1,e}}{x_{1,0} - y_{1,e}/\text{He}}$$

$$= \frac{\mathcal{G}_m \text{He} R}{c \mathcal{D}_{12}} \frac{x_{1,u} - y_{1,e}/\text{He}}{x_{1,0} - y_{1,e}/\text{He}}$$

and thus

$$-\left.\frac{\partial \phi}{\partial \eta}\right|_{\eta=1} = \text{Bi}_m \phi \mid_{\eta=1}; \quad \text{Bi}_m = \frac{\mathcal{G}_m \text{He} R}{c \mathcal{D}_{12}} \tag{1.50}$$

Table A.21 shows that Henry constants for sparingly soluble gases such as O_2 and CO_2 are very large: thus Bi_m for such gases also tend to be large. The convective resistance is then negligible, and the absorption process is controlled by diffusion in the liquid droplet. Clearly, a lumped capacity model is inappropriate for such situations. For a highly soluble gas such as ammonia, an effective Henry number calculated from the solubility data in Table A.22 is of order unity, and the convective resistance may not be negligible.

We have used gas absorption into a liquid to develop the concept of a mass transfer Biot number because the solubility data are conveniently represented by Henry's law. In the case of gas absorption into a solid, solubility data in the form of the solubility $S[\text{m}^3$ solute gas (STP)/m^3 solid atm] are given in Table A.23. If molar concentration is used as the measure of concentration in the solid, it can be easily shown that the appropriate definition of the Biot number is $\text{Bi} = \mathcal{G}_m (22.414/SP [\text{atm}]) R/\mathcal{D}_{12}$. Data for equilibrium between porous solids and a gas phase—for example, between wet wood and moist air, or between silica gel and moist air—are often given in graphical form or as analytical expressions. An effective Biot number at the moisture content of concern can be obtained from such data.

The Computer Program MCONV

MCONV calculates mass transfer rates according to the low mass transfer rate theory presented in Section 1.4.2, for the 12 flow configurations listed in Table 1.3. The transferred species is assumed to be in dilute concentration in air, so that the properties of the mixture can be approximated by pure air values at the mean film temperature. Calculations can be made on either a mass or a molar basis. Two options are available.

1. *Transfer of an inert chemical species i*. A menu of 16 chemical species is available, with Schmidt numbers taken from Table A.17b.

2. *Evaporation of water or condensation of water vapor*. In this option the s-surface composition is calculated as the equilibrium value corresponding to the input surface temperature. In addition to the mass transfer rate, the convective and evaporative heat transfer rates are also calculated (see Section 1.5.1).

Table 1.3 Menu of flow configurations in MCONV.

Item Number	Flow Configuration	Heat Transfer Table 4.10 Designation
1	Turbulent flow in smooth ducts	1
2	Laminar flow in a pipe	2
3	Laminar flow between parallel plates	3
4	Laminar boundary layer on a flat plate	4
5	Turbulent boundary layer on a smooth flat plate	5
6	Flow across a cylinder	6
7	Flow across a sphere	7
8	Laminar natural convection on a vertical wall	8
9	Turbulent natural convection on a vertical wall	9
10	Natural convection on a heated horizontal plate facing down	12
11	Natural convection on a heated horizontal plate facing up	13
12	Counterflow of gas and a falling water film	Eq. (1.47)

The use of option 1 is illustrated in Example 1.6, and the use of option 2 is illustrated in Example 1.8 of Section 1.5. Note that MCONV gives a warning if Δm_1 or Δx_1 is greater than 0.1, signifying that low mass transfer rate theory may be inadequate.

Example 1.5 Sublimation of a Naphthalene Model

A useful method for determining convective heat transfer coefficients on surfaces of complicated shape is to cast models of naphthalene and expose the models to an air flow for a specified duration: the local mass loss by sublimation is obtained by weighing, or from the surface recession. In a particular experiment, the recession in one hour at the location of concern was 112 μm. The free-stream air velocity was 1 m/s, the pressure 1 bar, and both the model and the air temperatures were 300 K. Determine the convective heat transfer coefficient under the same air flow conditions if it is known that the flow is laminar. Properties of naphthalene include:

Solid density, $\rho = 1075$ kg/m^3

Molecular weight, $M = 128.2$ kg/kmol

Schmidt number in air at 300 K, Sc$_i$ $= 2.35$

Vapor pressure, $P = 3.631 \times 10^{13} \exp(-8586/T \text{ [K]})$ Pa

Solution

Given: Rate of recession of a naphthalene surface in a laminar air flow.
Required: Convective heat transfer coefficient for a similar air flow.
Assumptions:

1. Low mass transfer rate theory.
2. Laminar flow.

The rate of sublimation is equal to the rate at which naphthalene vapor can diffuse away from the surface. Denoting naphthalene as species 1, the recession rate $\dot{s}$ is

$$\dot{s} = \frac{j_{1,s}}{\rho_{\text{solid}}}; \quad j_{1,s} = g_{m1}(m_{1,s} - m_{1,e}), \text{ where } m_{1,e} = 0$$

Solving,

$$g_{m1} = \frac{\rho_{\text{solid}}\dot{s}}{m_{1,s}}$$

The mass fraction $m_{1,s}$ is obtained from the partial pressure as

$$m_{1,s} \simeq \left(\frac{P_{1,s}}{P}\right)\left(\frac{M_1}{M_{\text{air}}}\right)$$

since species 1 is in very small concentration.

$$P_{1,s} = 3.631 \times 10^{13} \exp(-8586/300) = 13.51 \text{Pa}$$

Hence

$$m_{1,s} = \frac{13.51}{10^5}\frac{128.2}{29} = 5.97 \times 10^{-4}$$

and

$$g_{m1} = \frac{(1075 \text{ kg/m}^3)(112 \times 10^{-6} \text{ m/h})(1/3600 \text{ h/s})}{(5.97 \times 10^{-4})} = 0.0560 \text{ kg/m}^2 \text{ s}$$

The mass transfer Stanton number is then

$$\text{St}_m = \frac{g_{m1}}{\rho V} = \frac{0.0560 \text{ kg/m}^2 \text{ s}}{(1.163 \text{ kg/m}^3)(1\text{m/s})} = 0.0482$$

where ρ has been evaluated for pure air at 300 K and 1 bar.

Air flow

Solid naphthalene

We do not know the precise form of an appropriate heat or mass transfer correlation for this configuration; however, Chapter 4 of *Heat Transfer* shows that for forced-convection laminar boundary layer flows of air, $\text{St} \propto \text{Pr}^{-2/3}$. Thus, by analogy, $\text{St}_m \propto \text{Sc}^{-2/3}$, and hence

$$\text{St} = \text{St}_m(\text{Pr}/\text{Sc})^{-2/3} = (0.0482)(0.69/2.35)^{-2/3} = 0.109$$

$$h_c = \rho c_p V \text{St} = (1.163 \text{ kg/m}^3)(1005 \text{ J/kg K})(1\text{m/s})(0.109) = 127 \text{ W/m}^2 \text{ K}$$

where c_p has been evaluated for pure air at 300 K.

Comments

1. As a shortcut we could simply use $(h_c/c_p)/\mathcal{G}_{m1} = (\mathrm{Pr}\,/\mathrm{Sc})^{-2/3}$.

2. Some examples of the use of this method to determine convective transfer coefficients may be found in references [5] and [6].

3. In older literature concerning sublimation of naphthalene, a value of $\mathrm{Sc} = 2.5$ is often seen; also, there are slight variations in the vapor pressure data. ▲

Example 1.6   Scrubbing of Sulfur Dioxide from a Process Effluent

In a scrubber to remove SO_2 from a process effluent stream, the gas mixture flows countercurrent to a falling film of an aqueous alkaline solution in vertical tubes. The gas velocity is 1.8 m/s, the tube diameter is 5 cm, the liquid feed rate is 110 kg/h, and the column height is 4 m. The solubility of SO_2 in the alkaline solution is so high that the concentration of SO_2 in the gas phase adjacent to the liquid surface can be taken to be approximately zero, and the mass transfer process is "gas-side controlled." At a particular location in the column the pressure is 1 atm, the temperature is 300 K, and the bulk mole fraction of SO_2 in the gas stream is 0.03. Calculate the rate of removal of SO_2 per unit area of liquid-gas interface.

Solution

Given: Gas containing SO_2 flowing countercurrent to a falling film of aqueous alkaline solution.
Required: Rate of removal of SO_2 per unit area.
Assumptions:

1. Gas-side controlled absorption with $x_{SO_2,s} \simeq 0$.

2. Approximate liquid properties by those for pure water.

3. Approximate gas properties by those for pure air.

Since $x_{SO_2,s} \simeq O$, the rate of mass transfer is simply the rate at which SO_2 can be transported from the bulk gas stream to the s-surface; transport in the liquid phase need not be considered. For this gas-side controlled absorption, Eq. (1.37), $J_{1,s} = \mathcal{G}_{m1}\Delta x_1$, gives the rate of SO_2 absorption per unit area of interface. In order to calculate the mole transfer conductance $\mathcal{G}_{m1}$, we first check the liquid and gas flow Reynolds numbers. For water at 300 K, $\mu = 8.67 \times 10^{-4}$kg/m s. The water flow rate per unit perimeter is

$$\Gamma = \dot{m}/\pi D = (110)(1/3600)/(\pi)(0.05)$$
$$= 0.1945 \text{ kg/m s}$$

and the film Reynolds number is

$$4\Gamma/\mu = (4)(0.1945)/(8.67 \times 10^{-4}) = 897$$

The film is in the wavy laminar regime (see Section 7.2.2 of *Heat Transfer*).

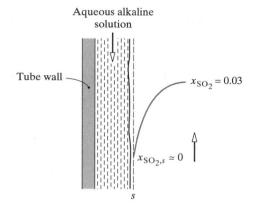

For air at 300 K and 1 atm, $\rho = 1.177 \text{ kg/m}^3$, $\nu = 15.66 \times 10^{-6} \text{ m}^2/\text{s}$, $c = \rho/M = 1.177/29 = 0.0406 \text{ kmol/m}^3$. The gas-phase Reynolds number based on tube diameter is

$$\text{Re}_D = (1.8)(0.05)/15.66 \times 10^{-6} = 5747$$

The flow is just turbulent. The effect of the waves on the film surface is to increase the gas-side convective transfer. From Eq. (1.47),

$$\text{Sh}_D = 0.00814 \text{Re}_D^{0.83} \text{Sc}^{0.44} (4\Gamma/\mu_l)^{0.15}$$

The diffusion coefficient of SO_2 in the mixture, $\mathcal{D}_{1m}$, is approximated by its value in air; from Table A.17a, $\mathcal{D}_{1m} = 12.6 \times 10^{-6} \text{ m}^2/\text{s}$, and the corresponding Schmidt number is

$$\text{Sc}_{1m} = \frac{\nu}{\mathcal{D}_{1m}} = \frac{15.66 \times 10^{-6}}{12.6 \times 10^{-6}} = 1.243$$

Thus

$$\text{Sh}_D = (0.00814)(5747)^{0.83}(1.243)^{0.44}(897)^{0.15} = 32.8$$

$$\mathcal{G}_{m1} = \frac{c\mathcal{D}_{1m}\text{Sh}_D}{D} = \frac{(0.0406 \text{ kmol/m}^3)(12.6 \times 10^{-6} \text{ m}^2/\text{s})(32.8)}{(0.05 \text{ m})}$$

$$= 3.36 \times 10^{-4} \text{ kmol/m}^2 \text{ s}$$

Then, from Eq. (1.37), with $\Delta x_1 = x_{1,s} - x_{1,b}$ for an internal flow,

$$J_{1,s} = \mathcal{G}_{m1}(x_{1,s} - x_{1,b}) = (3.36 \times 10^{-4})(0 - 0.03)$$

$$= -1.01 \times 10^{-5} \text{ kmol/m}^2 \text{ s}$$

where the negative sign indicates that the flux $J_{1,s}$ is toward the liquid surface— that is, SO_2 is being absorbed.

Solution using MCONV

The required input is:

$$\text{Configuration number} = 12$$

$$\text{Option} = 1 \text{ (transfer of chemical species } i)$$

$$\text{Chemical species number} = 7 \text{ (SO}_2)$$

1. Interface temperature, $T_s = 300$
2. Bulk temperature, $T_b = 300$
3. Pressure, $P = 1.013 \times 10^5$
4. Hydraulic diameter, $D_h = 0.05$

$$\text{Water flow rate per unit perimeter, } \Gamma = 0.1945$$

$$\text{Bulk velocity of air, } V = 1.8$$

5. Option 2 (mole fractions)

$$x_{1,s} = 0$$

$$x_{1,b} = 0.03$$

The output gives:

$$J_{1,s} = -1.01 \times 10^{-5} \text{ kmol/m}^2 \text{ s}$$

Comments

1. If Eq. (1.45) is used for the Sherwood number, the result is

$$\text{Sh} = 0.023 \text{Re}_D^{0.8} \text{Sc}^{0.4} = 0.023(5747)^{0.8}(1.243)^{0.4} = 25.5$$

which is 22% less than the value given by Eq. (1.47).

2. Check the complete output of MCONV.
3. Notice the similarity between gas-side controlled absorption and diffusion-controlled catalysis (Section 1.3.3). ▲

1.4.5 The Equivalent Stagnant Film Model

Figure 1.17 shows flow past a surface (for example, a salt, species 1, dissolving in water, species 2). The concentration of the salt at the s-surface is $m_{1,s}$, the equilibrium value obtained from solubility data; at the e-surface in the external flow, the salt concentration is $m_{1,e}$. From Eq. (1.36), the rate at which the salt is transferred into the water from a surface area A is

$$\dot{m}_1 = j_{1,s}A = g_{m1}A(m_{1,s} - m_{1,e})$$

If we now imagine the transfer process to be that of diffusion across an equivalent stagnant film of thickness δ_f, Eq. (1.21a) gives

$$\dot{m}_1 = \frac{\rho \mathcal{D}_{12}A}{\delta_f}(m_{1,s} - m_{1,e})$$

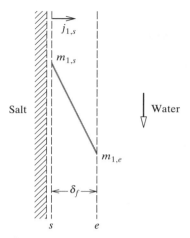

Figure 1.17 Equivalent
stagnant-film model for
convective mass transfer: a
salt dissolving in water.

Comparing these two expressions for $\dot{m}_1$ gives

$$\mathcal{g}_{m1} = \frac{\rho \mathcal{D}_{12}}{\delta_f} \tag{1.51}$$

and the corresponding diffusion resistance is $\delta_f / \rho \mathcal{D}_{12} A$.

Although the concept of an equivalent stagnant film is oversimplistic (there is, of course, some velocity profile in the fluid), it proves a useful model for the analysis of more complicated mass transfer problems. On a molar basis, the corresponding definition of an equivalent stagnant-film thickness is by the relation

$$\mathcal{G}_{m1} = \frac{c \mathcal{D}_{12}}{\delta_f} \tag{1.52}$$

As an example of the utility of the equivalent stagnant film concept, consider again the problem of heterogeneous catalysis analyzed in Section 1.3.3, except that now the reactants are flowing through a honeycomb matrix coated with the catalyst. If the actual process of convection and diffusion in the flow through the passages of the matrix is modeled as diffusion through a stagnant gas film of thickness δ_f, then the tube length L in the analysis of Section 1.3.3 can simply be replaced by δ_f to give

$$\dot{m}_1 = \frac{m_{1,e}}{1/\rho k'' A + \delta_f / \rho \mathcal{D}_{1m} A}$$

Since δ_f must be calculated from the mass transfer conductance using Eq. (1.51), this result is more conveniently written in terms of $\mathcal{g}_{m1}$ as

$$\dot{m}_1 = \frac{m_{1,e}}{1/\rho k'' A + 1/\mathcal{g}_{m1} A} \tag{1.53}$$

Notice how the equivalent stagnant-film thickness δ_f does not appear in the final result: the concept was used simply to relate the result of Section 1.3.3 to a convective situation. Equation (1.53) also can be derived directly as follows. Since the rate at which species CO is consumed equals the rate at which it is transported to the surface,

$$\dot{m}_1 = k'' \rho m_{1,s} A = -g_{m1}(m_{1,s} - m_{1,e})A$$

Solving for $m_{1,s}$ and substituting back to obtain $\dot{m}_1$ gives Eq. (1.53).

Example 1.7 Removal of Carbon Monoxide from an Automobile Exhaust

The packing for an automobile catalytic converter is in the form of a matrix with porous walls impregnated with catalyst. The passages are of square cross section with 1-mm sides, and the *effective* rate constant of the porous catalyst at the operating temperature of 800 K is 0.070 m/s. If the pressure in the reactor is 1.15×10^5 Pa and the mean molecular weight of the exhaust gases is 28, calculate the CO reduction rate at a location where the CO concentration is 0.187% by weight. Also calculate the equivalent stagnant-film thickness.

Solution

Given: Catalytic packing for CO removal from exhaust gases.
Required: CO reduction rate and equivalent stagnant-film thickness δ_f.
Assumptions:

1. Fully developed laminar flow in the matrix.

2. Approximate the CO diffusion coefficient by its value in air.

Equation (1.53) gives the rate at which CO is reduced:

$$\frac{\dot{m}_1}{A} = \frac{m_{1,e}}{1/\rho k'' + 1/g_{m1}}$$

$$g_{m1} = \frac{\rho \mathcal{D}_{1m}}{D_h} \mathrm{Sh}$$

Sh = 2.98 from Table 4.5 of *Heat Transfer* for fully developed laminar flow in a square duct.

$$D_h = 4A_c/\mathcal{P} = 4(1^2)/4(1) = 1 \text{ mm}$$

$$\rho = \frac{P}{(\mathcal{R}/M)T} = \frac{1.15 \times 10^5}{(8314/28)(800)} = 0.484 \text{ kg/m}^3$$

From Table A.17a, approximating $\mathcal{D}_{1m}$ as $\mathcal{D}_{CO_{air}}$,

$$\mathcal{D}_{1m} = (106 \times 10^{-6})(1.013 \times 10^5/1.15 \times 10^5) = 93.4 \times 10^{-6} \text{ m}^2/\text{s}$$

$$g_{m1} = (0.484)(93.4 \times 10^{-6})(2.98)/(1 \times 10^{-3}) = 0.135 \text{ kg/m}^2 \text{ s}$$

$$m_{1,e} = 0.00187$$

Substituting in Eq. (1.53),

$$\frac{\dot{m}_1}{A} = \frac{0.00187}{1/(0.484)(0.070) + 1/0.135} = 5.06 \times 10^{-5} \text{ kg/m}^2 \text{ s}$$

Also, from Eq. (1.51),

$$\delta_f = \frac{\rho \mathcal{D}_{1m}}{g_{m1}}, \quad \text{where } g_{m1} = \frac{\text{Sh}\rho \mathcal{D}_{1m}}{D_h}$$

Hence

$$\delta_f = \frac{D_h}{\text{Sh}} = \frac{1 \text{ mm}}{2.98} = 0.336 \text{ mm}$$

Comments

1. The effective rate constant takes into account the added surface area of the pores in a porous catalyst (see Section 1.6).

2. Physical chemists often simply guess a value for the equivalent stagnant-film thickness in a mass transfer process. This is a bad practice: it is relatively simple to obtain a reliable value from an appropriate correlation for the Sherwood number.   ▲

1.5 SIMULTANEOUS HEAT AND MASS TRANSFER

In many situations heat and mass transfer occur simultaneously at a phase interface. Whenever water evaporates into an air stream, the enthalpy of vaporization must be supplied by heat transfer from the bulk water, the air stream, or both. Thus, analysis of the performance of equipment such as cooling towers and humidifiers involves simultaneous consideration of heat and mass transfer processes. Combustion of a fuel oil droplet also requires simultaneous consideration of heat and mass transfer, since the rate of combustion depends both on the rate of diffusion of the fuel vapor and oxygen to the flame front, and on the rate of heat transfer from the flame to the droplet surface, which supplies the required enthalpy of vaporization. The rate at which an ablative heat shield loses mass depends on the rate of heat transfer to the shield. The combustion of carbon in the form of coke or coal also involves simultaneous heat and mass transfer: heat transfer considerations determine the temperature of the carbon and hence the mass transfer regime—that is, whether or not the reaction is rate or diffusion controlled—and the details of the kinetics mechanism and combustion products. In some simple situations engineering approximations can be made in order to *uncouple* the heat and mass transfer processes so that each can be analyzed separately: the problem of determining the heat loss from a well-stirred chemical dip tank is considered in Example 1.8. On the other hand, in most situations the heat and mass transfer processes are *coupled* in the sense that the equations governing the respective processes must be simultaneously solved. The wet- and dry-bulb psychrometer considered in Section 1.5.2 is such a situation, as are the combustion of a volatile hydrocarbon fuel droplet and the ablation of a heat shield.

Some simple examples of simultaneous heat and mass transfer involving low mass transfer rates are given in this section. Examples involving high mass transfer rates will be given in Chapter 2. The performance of exchangers that transfer both heat and mass is considered in Section 3.5. The use of the computer program MCONV, which was described in Section 1.4.4, is illustrated in Example 1.8.

1.5.1 Surface Energy Balances

Surface energy balances for heat transfer were introduced in Section 1.4 of *Heat Transfer*. We now extend this concept to situations where there is simultaneous mass transfer. An important problem is evaporation of a liquid, for example, evaporation of water into air, as shown in Fig. 1.18. With H_2O denoted as species 1, the steady-flow energy equation, Eq. (1.4) of *Heat Transfer*, applied to the control volume located between the u- and s-surfaces (Fig. 1.18a) requires that

$$\dot{m}\Delta h = \dot{Q}$$
$$\dot{m}(h_{1,s} - h_{1,u}) = A(q_{\text{cond}} - q_{\text{conv}} - q_{\text{rad}}) \tag{1.54}$$

where it has been recognized that only species 1 crosses the u- and s-surfaces. Also, the water has been assumed perfectly opaque so that all radiation is emitted or absorbed between the u-surface and the interface.[4]

If we restrict our attention to conditions for which low mass transfer rate theory is valid, we can write $\dot{m}/A \simeq j_{1,s} = g_{m1}(m_{1,s} - m_{1,e})$. Also, we can then calculate the convective heat transfer as if there were no mass transfer, and write $q_{\text{conv}} = h_c(T_s - T_e)$. Substituting in Eq. (1.54) with $q_{\text{cond}} = -k\partial T/\partial y\,|_u$, $h_{1,s} - h_{1,u} = h_{\text{fg}}$, and rearranging, gives

$$-k\left.\frac{\partial T}{\partial y}\right|_u = h_c(T_s - T_e) + g_{m1}(m_{1,s} - m_{1,e})h_{\text{fg}} + q_{\text{rad}} \tag{1.55}$$

It is common practice to refer to the convective heat flux $h_c(T_s - T_e)$ as the *sensible* heat flux, whereas the term $g_{m1}(m_{1,s} - m_{1,e})h_{\text{fg}}$ is called the *evaporative* or *latent* heat flux. Each of the terms in Eq. (1.55) can be positive or negative, depending on the particular situation. Also, the evaluation of the conduction heat flux at the u-surface, $-k\,\partial T/\partial y|_u$, depends on the particular situation. Four examples are shown in Fig. 1.18. For a water film flowing down a packing in a cooling tower (Fig. 1.18b), this heat flux can be expressed in terms of convective heat transfer from the bulk water at temperature T_L to the surface of the film, $-k\,\partial T/\partial y|_u = h_{cL}(T_L - T_s)$. If the liquid-side heat transfer coefficient h_{cL} is large enough, we can simply set $T_s \simeq T_L$, which eliminates the need to estimate h_{cL}. The evaporation process is then *gas-side controlled* (cf. Example 1.6). Figure 1.18c shows film condensation

---

[4] Strictly speaking, radiation is emitted or absorbed below the u-surface, since the u-surface is indefinitely close to the real interface. However, unless we are concerned about the detailed effect of emission and absorption on the temperature profile in the water very close to the interface, it is convenient to assume that the water is perfectly opaque so that heat is transferred across the u-surface by conduction only.

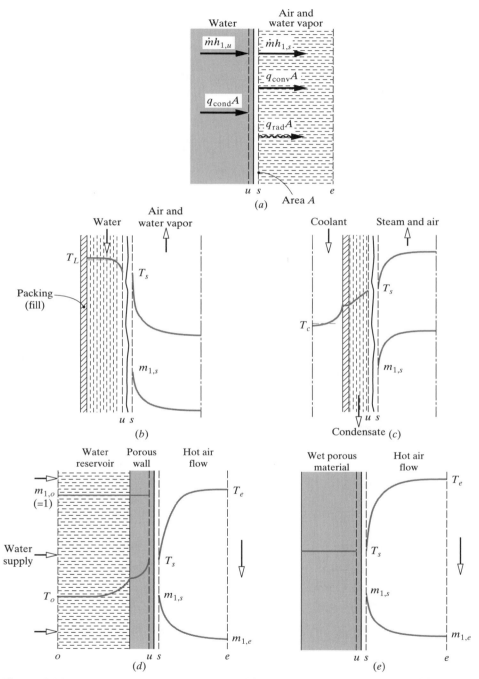

Figure 1.18   Evaporation of water into air. (a) The surface energy balance. (b) A falling water film in a cooling tower. (c) Film condensation from a steam–air mixture. (d) Sweat cooling. (e) Constant-rate drying of a porous material.

from a steam–air mixture on the outside of a vertical tube. In this case we can write $k\partial T/\partial y|_u = U(T_s - T_c)$, where T_c is the coolant bulk temperature. The overall heat transfer coefficient U includes the resistances of the condensate film, the tube wall, and the coolant. Sweat cooling is shown in Fig. 1.18d, with water from a reservoir (or *plenum chamber*) injected through a porous wall at a rate just sufficient to keep the wall surface wet. In this case, the conduction across the u-surface can be related to the reservoir conditions by application of the steady-flow energy equation to a control volume located between the o- and u-surfaces. Finally, Fig. 1.18e shows drying of a wet porous material (e.g., a textile or wood). During the constant-rate period of the process, evaporation takes place from the surface with negligible heat conduction into the solid; then $-k\,\partial T/\partial y|_u \simeq 0$. The term *adiabatic vaporization* is used to describe evaporation when $q_{cond} = 0$: constant-rate drying is one example, and the wet-bulb psychrometer analyzed in Section 1.5.2 is another.

Evaluation of Properties

Evaluation of composition-dependent properties, in particular the mixture specific heat and Prandtl number, poses a problem. In general, low mass transfer rates imply small composition variations across a boundary layer, and properties can be evaluated for a mixture of the free-stream composition at the mean film temperature. In fact, when dealing with evaporation of water into air, we will simply use the properties of dry air at the mean film temperature to give results of adequate engineering accuracy. If there are large composition variations across the boundary layer, as can occur in some catalysis problems, properties should be evaluated at the mean film composition and temperature. Methods for evaluating properties of mixtures and solutions are given in Section 1.7.

Example 1.8   Heat Loss from a Chemical Dip Bath

A bath 1 m square containing a dilute aqueous solution is maintained at 330 K in still air at 1 atm, 300 K, and 15% relative humidity. Estimate the convective (sensible) and evaporative (latent) heat losses from the surface, if the bath is well stirred so that its surface temperature is also approximately 330 K. Assume that the vapor pressure can be approximated as that of water.

Solution

Given: Dip bath at 330 K evaporating into air at 300 K.
Required: Convective and evaporative heat losses.
Assumptions:

1. The vapor pressure of the aqueous solution can be approximated by data for pure water.

2. The ambient air is still.

3. Negligible difference between the bath surface and bulk temperatures.

4. Moist air transport properties can be approximated by those for dry air.

• Ambient air at 1 atm, 300 K, 15% RH

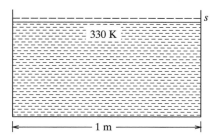

This is a natural-convection problem. To calculate the Grashof number $\mathrm{Gr} = (\Delta\rho/\rho)gL^3/\nu^2$, ρ_s and ρ_e are required. Denoting H_2O as species 1 and air as species 2, Table A.12a gives

$$P_{1,s} = P_{\mathrm{sat}}(T_s) = P_{\mathrm{sat}}(330\ \mathrm{K}) = 17,190\ \mathrm{Pa}$$

$$P_{1,e} = (\mathrm{RH})P_{\mathrm{sat}}(T_e) = (0.15)(3533) = 530\ \mathrm{Pa}$$

Assuming an ideal gas mixture,

$$\rho = \rho_1 + \rho_2 = \frac{P_1 M_1}{\mathcal{R}T} + \frac{P_2 M_2}{\mathcal{R}T}$$

$$\rho_s = \frac{(17,190)(18)}{(8314)(330)} + \frac{(101,330 - 17,190)(29)}{(8314)(330)}$$

$$= 0.1128 + 0.8894 = 1.0022\ \mathrm{kg/m^3}$$

$$\rho_e = \frac{(530)(18)}{(8314)(300)} + \frac{(101,330 - 530)(29)}{(8314)(300)}$$

$$= 0.003825 + 1.1720 = 1.1758\ \mathrm{kg/m^3}$$

$$\Delta\rho = \rho_e - \rho_s = 1.1758 - 1.0022 = 0.1736\ \mathrm{kg/m^3}$$

Also,

$$m_{1,s} = \frac{\rho_{1,s}}{\rho_s} = \frac{0.1128}{1.0022} = 0.1126; \quad m_{1,e} = \frac{\rho_{1,e}}{\rho_e} = \frac{0.003825}{1.1758} = 0.003253$$

The Grashof number can be calculated with ν approximated by the value for pure air at a mean film temperature of $(1/2)(300 + 330) = 315\ \mathrm{K}$, and ρ as the mean of ρ_s and ρ_e, $\rho = (1/2)(1.176 + 1.002) = 1.089\ \mathrm{kg/m^3}$.

$$\mathrm{Gr}_L = \frac{(\Delta\rho/\rho)gL^3}{\nu^2} = \frac{(0.1736/1.089)(9.81)(1)^3}{(16.99 \times 10^{-6})^2}$$

$$= 5.42 \times 10^9 > 10^9 \quad \text{(turbulent flow)}$$

Using Eqs. (4.96) in *Heat Transfer* and (1.46*b*) with the Prandtl number approximated as a pure air value of 0.69, and $Sc_{12} = 0.61$ from Table 1.2,

$$\overline{Nu}_L = 0.14(Gr_L\,Pr)^{1/3} = 0.14[(5.42 \times 10^9)(0.69)]^{1/3} = 217$$

$$\overline{Sh}_L = 0.14(Gr_L Sc)^{1/3} = 0.14[(5.42 \times 10^9)(0.61)]^{1/3} = 209$$

$$\overline{h}_c = \frac{k}{L}\overline{Nu}_L = \frac{(0.0277)(217)}{1} = 6.01 \text{ W/m}^2\text{ K}$$

$$\overline{g}_{m1} = \frac{\rho \mathcal{D}_{12}}{L}\overline{Sh}_L = \frac{\rho \nu}{Sc_{12}L}\overline{Sh}_L = \frac{(1.089)(16.99 \times 10^{-6})(209)}{(0.61)(1)}$$

$$= 6.34 \times 10^{-3} \text{ kg/m}^2\text{ s}$$

where k has been evaluated for pure air at the mean film temperature, and ρ again has been evaluated as the mean of ρ_s and ρ_e. Thus, the heat losses are

$$\dot{Q}_{\text{conv}} = \overline{h}_c A(T_s - T_e) = (6.01)(1)(330 - 300) = 180 \text{ W}$$

$$\dot{Q}_{\text{evap}} = \overline{g}_{m1} A(m_{1,s} - m_{1,e})h_{\text{fg}}(T_s)$$

$$= (6.34 \times 10^{-3})(1)(0.1126 - 0.0033)(2.365 \times 10^6) = 1640 \text{ W}$$

Solution using MCONV

The required input is:

Configuration number $= 11$

Option $= 2$ (evaporation of water into air)

1. Interface temperature, $T_s = 330$
2. Ambient temperature, $T_e = 300$
3. Pressure, $P = 1.0133 \times 10^5$
4. Length, $L = 1$
5. Relative humidity, RH $= 15$

The output gives:

$$q_{\text{evap}} = 1660 \text{ [W/m}^2]$$

$$q_{\text{conv}} = 180 \text{ [W/m}^2]$$

Comments

1. The latent heat loss is an order of magnitude larger than the sensible heat loss because the air is very dry.
2. This result should be viewed as an upper bound: in reality, the surface temperature will be a little lower than the bulk temperature, in order to have heat transfer from the bulk liquid to the interface. Of course, the interface temperature may not be uniform either (see Exercise 1–44).

3. For natural-convection simultaneous heat and mass transfer, the simple analogy between heat and mass transfer given in Section 1.4.4 is not quite true. More exact analysis of this problem shows that the true Nusselt number is somewhat higher than 217 and the true Sherwood number is somewhat lower than 209. However, our result is quite adequate for engineering purposes (see footnote 2).

4. MCONV gives a warning that $\Delta m_1 = 0.109 > 0.10$. Low mass transfer rate theory is inaccurate, but more exact analysis shows that the error is approximately $(1/2)\Delta m_1$ expressed as a percentage—that is, approximately 5%.

5. It is of interest to calculate the radiation heat flux. For a water emittance of 0.9 from Table A.5a, and black surroundings at 300 K,

$$q_{\text{rad}} = (0.9)(5.67 \times 10^{-8})(330^4 - 300^4) = 192 \text{ W/m}^2$$

Thus, the radiation heat loss is a little larger than the sensible heat loss (though still much smaller than the latent heat loss). ▲

1.5.2 The Wet- and Dry-Bulb Psychrometer

The wet- and dry-bulb psychrometer is widely used to measure the moisture content of air. In its simplest form, the air is made to flow over a pair of thermometers, one of which has its bulb covered by a wick whose other end is immersed in a small water reservoir. Evaporation of water from the wick causes the wet bulb to cool, and its steady-state temperature is a function of the air temperature measured by the dry bulb and the air humidity. Commonly, a *psychrometric chart* is used to deduce the air humidity from the two temperature readings. The *sling* psychrometer has thermometers mounted in a frame with a handle that allows the unit to be swung around the user's head. Alternatively, a fan is used to supply the air flow. Nowadays the common availability of digital readout thermocouples with built-in cold junctions has led to the increased use of thermocouple-based psychrometers. Since relatively small thermocouple junctions can be used, their response to changes in temperature and hence humidity is fast, and they are more appropriate than thermometer-based psychrometers in situations where the air state changes rapidly with time.

The principle of the wet- and dry-bulb psychrometer can be demonstrated using the mass convection theory developed in Section 1.4. Figure 1.19 shows a wet-bulb thermometer exposed to an air flow of temperature T_e (as measured by the dry bulb) and mass fraction of water vapor $m_{1,e}$. The surface of the moist wick is at temperature T_s, and the mass fraction of water vapor adjacent to the wick is $m_{1,s}$, the saturation value corresponding to temperature T_s. When the wet-bulb temperature is steady, there is negligible heat conduction into the thermometer, and from Eq. (1.55) the surface energy balance on the wet bulb requires that the heat transfer from the air equal the evaporation rate times the enthalpy of vaporization:

$$\overline{h}_c(T_e - T_s)A = \overline{g}_{m1}(m_{1,s} - m_{1,e})h_{\text{fg}}A$$

Radiation heat transfer from the surroundings has been ignored since, if the air velocity is sufficiently high, radiation can be neglected compared with convection.

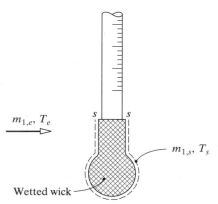

$m_{1,e}, T_e$

$m_{1,s}, T_s$

Wetted wick

Figure 1.19 Wet bulb of a wet- and dry-bulb psychrometer.

Also, the very small amount of heat given up by the water as it cools from the reservoir temperature to T_s is neglected in taking $-k\,\partial T/\partial y|_u = 0$. Rearranging gives

$$m_{1,e} = m_{1,s} - \frac{\overline{h}_c}{\overline{g}_{m1} h_{\text{fg}}}(T_e - T_s) \tag{1.56}$$

For laminar forced flow over a cylinder, sphere, or similar shape, Stanton numbers for heat and mass transfer in gases can be approximated by Chilton–Colburn analogy power law relations,

$$\overline{\text{St}} = C\text{Re}^{-1/2}\text{Pr}^{-2/3} = \frac{\overline{h}_c}{\rho c_p V}$$

$$\overline{\text{St}}_m = C\text{Re}^{-1/2}\text{Sc}^{-2/3} = \frac{\overline{g}_{m1}}{\rho V}$$

Thus

$$\frac{\overline{h}_c/c_p}{\overline{g}_{m1}} = \left(\frac{\text{Pr}}{\text{Sc}}\right)^{-2/3}$$

and substituting into Eq. (1.56) gives

$$m_{1,e} = m_{1,s} - \frac{c_p}{h_{\text{fg}}}\left(\frac{\text{Pr}}{\text{Sc}_{12}}\right)^{-2/3}(T_e - T_s) \tag{1.57}$$

Strictly speaking, the specific heat and Prandtl number should be evaluated at an appropriate reference temperature and composition. However, if the water vapor is in small concentrations, it is sufficient to use pure air values at the mean film temperature. Also, it is sufficiently accurate to take $\text{Pr}/\text{Sc}_{12} \simeq (0.69/0.61) = 1.13$ and $(\text{Pr}/\text{Sc}_{12})^{-2/3} = 1/1.08$. The ratio of Prandtl to Schmidt number is a new dimensionless group that is relevant to simultaneous convective heat and mass transfer: it is called the **Lewis number**.[5] Equation (1.57) can thus be written as

$$m_{1,e} = m_{1,s} - \frac{c_{p,\text{air}}}{1.08 h_{\text{fg}}}(T_e - T_s) \tag{1.58}$$

Temperatures T_s and T_e are, respectively, the known measured wet- and dry-bulb temperatures, and $m_{1,e}$ is the unknown air moisture content. For T_s known, $m_{1,s}$ can be obtained from steam tables in the usual way: such data can be represented by the relation

$$m_{1,s} = m_{1,s}(T_s, P) \tag{1.59}$$

where P is the total pressure. Equations (1.58) and (1.59) can be viewed as two equations in the unknowns $m_{1,e}$ and $m_{1,s}$.

Equation (1.58) is accurate only for humidity measurements in water vapor–air mixtures at low temperatures, that is, the usual conditions encountered in air-conditioning practice. The effect of ignoring radiation heat transfer in the analysis is explored in Exercise 1–41. The wet-bulb temperature recorded by a psychrometer is termed the *psychrometric* wet-bulb temperature and is in general not the same as the adiabatic saturation temperature or *thermodynamic* wet-bulb temperature.

The thermodynamic wet-bulb temperature is the temperature at which water, by evaporating into air in a constant pressure-mixing process, can bring the air to saturation adiabatically at the same temperature. The essential point is that the required enthalpy of vaporization exactly equals the enthalpy given up by the air as it cools down from its initial temperature to the wet-bulb temperature. The psychrometric and thermodynamic wet-bulb temperatures can be taken to be equal when the Lewis number is unity.[6] The Lewis number for water vapor–air mixtures is somewhat greater than unity, so that the two temperatures are never exactly equal; however, the difference is usually small enough to be ignored. The psychrometric charts that are widely used by air-conditioning engineers are based on the thermodynamic wet-bulb temperature.

If the wet- and dry-bulb temperatures are known, the calculation of moisture content using Eqs. (1.58) and (1.59) is straightforward, as will be seen in Example 1.9. If the unknown is the wet-bulb temperature, however, an iterative calculation is required. Such calculations must be made accurately to obtain a satisfactory result and are tedious. Air-conditioning engineers use psychrometric charts to avoid the problem. Alternatively, the computer program PSYCHRO can be used for this purpose. The principles of psychrometry and psychrometric charts may be found in various textbooks and handbooks, for example, [7, 8, 9, 10].

---

[5] Other definitions of the Lewis number are found in the literature, for example, the ratio of Schmidt to Prandtl number and the ratio of the heat to mass transfer conductance.

[6] Proof of this statement is left to more advanced texts.

The Computer Program PSYCHRO

PSYCHRO calculates psychrometric and thermodynamic properties of water vapor and air mixtures. Values of three properties are required to specify the thermodynamic state of the mixture. In PSYCHRO these are total pressure and two of the following: dry-bulb temperature, wet-bulb temperature, dewpoint temperature, relative humidity, vapor mass fraction, and humidity ratio. Temperatures may be input in either kelvins or degrees Celsius. Upon input of three property values, PSYCHRO calculates and displays the remaining properties, together with the mixture enthalpy, density, and specific volume, as well as the vapor mole fraction (see Table 1.4).

Two options for making subsequent changes are available:

1. Fixed total pressure and dry-bulb temperature

2. Fixed total pressure and water vapor concentration (mass fraction and humidity ratio)

PSYCHRO is based on data given in the ASHRAE *Handbook of Fundamentals* [10].

Table 1.4 Psychrometric and thermodynamic properties of water vapor–air mixtures in PSYCHRO.

Symbol	Property	Units	Definition
P	Total pressure	Pa	$P = P_1 + P_2$, where P_1 and P_2 are the partial pressures of water vapor and dry air, respectively
T_{dry}	Dry-bulb temperature	K or $^\circ$C	Mixture temperature
T_{wet}	Wet-bulb temperature	K or $^\circ$C	Thermodynamic wet-bulb temperature
T_{dew}	Dewpoint temperature	K or $^\circ$C	Saturation temperature corresponding to the partial pressure of water vapor in the mixture
RH	Relative humidity	%	$P_1/P_{\text{sat}}(T_{\text{dry}})$
m_1	Mass fraction of water vapor		$m_1 = \dfrac{\rho_1}{\rho} = \dfrac{\rho_1}{\rho_1 + \rho_2}$
ω	Humidity ratio		$\omega = \dfrac{\rho_1}{\rho_2} = \dfrac{m_1}{1 - m_1}$; also called the specific humidity ϕ
h	Mixture enthalpy	J/kg	$h = m_1 h_1 + m_2 h_2$ Datum states: $h_1 = 0$ for liquid water at 0°C $h_2 = 0$ for air at 0°C
ρ	Mixture density	kg/m^3	$\rho = \rho_1 + \rho_2$; $\rho_1 = \dfrac{P_1 M_1}{\mathcal{R} T_{\text{dry}}}, \rho_2 = \dfrac{P_1 M_2}{\mathcal{R} T_{\text{dry}}}$
v	Mixture specific volume	m^3/kg	$v = \dfrac{1}{\rho}$
x_1	Mole fraction of water vapor		$x_1 = \dfrac{c_1}{c} = \dfrac{m_1/M_1}{m_1/M_1 + m_2/M_2}$

Example 1.9   Humidity of an Air Flow

Air at 1010 mbar flows over a wet- and dry-bulb psychrometer. The dry bulb measures 31.8°C, and the wet bulb measures 26.8°C. Determine the relative humidity and humidity ratio.

Solution

Given: Wet- and dry-bulb temperatures of an air stream.
Required: Relative humidity and humidity ratio.
Assumptions: The air velocity is sufficiently high for radiation effects to be negligible.
Equation (1.58) gives the mass fraction of the water vapor in the air stream:

$$m_{1,e} = m_{1,s} - \frac{c_{p,\text{air}}}{1.08 h_{\text{fg}}} (T_e - T_s)$$

$$T_e = 305.0 \text{ K}; \quad T_s = 300.0 \text{ K}$$

From Table A.12a, $P_{1,s} = P_{\text{sat}}(300 \text{ K}) = 3533$ Pa.

$$x_{1,s} = P_{1,s}/P = 3533/101,000 = 0.0350$$

$$m_{1,s} = \frac{18 x_{1,s}}{18 x_{1,s} + 29 x_{2,s}} = \frac{x_{1,s}}{x_{1,s} + (29/18)(1 - x_{1,s})}$$

$$= \frac{0.0350}{0.0350 + 1.61(1 - 0.0350)} = 0.0220$$

Also from Table A.12a, $h_{\text{fg}}(T_s) = 2.437 \times 10^6$ J/kg, and from Table A.7, $c_{p,\text{air}} = 1005$ J/kg K. Notice that the specific heat of air is almost constant at normal temperatures, and it is unnecessary to evaluate it at a precise reference temperature. Substituting in Eq. (1.58),

$$m_{1,e} = 0.0220 - \frac{1005}{(1.08)(2.437 \times 10^6)} (305 - 300) = 0.0220 - 0.0019 = 0.0201$$

The mole fraction of water vapor in the air is then

$$x_{1,e} = \frac{m_{1,e}/18}{m_{1,e}/18 + m_{2,e}/29} = \frac{m_{1,e}}{m_{1,e} + (18/29)(1 - m_{1,e})}$$

$$= \frac{0.0201}{0.0201 + 0.621(1 - 0.0201)} = 0.0320$$

and the partial pressure is

$$P_{1,e} = x_{1,e} P = (0.0320)(101,000) = 3230 \text{ Pa}$$

By definition, the relative humidity is RH $= P_{1,e}/P_{\text{sat}}(T_e)$; from Table A.12a, $P_{\text{sat}}(305.0 \text{ K}) = 4714$ Pa. Thus,

$$\text{RH} = P_{1,e}/P_{\text{sat}}(T_e) = 3230/4714 = 68.5\%$$

The humidity ratio ω is the mass of water vapor per mass of dry air:

$$\omega = \frac{\text{kg water vapor}}{\text{kg dry air}} = \frac{m_{1,e}}{1 - m_{1,e}} = \frac{0.0201}{1 - 0.0201} = 0.0205$$

Solution using PSYCHRO

The required input is:

$$\text{Pressure, } P = 1.01 \times 10^5$$
$$\text{Dry-bulb temperature, } T_{\text{dry}} = 31.8°C$$
$$\text{Wet-bulb temperature, } T_{\text{wet}} = 26.8°C$$

The output includes:

$$\text{Relative humidity, RH} = 68.5\%$$
$$\text{Humidity ratio, } \omega = 0.0204$$

Comments

1. Use MCONV to check $m_{1,e}$ and the energy balance at the s-surface of the wet bulb.
2. Compare these results to values given in a psychrometric chart. ▲

Example 1.10   Evaporation of a Water Droplet

A 50 μm–diameter water droplet initially at 315 K is injected into an air stream at 315 K, 1.050×10^5 Pa, and 50.5% RH. Estimate the droplet lifetime.

Solution

Given: A small water droplet.
Required: Lifetime after injection into an air stream.
Assumptions:

1. After a short initial transient the droplet temperature attains a constant value.
2. Radiation heat transfer is negligible.
3. The droplet is entrained into the air stream.

The above assumptions may appear to be rather bold! We will assume that they are valid and check later. Using Eq. (1.36), the evaporation rate of the droplet is

$$j_{1,s} A = g_{m1}(m_{1,s} - m_{1,e})A; \quad A = \pi D^2$$

If the droplets are entrained in the air flow, from Eq. (4.76) in *Heat Transfer* for Re $\rightarrow$ 0, or from Eq. (4.78), the Sherwood number can be approximated by its lower-limit value of 2.

$$\text{Sh} = \frac{g_{m1} D}{\rho \mathcal{D}_{12}} \simeq 2; \text{ hence } g_{m1} = \frac{2\rho \mathcal{D}_{12}}{D} \text{ and } j_{1,s} A = \frac{2\rho \mathcal{D}_{12} A}{D}(m_{1,s} - m_{1,e})$$

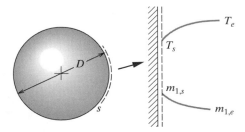

A mass balance on a droplet requires that the rate of mass loss equal the evaporation rate:

$$\frac{d}{dt}\left(\frac{1}{6}\pi D^3 \rho_l\right) = -j_{1,s} A$$

where ρ_l is the water density. Substituting for $j_{1,s}$ with $A = \pi D^2$, and differentiating,

$$\frac{dD}{dt} = -\frac{4\rho \mathcal{D}_{12}(m_{1,s} - m_{1,e})}{D\rho_l}$$

If D_0 is the initial diameter and τ is the droplet lifetime, then integrating with T_s constant gives

$$\int_{D_0}^{0} D\,dD = -\frac{4\rho \mathcal{D}_{12}(m_{1,s} - m_{1,e})}{\rho_l} \int_{0}^{\tau} dt$$

$$\frac{D_0^2}{2} = \frac{4\rho \mathcal{D}_{12}(m_{1,s} - m_{1,e})\tau}{\rho_l}$$

$$\tau = \frac{\rho_l D_0^2}{8\rho \mathcal{D}_{12}(m_{1,s} - m_{1,e})}$$

First we evaluate $m_{1,e}$; at 315 K, Table A.12a gives

$$P_{\text{sat}} = 0.08135 \times 10^5 \text{ Pa}$$

$$P_{1,e} = (\text{RH})P_{1,\text{sat}}(T_e) = (0.505)(0.08135 \times 10^5) = 0.0411 \times 10^5 \text{ Pa}$$

$$x_{1,e} = P_{1,e}/P = 0.0411/1.050 = 0.0391$$

$$m_{1,e} = \frac{0.0391}{0.0391 + (29/18)(1 - 0.0391)} = 0.0246$$

In order to evaluate $m_{1,s}$ we need to know the droplet surface temperature. After injection, the droplet will cool until the latent heat required for evaporation just balances the rate of heat transfer from the air. For a precise result we should not use Eq. (1.58) to obtain T_s since the psychrometric wet-bulb temperature assumes a convective situation, whereas for an entrained droplet we have a purely diffusive situation with $\text{Nu} = \text{Sh} = 2$ and $(h_c/c_p)/g_{m1} = (\text{Pr/Sc})^{-1} \sim 1.13^{-1}$.

Thus, Eq. (1.58) is replaced by

$$m_{1,s} = m_{1,e} + \frac{c_{p,\text{air}}}{1.13 h_{\text{fg}}}(T_e - T_s) = 0.0246 + \frac{1006}{1.13 h_{\text{fg}}(T_s)}(315 - T_s)$$

$$m_{1,s} = m_{1,s}(T_s, P)$$

Using Table A.12a and solving by iteration gives $T_s = 305.0$ K, $m_{1,s} = 0.0283$.

The properties required are as follows. At 305 K the density of water is $\rho_l = 995$ kg/m^3. Gas-phase properties are evaluated at the mean film temperature, $T_r = (1/2)(T_s + T_e) = (1/2)(305 + 315) = 310$ K. Using $\rho\mathcal{D}_{12} = \mu/\text{Sc}_{12}$, approximating μ as the pure air value of 18.87×10^{-6} kg/m s, and taking Sc $= 0.61$ for dilute water vapor–air mixtures gives $\rho\mathcal{D}_{12} = (18.87 \times 10^{-6})/0.61 = 3.09 \times 10^{-5}$ kg/m s.

$$\tau = \frac{\rho_l D_0^2}{8\rho\mathcal{D}_{12}(m_{1,s} - m_{1,e})} = \frac{(995)(50 \times 10^{-6})^2}{(8)(3.09 \times 10^{-5})(0.0283 - 0.0246)} = 2.72 \text{ s}$$

The various assumptions should now be checked.

1. *Duration of initial temperature transient.* The time required for the droplet to cool from its injection temperature of 315 K to 305 K can be estimated by calculating dT/dt from an energy balance at the average droplet temperature of 310 K.

Rate of decrease in   =   Rate at which latent   +   Sensible heat
droplet internal energy   heat is supplied   transfer rate

$$-\rho_l(\pi D^3/6)c_v \frac{dT}{dt} = \pi D^2(q_{\text{evap}} + q_{\text{conv}})$$

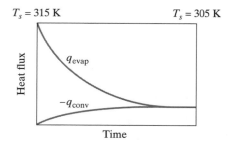

The latent heat transfer is

$$q_{\text{evap}} = j_{1,s} h_{\text{fg}} = 2(\rho\mathcal{D}_{12}/D)(m_{1,s} - m_{1,e})h_{\text{fg}}$$

At 310 K, $P_{1,s} = 0.06224 \times 10^5$ Pa, $x_{1,s} = 0.0593$, $m_{1,s} = 0.0377$; $h_{\text{fg}} = 2.414 \times 10^6$ J/kg. Approximating $\rho\mathcal{D}_{12}$ as 3.09×10^{-5} kg/m s, as previously calculated,

$$q_{\text{evap}} = 2[3.09 \times 10^{-5}/(50 \times 10^{-6})](0.0377 - 0.0246)(2.414 \times 10^6)$$

$$= 3.91 \times 10^4 \text{ W/m}^2$$

The sensible heat transfer is

$$q_{\text{conv}} = h_c(T_s - T_e)$$

$h_c = (k/D)\text{Nu}$, $\text{Nu} = 2$, $k = 0.0274$ W/m K for air at 310 K. Hence,

$$q_{\text{conv}} = 2[0.0274/(50 \times 10^{-6})](310 - 315) = -5.48 \times 10^3 \text{ W/m}^2$$

$$\frac{dT}{dt} = -(6/\rho_l c_v D)(q_{\text{evap}} + q_{\text{conv}})$$

$$= -[6/(995)(4174)(50 \times 10^{-6})](10^4)(3.91 - 0.548) = -971 \text{ K/s}$$

The duration of the 10 K temperature transient is of the order $10/971 \simeq 10^{-2}$ s, which is negligible compared to the lifetime of 2.7 s.

2. *Radiation heat transfer.* We will estimate the radiation heat transfer when the droplet is at 310 K. Assuming black surfaces for an upper bound,

$$q_{\text{rad}} = \sigma\epsilon(T_s^4 - T_e^4) = (5.67 \times 10^{-8})(310^4 - 315^4) = -34.6 \text{ W/m}^2$$

which is two orders of magnitude less than the convection heat transfer, and is negligible.

3. *Droplet acceleration.* To estimate the time required for the droplet to be entrained, we apply Newton's second law of motion to the droplet:

$$\rho_l(\pi D^3/6)\frac{dV}{dt} = -C_D\left(\frac{1}{2}\rho V^2\right)(\pi D^2/4)$$

where V is the droplet velocity relative to the air stream. The drag coefficient C_D can be estimated from Stokes' law, Eq. (4.73) in *Heat Transfer*, that is, $C_D = 24/\text{Re}_D$. Substituting and rearranging,

$$\frac{dV}{dt} = -\frac{18\mu}{\rho_l D^2}V$$

Integrating with $V = V_0$ at $t = 0$ and neglecting the change in D gives

$$V = V_0 e^{-t/t_c}; \quad t_c = \rho_l D^2/18\mu$$

where t_c is a time constant for the acceleration process. Using the same property values as before,

$$t_c = (995)(50 \times 10^{-6})^2/(18)(18.87 \times 10^{-6}) = 7.32 \times 10^{-3} \text{ s}$$

which is two orders of magnitude less than the evaporation time.

Comments

1. The fact that radiation has a negligible effect is due to the small size of the droplet. The convective heat transfer is inversely proportional to droplet size ($h_c = 2k/D$), while the radiation is independent of droplet size.

2. The small time constants for droplet cool-down and acceleration are also due to the small droplet size.

3. Use PSYCHRO to calculate the thermodynamic wet-bulb temperature and compare it to the value of T_s used in this example. Would it be an adequate approximation?

4. Use MCONV to check q_{conv} and q_{evap}.   ▲

1.5.3 Heterogeneous Combustion

Combustion is an important engineering problem in which the heat and mass transfer processes are coupled, and the equations governing each process often must be solved simultaneously. In diffusion-controlled combustion, the chemical kinetics are so fast that mass transfer considerations control the rate of combustion, while heat transfer considerations control the rate at which the heat of combustion can be removed, and hence the temperature of combustion. Since the transport properties, such as the diffusion coefficient, are temperature dependent, the equations governing heat and mass transfer must be solved simultaneously. An example for a solid fuel will be analyzed below. More complex problems are the interesting phenomena of ignition and extinction of combustion, which are results of conflicting requirements of the mass transfer, chemical kinetics, and heat transfer involved. Such problems are dealt with in combustion texts.

We will consider the combustion of carbon particles in an air stream in the temperature range 1000–1600 K at 1 atm, for which the reaction is

$$C + O_2 \rightarrow CO_2$$

The combustion is diffusion controlled; that is, the kinetics are so fast that the gas mixture at the s-surface is in chemical equilibrium with solid carbon. Equilibrium data for the reaction indicate that the resulting concentration of oxygen at the s-surface is essentially zero. Figure 1.20 shows a schematic of the situation. We will analyze the problem on a molar basis. The rate at which oxygen, species 1, diffuses across the s-surface is, from Eq. (1.37),

$$J_{1,s} = G_{m1}(x_{1,s} - x_{1,e}) = -G_{m1}x_{1,e} \tag{1.60}$$

If the particles are spherical and small enough to be entrained in the air stream, the Sherwood number can be taken to be its lower-limit value of 2.

$$Sh = \frac{G_{m1}D}{c\mathcal{D}_{1m}} = 2$$

Hence

$$G_{m1} = \frac{2c\mathcal{D}_{1m}}{D}$$

and

$$J_{1,s} = -\frac{2c\mathcal{D}_{1m}x_{1,e}}{D} \tag{1.61}$$

Since one mole of carbon is consumed for each mole of oxygen crossing the s-surface, Eq. (1.61) also gives the rate at which carbon is consumed in $kmol/m^2\, s$. A mass

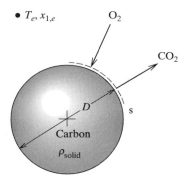

Figure 1.20 Schematic of a carbon particle oxidizing in an air stream.

balance on the carbon particle requires that

$$\frac{d}{dt}\left(\frac{1}{6}\pi D^3 \rho_{\text{solid}}\right) = J_{1,s} M_C (\pi D^2) \tag{1.62}$$

where ρ_{solid} is the density of solid carbon and M_C is the molecular weight of element carbon. Differentiating and substituting from Eq. (1.61) then gives the rate of decrease of particle diameter as

$$\frac{dD}{dt} = -\frac{4c\mathcal{D}_{1m} M_C x_{1,e}}{\rho_{\text{solid}} D} \tag{1.63}$$

Since both c and $\mathcal{D}_{1m}$ depend on temperature, our next task is to obtain the particle temperature. If conduction into the particle is negligible, the surface energy balance requires that the rate at which heat is lost by convection and radiation equal the rate at which heat is liberated by the combustion reaction:

$$\dot{Q}_{\text{conv}} + \dot{Q}_{\text{rad}} = \dot{Q}_{\text{comb}}$$

or

$$h_c A(T_s - T_e) + \epsilon \sigma A(T_s^4 - T_e^4) = (-J_{1,s})\Delta H_c A \tag{1.64}$$

where ΔH_c is the heat released per kmol of oxygen (or carbon) consumed. The expression for $\dot{Q}_{\text{rad}}$ is Eq. (1.18) in *Heat Transfer* for a small gray body in large surroundings, and assumes that the surroundings are at the same temperature as the gas. For an entrained spherical particle, the Nusselt number is also 2; thus $h_c = 2k/D$. Substituting and using Eq. (1.61) gives

$$\frac{2k}{D}(T_s - T_e) + \sigma\epsilon(T_s^4 - T_e^4) = \frac{2c\mathcal{D}_{1m} x_{1,e}}{D}\Delta H_c \tag{1.65}$$

For a given particle diameter D, Eq. (1.65) can be solved for T_s by iteration. Alternatively, it makes more sense to specify a range of T_s values and solve directly for the corresponding values of D, as will be done in Example 1.11. Equation (1.65) shows that both q_{conv} and q_{comb} are inversely proportional to particle size, while q_{rad} is independent of particle size. Thus, as the particle burns and its size decreases, the radiation loss becomes relatively smaller and the combustion temperature increases. This feature is common to many types of particle combustion: the color brightness is seen to increase toward the end of the particle life.

The $c\mathcal{D}_{1m}$ product depends on temperature and hence particle size, so that Eq. (1.63) should be integrated numerically to obtain the particle lifetime. An approximate estimate can be obtained if a constant value of $c\mathcal{D}_{1m}$ is used. Rearranging Eq. (1.63) and integrating,

$$\int_{D_0}^{0} \frac{D\,dD}{c\mathcal{D}_{1m}} = -\frac{4M_C x_{1,e}}{\rho_{\text{solid}}} \int_0^{\tau} dt \qquad (1.66)$$

where D_0 is the particle initial diameter and τ is its lifetime. Taking $c\mathcal{D}_{1m} = (c\mathcal{D}_{1m})_r$ at an appropriate reference temperature yields

$$\frac{1}{(c\mathcal{D}_{1m})_r}\frac{D_0^2}{2} = \frac{4x_{1,e}M_C\tau}{\rho_{\text{solid}}}$$

or

$$\tau = \frac{\rho_{\text{solid}}D_0^2}{8(c\mathcal{D}_{1m})_r M_C x_{1,e}} \qquad (1.67)$$

Since $(1/c\mathcal{D}_{1m})_r$ is weighted with D in the integral, it should be evaluated at the temperature corresponding to a diameter somewhat larger than the mean value, perhaps $(0.5)^{1/2}D_0 \simeq 0.70D_0$.

Two comments close this section:

1. We chose to perform this analysis on a molar basis because the net rate of transfer of moles across the s-surface was zero: the molar fluxes of CO_2 and O_2 balance due to the stoichiometry of the reaction. Thus, low *mole* transfer rate theory applies and, in fact, is exact (as is the case of low *mass* transfer rate theory applied to catalysis). Had we analyzed the problem on a mass basis, low mass transfer rate theory would have been of adequate accuracy for most purposes, but not exact like the molar result.

2. In considering a small entrained particle so that the Sherwood and Nusselt numbers could be taken equal to 2, we have, in fact, analyzed a diffusion rather than a convection problem. For larger particles there would be a relative velocity between the particle and air stream, and appropriate correlations should be used for Sh and Nu [e.g., Eq. (4.76) in *Heat Transfer*]. Then a numerical integration may be required to obtain the particle lifetime.

Example 1.11   Low-Temperature Oxidation of a Carbon Particle

Hot carbon particles of initial diameter 300 μm are entrained into a gas stream containing 10% oxygen by volume at 500 K and 1 atm. Determine the particle temperature as a function of particle size, and hence estimate the time for a particle to be consumed. Take $\epsilon = 0.9$ for the carbon and $\Delta H_c = 3.94 \times 10^8$ J/kmol carbon consumed.

Solution

Given: Carbon particles oxidizing in a gas stream containing 10% oxygen by volume.

Required: Particle temperature and lifetime.

Assumptions:

1. Diffusion-controlled oxidation, $C + O_2 \rightarrow CO_2$.
2. Gas properties can be approximated with air values.

We will assume that the particle temperature is in the range 1000–1600 K so that the analysis of Section 1.5.3 is valid. To avoid iteration, the particle energy balance, Eq. (1.65), is solved for particle diameter D:

$$D = \frac{(c\mathcal{D}_{1m}/k)x_{1,e}\Delta H_c - (T_s - T_e)}{(\sigma\epsilon/2k)(T_s^4 - T_e^4)}$$

$T_e = 500\,\text{K}, x_{1,e} = 0.10, \epsilon = 0.9, \sigma = 5.67\times10^{-8}\,\text{W/m}^2\,\text{K}^4$, and $\Delta H_c = 3.94\times10^8$ J/kmol. Substituting,

$$D = \frac{3.94 \times 10^7(c\mathcal{D}_{1m}/k) - (T_s - 500)}{(2.552/k)[(T_s \times 10^{-2})^4 - 625]}$$

Using $c = P/\mathcal{R}T = (1.0133 \times 10^5/8314T)$, and approximating $\mathcal{D}_{1m}$ as the value for O_2 in air from Table A.17a and k as the value for air from Table A.7, allows D to be obtained as a function of T_s, as given below. Properties are evaluated at the mean film temperature.

T_s, K	1000	1100	1200	1300	1400	1500	1600
D,μm	1420	824	470	252	128	45	—

The particle lifetime is given by Eq. (1.67) as

$$\tau = \frac{\rho_{\text{solid}}D_0^2}{8(c\mathcal{D}_{1m})_r M_C x_{1,e}}$$

where $(c\mathcal{D}_{1m})_r$ will be evaluated (somewhat arbitrarily) at a temperature corresponding to $0.7D_0 = 210\,\mu$m. From the table, the required values are $T_s \simeq 1320\,\text{K}$ and $T_r = (500+1320)/2 = 910\,\text{K}$: $c = 0.0134\,\text{kmol/m}^3$, $\mathcal{D}_{1m} = 129\times10^{-6}\,\text{m}^2/\text{s}$. Then, with $\rho_{\text{solid}} = 1810\,\text{kg/m}^3$ from Table A.1,

$$\tau = \frac{(1810)(300 \times 10^{-6})^2}{8(0.0134)(129 \times 10^{-6})(12)(0.10)} = 9.82\,\text{s}$$

Comments

1. The particle temperature is in the range 1000–1600 K, so that the reaction regime is diffusion-controlled, as assumed.
2. By performing additional calculations, investigate the sensitivity of particle temperature to parameters such as T_e and $x_{1,e}$.
3. Notice that the particle temperature is independent of pressure.   ▲

1.6 MASS TRANSFER IN POROUS CATALYSTS

Porous solids are widely used in mass transfer operations to allow a large surface area per unit volume for transfer. An example is the engineering application of catalysis, such as hydrodesulfurization in petroleum refining or removal of CO, NO, and unburnt hydrocarbons from automobile exhausts. In common use are fixed-bed reactors where the catalyst is in the form of porous pellets of size ranging from 1 to 15 mm in diameter. For CO oxidation, a pellet of copper oxide on alumina may be used; the alumina provides a suitable support for the catalyst, as it is relatively inert and structurally stable at high temperatures. In one manufacturing procedure the pellet is prepared by first impregnating alumina powder with copper nitrate solution. When the powder is heated to 700 K, the nitrate decomposes, evolving gases and leaving cupric oxide. Finally, the pellet is formed by molding at high pressure. The resulting pellet has a density in the range 600–1200 kg/m^3, and a surface area ranging from 100 to 400 m^2/g. A pore diameter of 1 μm is typical.

1.6.1 Diffusion Mechanisms

Three mechanisms of diffusion can occur in porous solids: *ordinary diffusion, Knudsen diffusion*, and *surface diffusion*. Ordinary diffusion of gaseous species, as described by Fick's law, dominates when the pores are large and the gas relatively dense. When the pores are small or the gas density low, the molecules collide with pore walls more frequently than with each other. Then diffusion of molecules along the pore is described by the equations for free molecule flow and is called Knudsen diffusion. At intermediate pressures and pore sizes both types of collisions play an important role. Molecules are also *adsorbed* on the walls of the pores: these molecules are mobile and will diffuse along the walls in the direction of decreasing *surface concentration*. Surface diffusion is the dominant mechanism of transport for the smallest pores, for which ordinary diffusion and Knudsen diffusion rates are very small. Mass transport can be further complicated by the presence of gradients of total pressure and temperature in the porous solid.

Notwithstanding the complexity just described, relative success has been achieved in the analysis of mass transport within catalyst pellets by assuming that transport is governed by a linear relation,

$$J_1 \simeq -c\mathcal{D}_{1,\text{eff}}\nabla x_1 \tag{1.68}$$

where the subscript "eff" denotes an effective diffusivity that accounts for the presence of the solid material. The phenomena of ordinary diffusion and Knudsen diffusion are accounted for in an approximate manner, by simply assuming additive resistances,

$$\frac{1}{\mathcal{D}_{1,\text{eff}}} = \frac{1}{\mathcal{D}_{12,\text{eff}}} + \frac{1}{\mathcal{D}_{K1,\text{eff}}} \tag{1.69}$$

Surface diffusion can be ignored for catalyst pellets since the pores are relatively large.[7] The first of the two effective diffusivities is taken to be

$$\mathcal{D}_{12,\text{eff}} = \frac{\epsilon_v}{\tau} \mathcal{D}_{12} \tag{1.70}$$

where $\mathcal{D}_{12}$ is the binary diffusion coefficient of Fick's law; ϵ_v is the *porosity*, or volume void fraction, of the pellet and accounts for the reduction in cross-sectional area for diffusion posed by the solid material; and τ is the *tortuosity factor* and accounts for the increased diffusion length due to the tortuous paths of real pores, and for the effects of constrictions and dead-end pores. The second is

$$\mathcal{D}_{K1,\text{eff}} = \frac{\epsilon_v}{\tau} \mathcal{D}_{K1} \tag{1.71}$$

where $\mathcal{D}_{K1}$ is the Knudsen diffusion coefficient for species 1, which is given by free molecule flow theory [11] as

$$\mathcal{D}_{K1} = \frac{2}{3} r_e \bar{v}_1 \tag{1.72}$$

where r_e is the effective pore radius and $\bar{v}_1$ is the average molecular speed of species 1,[8]

$$\bar{v}_1 = \left(\frac{8\mathcal{R}T}{\pi M_1} \right)^{1/2} \tag{1.73}$$

Substituting in the value for the gas constant gives a dimensional equation,

$$\mathcal{D}_{K1} = 97 r_e \left(\frac{T}{M_1} \right)^{1/2} \text{m}^2/\text{s} \tag{1.74}$$

for r_e in meters and T in kelvins.

At atmospheric pressure, ordinary diffusion is dominant for pore radii larger than about 2 μm and Knudsen diffusion is dominant for pores smaller than about 0.2 μm. Notice that Eq. (1.69) ensures that $\mathcal{D}_{1,\text{eff}} \to \mathcal{D}_{12,\text{eff}}$ in large pores for which $\mathcal{D}_{K1,\text{eff}}$ is large, and $\mathcal{D}_{1,\text{eff}} \to \mathcal{D}_{K1,\text{eff}}$ in small pores for which $\mathcal{D}_{K1,\text{eff}}$ is small.

1.6.2 Effectiveness of a Catalyst Pellet

When a chemical reaction takes place within a porous pellet, a concentration gradient is set up, and surfaces on pores deep within the pellet are exposed to lower reactant concentrations than surfaces near the pore openings. Since the rate at which a chemical reaction proceeds is dependent on reactant concentration, the average reaction rate throughout a catalyst pellet will be less than if all the available catalyst surface were exposed to reactant at the concentration prevailing at the exterior of the pellet. Figure 1.21 shows a spherical catalyst pellet of a fixed-bed reactor, where it is

---

[7] Surface diffusion plays an important role when water vapor is adsorbed by regular-density silica gels in air-drying operations.

[8] Both c and $\bar{v}$ are commonly used symbols for the average (mean) molecular speed. In the present chapter we will use $\bar{v}$ to avoid confusion, since c is used for molar concentration.

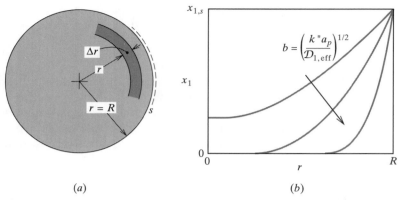

(a)   (b)

Figure 1.21   A spherical catalyst pellet. (a) Elemental control volume for application of the species conservation principle. (b) Concentration profiles.

immersed in a gas flow containing species 1 in dilute concentration. The catalyst promotes a reaction that consumes species 1; for example, the catalyst might be cupric oxide or platinum, which promotes the reaction

$$2CO + O_2 \rightarrow 2CO_2$$

in oxygen-rich automobile exhaust gases; CO can be species 1. The pellet has radius R and catalytic surface area per unit volume of pellet a_p [m^{-1}]. The reaction will be taken to be of first order; thus, the reaction consumes $k'' c x_1$ moles of species 1 per unit area–unit time. The gas flow maintains the concentration of species 1 at the exterior of the pellet at the value $x_{1,s}$.

The Governing Equation and Boundary Conditions

In order to set up the differential equation governing the concentration distribution of species 1, we apply the principle of conservation of species to an elemental control volume located between spherical surfaces at radii r and $r + \Delta r$. At steady state

$$\text{Net inflow of species 1} + \text{Rate of production of species 1} = 0$$

$$[(4\pi r^2 J_1)_r - (4\pi r^2 J_1)_{r+\Delta r}] - k'' c x_1 a_p 4\pi r^2 \Delta r = 0$$

Dividing by $4\pi \, \Delta r$ and letting $\Delta r \rightarrow 0$ gives

$$-\frac{d}{dr}(r^2 J_1) - r^2(k'' c x_1 a_p) = 0$$

Substituting for J_1 from Eq. (1.68) and rearranging yields

$$\frac{1}{r^2}\frac{d}{dr}\left(r^2 c \mathcal{D}_{1,\text{eff}}\frac{dx_1}{dr}\right) - k'' c x_1 a_p = 0$$

If the pellet is nearly isothermal and is at nearly constant total pressure, $c\mathcal{D}_{1,\text{eff}}$ may be assumed constant; then

$$\frac{1}{r^2}\frac{d}{dr}\left(r^2\frac{dx_1}{dr}\right) - \frac{k''a_p}{\mathcal{D}_{1,\text{eff}}}x_1 = 0 \tag{1.75}$$

This differential equation governs the concentration distribution of species 1; it must be solved subject to the boundary conditions

$$r = 0: \quad x_1 \text{ bounded (not infinite)}$$
$$r = R: \quad x_1 = x_{1,s}$$

Solution for the Concentration Distribution and Reaction Rate

In order to simplify notation, we define $b = (k''a_p/\mathcal{D}_{1,\text{eff}})^{1/2}$ and then introduce a new dependent variable $\theta = rx_1$. Equation (1.75) becomes

$$\frac{d^2\theta}{dr^2} - b^2\theta = 0 \tag{1.76}$$

which has the general solution

$$\theta = C_1 \sinh br + C_2 \cosh br$$

or, since $\theta = rx_1$,

$$x_1 = \frac{C_1}{r}\sinh br + \frac{C_2}{r}\cosh br \tag{1.77}$$

Since $\sinh 0 = 0$ and $\cosh 0 = 1$, from the first boundary condition $C_2 = 0$. The constant C_1 is evaluated from the second boundary condition, and the resulting concentration distribution is

$$\frac{x_1}{x_{1,s}} = \frac{R\,\sinh br}{r\,\sinh bR}; \quad b = \left(\frac{k''a_p}{\mathcal{D}_{1,\text{eff}}}\right)^{1/2} \tag{1.78}$$

The rate at which species 1 is *consumed* within the catalyst pellet is given by $-4\pi r^2 J_1\big|_{r=R}$, where

$$4\pi r^2 J_1\big|_{r=R} = -4\pi R^2 c\mathcal{D}_{1,\text{eff}}\frac{dx_1}{dr}\bigg|_{r=R}$$

$$= 4\pi Rc\mathcal{D}_{1,\text{eff}}x_{1,s}\left(1 - \frac{bR}{\tanh bR}\right) \tag{1.79}$$

The Pellet Effectiveness

If the diffusion coefficient were infinite, the complete internal surface would be exposed to reactant at concentration $x_{1,s}$; the rate of consumption would be

$$(4/3)\pi R^3 a_p k'' c x_{1,s}$$

The effectiveness $\eta_p (0 < \eta_p < 1)$ of the catalyst pellet is the ratio of the actual consumption rate divided by that for a pellet with an infinite diffusion coefficient:

$$\eta_p = \frac{-4\pi Rc\mathcal{D}_{1,\text{eff}} x_{1,s} \left(1 - \dfrac{bR}{\tanh bR} \right)}{(4/3)\pi R^3 a_p k'' c x_{1,s}}$$

$$= \frac{3}{bR} \left(\frac{1}{\tanh bR} - \frac{1}{bR} \right) \tag{1.80}$$

Notice that this effectiveness is used in the same sense as the efficiency of a cooling fin, which is the ratio of actual heat transfer to that for a fin of infinite thermal conductivity. Following chemical engineering practice, we recast this result in final form by introducing the Thiele modulus Λ,

$$\Lambda = \frac{V_p}{S_p} \left(\frac{k'' a_p}{\mathcal{D}_{1,\text{eff}}} \right)^{1/2} = \frac{V_p}{S_p} b \tag{1.81}$$

where V_p is the pellet volume and S_p is the area of the pellet exterior. For a sphere, $V_p/S_p = (4/3)\pi R^3 / 4\pi R^2 = R/3$. Substituting in Eq. (1.80),

$$\eta_p = \frac{1}{\Lambda} \left(\frac{1}{\tanh 3\Lambda} - \frac{1}{3\Lambda} \right) \tag{1.82}$$

We know from the definition of effectiveness that when the Thiele modulus is small compared with unity, η_p approaches unity. When the Thiele modulus is large compared with unity, Eq. (1.82) reduces to

$$\eta_p \simeq \frac{1}{\Lambda} \tag{1.83}$$

For nonspherical pellets Eq. (1.82) may be applied approximately by using the appropriate value for V_p/S_p in the Thiele modulus. For example, a cylindrical pellet of radius R and length L will have

$$\frac{V_p}{S_p} = \frac{\pi R^2 L}{2\pi RL + 2\pi R^2} = \frac{RL}{2(L + R)}$$

Figure 1.22 shows a plot of pellet effectiveness versus Thiele modulus, as given by Eq. (1.82).

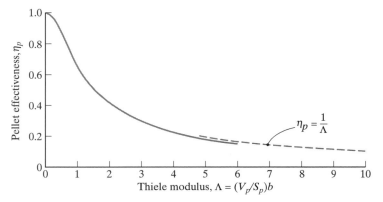

Figure 1.22   Pellet effectiveness as a function of the Thiele modulus, according to Eq. (1.82).

Example 1.12   Effectiveness of a CuO-on-Alumina Pellet

The catalyst bed of an oxidation converter for an automobile is packed with 0.5 cm–diameter spherical pellets of CuO on alumina. The pellets have a volume void fraction $\epsilon_v = 0.80$, tortuosity factor $\tau = 4.0$, average pore radius $r_e = 1\ \mu$m, and catalytic surface per unit volume $a_p = 7.12 \times 10^5$ cm^2/cm^3. Calculate the effectiveness of the pellet in promoting carbon monoxide oxidation at 800 K and 1 atm. The rate constant of the reaction is obtained from the chemical kinetics literature as $k'' = Ae^{-E_a/\mathcal{R}T}$, with $A = 50.6$ m/s and $E_a = 18.6$ kcal/mol.

Solution

Given: A 0.5 cm–diameter spherical catalyst pellet.
Required: Effectiveness for promoting CO oxidation at 800 K and 1 atm.
Assumptions:

1. The pellet is isothermal.
2. The reaction is of first order.
3. Equation (1.68) adequately describes diffusion inside the pellet.

Equation (1.82) gives the pellet effectiveness as

$$\eta_p = \frac{1}{\Lambda}\left[\frac{1}{\tanh 3\Lambda} - \frac{1}{3\Lambda}\right]$$

where $\Lambda = (V_p/S_p)(k''a_p/\mathcal{D}_{1,\text{eff}})^{1/2}$ is the Thiele modulus. We first calculate the effective diffusion coefficient $\mathcal{D}_{1,\text{eff}}$. From Eq. (1.70),

$$\mathcal{D}_{12,\text{eff}} = \frac{\epsilon_v}{\tau}\mathcal{D}_{12} = \frac{0.8}{4.0}(1.06 \times 10^{-4}) = 2.12 \times 10^{-5}\ \text{m}^2/\text{s}$$

where $\mathcal{D}_{12}$ is taken from Table A.17a as the value for a CO-air mixture at 800 K, 1 atm. From Eqs. (1.71) and (1.72),

$$\mathcal{D}_{K1,\text{eff}} = \frac{2}{3}\frac{\epsilon_v}{\tau}r_e\bar{v}_1$$

$$r_e = 1\,\mu\text{m} = 10^{-6}\,\text{m}$$

$$\bar{v}_1 = \left(\frac{8\mathcal{R}T}{\pi M_1}\right)^{1/2}$$

$$= \left\{\frac{(8)(8.314 \times 10^3\,\text{J/kmol K})(800\,\text{K})(1\,\text{N m/J})(1\,\text{kg m s}^{-2}/\text{N})}{(\pi)(28\,\text{kg/kmol})}\right\}^{1/2}$$

$$= 778\,\text{m/s}$$

$$\mathcal{D}_{K1,\text{eff}} = \frac{2}{3}(0.8/4.0)(10^{-6}\,\text{m})(778\,\text{m/s}) = 1.04 \times 10^{-4}\,\text{m}^2/\text{s}$$

Then, from Eq. (1.69), the effective diffusion coefficient is

$$\frac{1}{\mathcal{D}_{1,\text{eff}}} = \frac{1}{\mathcal{D}_{12,\text{eff}}} + \frac{1}{\mathcal{D}_{K1,\text{eff}}} = \frac{1}{2.12 \times 10^{-5}} + \frac{1}{1.04 \times 10^{-4}}$$

$$\mathcal{D}_{1,\text{eff}} = 1.76 \times 10^{-5}\,\text{m}^2/\text{s}$$

Next the rate constant is calculated:

$$k'' = Ae^{-E_a/\mathcal{R}T}\,\text{m/s}$$

$$= 50.6\exp\left(-\frac{(18.6 \times 10^3\,\text{cal/mol})}{(1.987\,\text{cal/mol K})(800\,\text{K})}\right)$$

$$= 50.6e^{-11.70} = 4.19 \times 10^{-4}\,\text{m/s}$$

The product of the rate constant and catalytic surface area per unit volume is

$$k''a_p = (4.19 \times 10^{-4}\,\text{m/s})(7.12 \times 10^5\,\text{cm}^2/\text{cm}^3)(10^2\,\text{cm/m}) = 2.98 \times 10^4\,\text{s}^{-1}$$

The Thiele modulus Λ may now be calculated:

$$\Lambda = \frac{R}{3}\left(\frac{k''a_p}{\mathcal{D}_{1,\text{eff}}}\right)^{1/2} = \frac{(0.25 \times 10^{-2}\,\text{m})}{3}\left(\frac{2.98 \times 10^4\,\text{s}^{-1}}{1.76 \times 10^{-5}\,\text{m}^2/\text{s}}\right)^{1/2} = 34.3$$

Since Λ is large, we may use Eq. (1.83) to obtain the effectiveness:

$$\eta_p \simeq \frac{1}{\Lambda} = \frac{1}{34.3} = 2.92 \times 10^{-2}\quad\text{or } 2.92\%$$

Comments

1. Pellets are usually designed to have a low effectiveness, because in practice catalysts become "poisoned," reducing k''. For a new pellet, the reaction zone will be a thin spherical shell just below the surface. As time goes on,

the zone will move toward the center of the pellet as the outer surface area becomes poisoned.

2. An alternative catalyst support used in automobile catalytic converters is a ceramic matrix. The catalyst is impregnated into a thin porous alumina layer (*washcoat*) that is applied to the passage walls. A typical matrix has passages of hydraulic diameter about 1 mm, and the washcoat may be about 20 μm thick. Diffusion into the porous layer can then be analyzed in Cartesian coordinates (see Exercise 1–54). ▲

1.6.3 Mass Transfer in a Pellet Bed

In Section 1.6.2 we were concerned only with diffusion inside the catalyst pellet. We now consider coupling between the transport of reactant to the exterior surface of the pellet on the one hand and transport within the pellet itself on the other, as occurs in a catalytic converter. We shall not be concerned with the overall performance of the converter; such matters belong more properly in Chapter 3, which deals with mass exchangers. Each pellet may be viewed as being surrounded by a gas flow containing a free-stream reactant concentration $x_{1,e}$. Mass transfer to the exterior surface of the pellet is described by Eq. (1.37), $J_{1,s} = \mathcal{G}_{m1}(x_{1,s} - x_{1,e})$, where $\mathcal{G}_{m1}$ is the mole transfer conductance and $x_{1,s}$ is the concentration of reactant at the s-surface shown in Fig. 1.23. As before, the pellet volume is V_p and the external surface is S_p; the catalytic area per unit volume is a_p. For a first-order reaction characterized by rate constant k'', the effectiveness at the specified temperature and pressure is η_p. We have two diffusive resistances in series, as shown in Fig. 1.24:

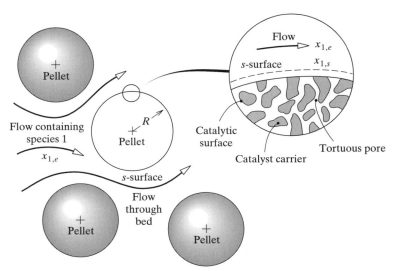

Figure 1.23 A porous catalyst pellet in a packed bed for the removal of species 1.

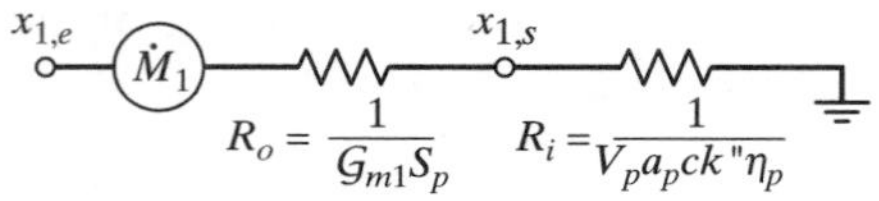

Figure 1.24 Mass flow circuit for porous catalyst pellets in a fixed-bed catalytic reactor.

1. An outside convective resistance $R_o = 1/\mathcal{G}_{m1}S_p$ (based on the pellet exterior surface area S_p).

2. An inside resistance that accounts for both diffusion through the pores and the chemical reaction. The potential difference across the inside resistance is $x_{1,s}(= x_{1,s} - 0)$, and the molar current is $\dot{M}_1 = -J_{1,s}S_p = V_p a_p k'' c x_{1,s} \eta_p$. Since $R_i = x_{1,s}/\dot{M}_1$, we obtain $R_i = 1/V_p a_p c k'' \eta_p$.

The rate of reaction is

$$\dot{M}_1 = \frac{(x_{1,e} - 0)}{R_o + R_i} \tag{1.84}$$

Substituting for the resistances R_o and R_i and rearranging gives

$$\dot{M}_1 = \frac{x_{1,e}}{\dfrac{1}{V_p a_p \eta_p k'' c} + \dfrac{1}{\mathcal{G}_{m1} S_p}} \tag{1.85}$$

The equivalence of Eqs. (1.85) and (1.35) is made clear if $\mathcal{G}_{m1}$ is replaced by $c\mathcal{D}_{12}/\delta$, where δ is the thickness of an equivalent stagnant film. Then

$$\dot{M}_1 = \frac{x_{1,e}}{\dfrac{1}{(V_p a_p/S_p)\eta_p k'' c S_p} + \dfrac{\delta}{c\mathcal{D}_{12}S_p}} \tag{1.86}$$

The quantity $(V_p a_p/S_p)$ is the area of catalyst per unit area of pellet exterior—that is, of s-surface—and η_p is the fraction of this area that is effective in promoting the reaction.

Equation (1.85) gives the rate of reaction for a single pellet. The rate of reaction per unit exterior surface area of the pellet is $\dot{M}_1/S_p$. When the pellets are packed in a bed, the surface area of the pellets can be obtained from either the specific surface area of the bed a or the transfer perimeter $\mathcal{P}$, both of which are defined in Section 4.5.2 in *Heat Transfer*. Notice that in deriving Eq. (1.85), we have assumed $\mathcal{G}_{m1}$ and $x_{1,s}$ to be uniform around the pellet, in order to use the result for η_p obtained in Section 1.6.2. In a packed bed, both $\mathcal{G}_{m1}$ and $x_{1,s}$ will vary somewhat around the pellet; however, our result is adequate for engineering purposes.

Example 1.13 A Pellet Bed for an Automobile Catalytic Converter

A catalytic converter attached to the exhaust system of an automobile is packed with 0.5 cm–diameter spherical CuO-on-alumina catalyst pellets. At a specific location in the bed the mole transfer conductance $\mathcal{G}_{m1}$ is 0.030 kmol/m² s, the pellet temperature is 800 K, and the free-stream concentration of CO is 5.0% by mass. Calculate the CO oxidation rate per unit volume if the bed void fraction is 30%.

Solution

Given: A packed-pellet-bed catalytic reactor to remove carbon monoxide from automobile exhaust.
Required: CO oxidation rate per unit volume of bed.
Assumptions:

1. Kinetics data for CO oxidation of CuO is as given in Example 1.12.
2. $\mathcal{G}_{m1}$ is uniform around the pellet.

The CO oxidation rate per unit volume is equal to the oxidation rate per unit pellet exterior area times the total surface area of pellets per unit volume. From Eq. (1.85) we have the oxidation rate of CO per unit pellet exterior area as

$$\frac{\dot{M}_1}{S_p} = \frac{x_{CO,e}}{\dfrac{1}{(V_p a_p/S_p)\eta_p k'' c} + \dfrac{1}{\mathcal{G}_{m1}}}$$

For a spherical pellet,

$$\frac{V_p}{S_p} = \frac{R}{3} = \frac{0.25 \times 10^{-2}}{3} = 0.833 \times 10^{-3} \text{ m}$$

From Example 1.12,

$$\eta_p = 0.0292 \quad \text{and} \quad k'' a_p = 2.98 \times 10^4 \text{ s}^{-1}$$

At 800 K and 1 atm,

$$c = \frac{(1.013 \times 10^5)}{(8.314 \times 10^3)(800)} = 1.523 \times 10^{-2} \text{ kmol/m}^3$$

Since N_2 predominates, we assume a mean molecular weight of 28 for the exhaust gases, $M_{CO} = 28$; thus, $x_{CO,e} = m_{CO,e} = 0.05$. Substituting above,

$$(V_p/S_p)\eta_p k'' a_p c = (0.833 \times 10^{-3} \text{ m})(0.0292)(2.98 \times 10^4 \text{ s}^{-1}) \times$$
$$\times (1.523 \times 10^{-2} \text{ kmol/m}^3)$$
$$= 0.0110 \text{ kmol/m}^2 \text{ s}$$

$$\frac{\dot{M}_1}{S_p} = \frac{0.05}{\dfrac{1}{0.0110} + \dfrac{1}{0.030}} = 4.02 \times 10^{-4} \text{ kmol/m}^2 \text{ s}$$

To find the total rate at which CO is oxidized per unit volume of the reactor, we multiply by the area-to-volume ratio A_s/V_{bed}. For N_p pellets in the bed,

$$\frac{A_s}{V_{bed}} = \frac{N_p S_p}{V_{bed}} = \left(\frac{N_p V_p}{V_{bed}}\right)\left(\frac{S_p}{V_p}\right) = (1 - \epsilon_{v,bed})\left(\frac{3}{R}\right)$$

$$= (0.70)/(0.833 \times 10^{-3})$$

$$= 840 \text{ m}^2/\text{m}^3$$

The CO consumed per unit volume is then

$$\frac{\dot{M}_1}{S_p}\frac{A_s}{V_{bed}} = (4.02 \times 10^{-4})(840) = 0.338 \text{ kmol/m}^3\text{ s}$$

Comments

1. Correlations for convective transport and pressure drop in packed particle beds are given in Section 4.5.2 in *Heat Transfer*.
2. The performance of the bed as a mass exchanger is analyzed in Section 3.3.1.   ▲

1.7 TRANSPORT PROPERTIES

The tables of transport property data given in Appendix A have proven useful for a wide range of engineering problem solving. However, in many situations the data are inadequate. For example, we often need binary diffusion coefficients for species pairs not listed in Tables A.17 and A.18 (which are restricted to air mixtures and aqueous solutions, respectively). We also often need the viscosity or thermal conductivity of a gas mixture of arbitrary composition. In this section we develop formulas that will be useful for such purposes. Very small particles (called aerosols when suspended in a gas) behave very much like large molecules. It is also useful to be able to estimate diffusion coefficients of such particles so as to exploit analogies between particle deposition and mass or heat transfer.

In Section 1.7.1 gas mixtures are considered. First, a kinetic theory model involving crude rigid spheres is used to give simple formulas for transport properties of pure species, and to give an understanding of the transport processes. Subsequently, the rigorous Chapman–Enskog kinetic theory and Lennard–Jones potential model are explained. Formulas are given for the viscosity and thermal conductivity of pure species, and for the binary diffusion coefficient. Then Wilke's mixture rules are given, which allow the viscosity and thermal conductivity of gaseous mixtures to be calculated. In Section 1.7.2, models for diffusion in dilute liquid solutions are briefly discussed, and selected empirical formulas for the binary diffusion coefficients are given. In Section 1.7.3, the phenomenon of Brownian motion of small particles is described, and formulas for the calculation of Brownian diffusion coefficients are given.

1.7.1 Gas Mixtures

Rigid-Sphere Kinetic Theory Model

We first consider a simple kinetic theory of gases in which the gas is uniform and the molecules are rigid spheres of diameter d. If molecule 1 is viewed as stationary, a collision will occur when molecule 2 moves into a sphere of radius d around molecule 1, as shown in Fig. 1.25. The area of the circle is termed the *collision cross section* $\sigma_c = \pi d^2$. The relative velocity of the molecules, $\Delta \mathbf{v} = \mathbf{v}_1 - \mathbf{v}_2$, is of the same order of magnitude as the average molecular speed, which may be obtained from a physics or thermodynamics text [11–15] as

$$\bar{v} = \left(\frac{8kT}{\pi m} \right)^{1/2} \tag{1.87}$$

where k is the Boltzmann constant and m is the mass of a molecule. It is quite simple to show that the average magnitude of the relative velocity between two colliding molecules is

$$\overline{\Delta v} = \sqrt{2}\bar{v} \tag{1.88}$$

During a time t, molecule 2 moves a distance $\overline{\Delta v}t$ relative to molecule 1. The number of collisions during this time is equal to the volume swept out, $\overline{\Delta v}t\pi d^2$, times the average number of molecules per unit volume, which is the number density $\mathcal{N}$. The average time between collisions is then

$$t_c = \frac{1}{\overline{\Delta v}\pi d^2 \mathcal{N}} = \frac{1}{\sqrt{2}\bar{v}\pi d^2 \mathcal{N}} \tag{1.89}$$

and during this time a molecule travels an average distance of

$$\ell = \bar{v}t_c = \frac{1}{\sqrt{2}\pi d^2 \mathcal{N}} = \frac{1}{\sqrt{2}\sigma_c \mathcal{N}} \tag{1.90}$$

which is the *collision mean free path*. Since $\mathcal{N} = P/kT$, ℓ is inversely proportional to pressure and to the collision cross section.

Detailed consideration of the flux of molecules per unit solid angle, and changes in the flux due to collisions, shows that the effective cross section affecting transport rates is

$$\sigma_t = (1 - \overline{\cos\theta})\sigma_c \tag{1.91}$$

where θ is the angle that a molecule is deflected by a collision. If there is no deflection, $\theta = 0$, and we do not count the collision. If a collision resulted in perfect backscattering with $\theta = \pi$, then such a collision would have twice the effect of one that merely turned the molecule through $\pi/2$. For molecules of the same mass, $\overline{\cos\theta} = 2/3$, and if a light molecule of mass m_1 is scattered from a heavy molecule of mass m_2, $\overline{\cos\theta} = (2/3)(m_1/m_2)$ [16]. The *transport mean free path* is defined in

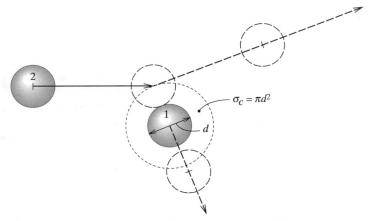

Figure 1.25 A molecule collision showing the collision cross section $\sigma_c = \pi d^2$.

terms of σ_t rather than σ_c,

$$\ell_t = \frac{1}{\sqrt{2}\sigma_t \mathcal{N}} \tag{1.92}$$

and is somewhat longer than the collision mean free path.

Self-Diffusion Coefficient

If the gas is a stationary isothermal mixture of two isotopes, say of uranium fluoride, the molecules have essentially the same mass m and diameter d. We wish to calculate the net transport of isotope 1 down its concentration gradient as shown in Fig. 1.26. The kinetic theory of gases gives the one-way flux of molecules across a plane in a uniform gas as (see also Eq. (7.103) in *Heat Transfer*):

$$J = \frac{1}{4}\bar{v}\mathcal{N} \tag{7.103}$$

The molecules of isotope 1 crossing the plane in the positive z-direction come, on average, from a distance $b\ell_t$ below the plane, where b is of order magnitude unity; thus

$$J_1^+ = \frac{1}{4}\bar{v}\mathcal{N}_1|_{z-b\ell_t} \tag{1.93a}$$

The molecules of species 1 crossing the plane in the negative z-direction come from an average distance $b\ell_t$ above the plane,

$$J_1^- = \frac{1}{4}\bar{v}\mathcal{N}_1|_{z+b\ell_t} \tag{1.93b}$$

The net flux of molecules of species 1 crossing the plane is

$$J_1 = J_1^+ - J_1^- = \frac{1}{4}\bar{v}\left[\mathcal{N}_1|_{z-b\ell_t} - \mathcal{N}_1|_{z+b\ell_t}\right] \tag{1.94}$$

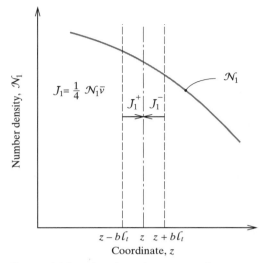

Figure 1.26 Diffusion of a gaseous isotope down a concentration gradient.

If the concentration gradient is small, a Taylor series expansion gives

$$\mathcal{N}_1|_{z-b\ell_t} = \mathcal{N}_1|_z + \frac{d\mathcal{N}_1}{dz}(-b\ell_t) + \cdots$$

$$\mathcal{N}_1|_{z+b\ell_t} = \mathcal{N}_1|_z + \frac{d\mathcal{N}_1}{dz}(+b\ell_t) + \cdots$$

Substituting in Eq. (1.94),

$$J_1 = -\frac{b}{2}\bar{v}\ell_t\frac{d\mathcal{N}_1}{dz} \tag{1.95}$$

Dividing by Avogadro's number $\mathcal{A}$ gives the molar flux of species 1 as

$$J_1 = -\frac{b}{2}\bar{v}\ell_t c\frac{dx_1}{dz} \tag{1.96}$$

where the molar concentration c has been taken to be constant for the isothermal isotope mixture. Comparing Eq. (1.96) with Fick's law, Eq. (1.18), gives the self-diffusion coefficient

$$\mathcal{D}_{11} = \frac{b}{2}\bar{v}\ell_t \tag{1.97}$$

Since molecules crossing the plane come from all angles, it is clear that $b < 1$; more careful analysis shows that an appropriate value of b for isotopes is 2/3 [15]. Then

$$\mathcal{D}_{11} = \frac{1}{3}\bar{v}\ell_t = \frac{1}{3}\left(\frac{8kT}{\pi m}\right)^{1/2}\left(\frac{1}{\sqrt{2}\sigma_t\mathcal{N}}\right) \tag{1.98}$$

Substituting $\bar{k}/m = \mathcal{R}/M$, $\mathcal{N} = c\mathcal{A} = P\mathcal{A}/RT$, $\sigma_t = (1 - \overline{\cos\theta})\pi d^2$, $\overline{\cos\theta} = 2/3$, gives

$$\mathcal{D}_{11} = 2\left(\frac{\mathcal{R}}{\pi}\right)^{3/2} \frac{T^{3/2}}{\mathcal{A}M^{1/2}d^2 P} \tag{1.99}$$

Equation (1.99) shows that the self-diffusion coefficient is proportional to temperature to the three-halves power, is inversely proportional to pressure, and decreases as the molecular weight and molecular diameter increase. However, the effective molecular diameter as it affects the collision process actually decreases with increasing temperature before becoming constant at high temperatures: thus the $T^{3/2}$ dependence is valid only at high temperatures.

Viscosity

If the gas is in simple shear flow, as shown in Fig. 1.27, molecules coming from below the plane at z have, on average, the velocity u at location $z - b\ell_t$. The momentum flux upward is

$$J^+ m \left(u|_z - \frac{du}{dz}b\ell_t + \cdots \right)$$

The momentum flux downward is similarly

$$J^- m \left(u|_z + \frac{du}{dz}b\ell_t + \cdots \right)$$

There is no net flow in the z direction, so $J^+ = J^- = (1/4)\bar{v}\mathcal{N}$. Then, consistent with the sign convention introduced in Section 5.2.1 in *Heat Transfer*, the shear stress

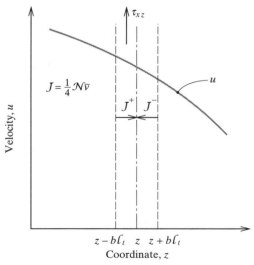

Figure 1.27 Transport of momentum in a simple shear flow of a gas.

at the plane z is the difference between the upward and downward momentum fluxes:

$$\tau_{xz} = \frac{1}{4}\bar{v}\mathcal{N}m\left[-2b\ell_t\frac{du}{dz}\right] = -\frac{b}{2}\bar{v}\rho\ell_t\frac{du}{dz} \tag{1.100}$$

Comparison with Newton's law of viscosity, $\tau = -\mu(du/dz)$, gives the viscosity as

$$\mu = \frac{b}{2}\bar{v}\rho\ell_t \tag{1.101}$$

Comparison with Eq. (1.97) shows that $\mu = \rho\mathcal{D}_{11}$, that is, the Schmidt number $\mathrm{Sc}_{11} = \mu/\rho\mathcal{D}_{11} = 1$. It follows that viscosity has the same pressure and temperature dependence as the $\rho\mathcal{D}$ product. If we again take $b = 2/3$, since the gas is uniform,

$$\mu = \frac{1}{3}\bar{v}\rho\ell_t \tag{1.102}$$

Equation (1.102) is the basis of the most convenient method for estimating the transport mean path; solving for ℓ_t,

$$\ell_t = 3v/\bar{v} = v\left(\frac{9\pi M}{8\mathcal{R}T}\right)^{1/2} \tag{1.103}$$

Thermal Conductivity

To complete this simple kinetic theory, we consider a uniform gas of monatomic molecules with an imposed temperature gradient, as shown in Fig. 1.28. Each molecule has, on an average, a kinetic energy of $(3/2)\,kT$; however, the more energetic molecules cross the plane more frequently so that the energy transport per

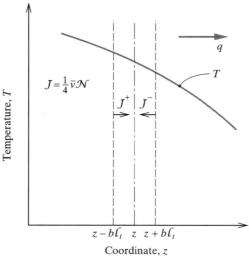

Figure 1.28 Transport of energy in a gas by monatomic molecules.

molecule proves to be $2kT$ [15]. The net energy transport across the plane is

$$q = \frac{1}{4}\bar{v}\mathcal{N}\left[(2kT)_{z-b\ell_t} - (2kT)_{z+b\ell_t}\right] = -\frac{b}{2}\bar{v}\mathcal{N}\ell_t(2k)\frac{dT}{dz} \qquad (1.104)$$

Comparing Eq. (1.104) with Fourier's law of heat conduction, $q = -k(dT/dz)$ with $b = 2/3$ gives the thermal conductivity as

$$k = \frac{1}{3}\bar{v}\rho\ell_t(2R/M) \qquad (1.105)$$

where the relations $\mathcal{N}m = \rho$ and $R/M = k/m$ have been used. The ratio of conductivity to viscosity is then $2R/M$. However, this rigid-sphere model result is not very accurate: more rigorous kinetic theory of gases gives this ratio as $5/2 R/M$, which is higher because the faster, more energetic molecules have a longer-than-average mean free path. In the case of a polyatomic gas, transport of energy associated with the internal degrees of freedom must also be considered.

The Chapman–Enskog Kinetic Theory

The simple kinetic theory presented above provides a model, on a molecular scale, for physical reasoning about transport mechanisms in gases. The Chapman–Enskog kinetic theory of gases [17, 18] is based both on a more realistic physical model and a rigorous mathematical formulation and solution. Some features of the theory that give an indication of the range of validity of the results are as follows. The density of the mixture must be low enough for three-body collisions to occur with negligible frequency. Except at very low temperatures, pressures less than 100 atm are low enough to ensure that this condition is met. The model assumes monatomic molecules, but little error is introduced by applying the results to polyatomic gases. The viscosity and mass diffusivity are not appreciably affected by internal degrees of freedom. The thermal conductivity depends on the energy of both the translational and internal degrees of freedom: the so-called *Eucken correction* is introduced to account for the additional contribution of rotational and vibrational modes. Finally, the theory neglects higher than first-order spatial derivatives of temperature, pressure, concentration, and so on. Thus, the results are inapplicable when gradients change abruptly, for example, within a shock wave.

The forces acting between a pair of molecules during a collision are characterized by a potential energy of interaction ϕ: the functional form of ϕ is illustrated in Fig. 1.29. An empirical representation of the potential energy function that has proven fairly successful is the Lennard–Jones 6–12 potential model,

$$\phi(r) = 4\epsilon\left[\left(\frac{\sigma}{r}\right)^{12} - \left(\frac{\sigma}{r}\right)^{6}\right] \qquad (1.106)$$

where σ, the *collision diameter*, is the value of r for which $\phi(r) = 0$, and ϵ is the maximum energy of attraction between a pair of molecules. The model exhibits weak attraction, due to London dispersion forces at large separations (like r^{-6}), and strong repulsion due to electron cloud overlapping, at small separations (nearly like r^{-12}). Table A.26 lists values of σ and ϵ for a number of chemical species. In cases where the Lennard–Jones model parameters are not given in Table A.26, they may

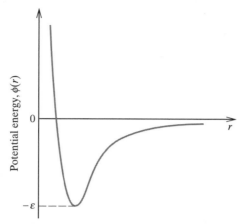

Figure 1.29 Potential energy for interaction as given by the Lennard–Jones 6–12 model.

be estimated using the following rules:

$$\sigma \simeq 8.33V_c^{1/3} \simeq 11.8V_b^{1/3} \tag{1.107}$$

$$\epsilon/k \simeq 0.75T_c \simeq 1.21T_b \tag{1.108}$$

where subscripts c and b refer to the critical state and boiling point, respectively, and V has units $[\text{m}^3/\text{kmol}]$. Table A.27 lists boiling point data for selected species.

The Lennard–Jones model describes a spherically symmetrical force field and hence is intended for use with nonpolar, nearly symmetrical molecules (for example, O_2, He, CO). Indeed, the Chapman–Enskog theory is, strictly speaking, valid only for molecules with spherically symmetrical force fields. Molecules with appreciable dipole moments (for example, H_2O, NH_3) or that are highly elongated (for example, C_3H_6, n-C_6H_{14}) interact with potentials that are angle-dependent. For polar molecules the Stockmayer potential model, which adds an angle-dependent factor to the Lennard–Jones expression, has been successfully used. However, for many practical purposes, the Lennard–Jones potential has been found adequate even for polar and elongated molecules. The usual practice is to *determine* the parameters σ and ϵ by matching theoretical viscosity predictions with experimental data. In this way experimental viscosity data are extrapolated outside the original temperature range; also, the same values of σ and ϵ usually (but not always) prove to be the best available for the estimation of thermal conductivity and mass diffusivity.

When the Lennard–Jones potential is used in the Chapman–Enskog kinetic theory of gases, the viscosity of a pure monatomic gas is given by

$$\mu = 2.67 \times 10^{-6}\frac{\sqrt{MT}}{\sigma^2\Omega_\mu} \text{ kg/m s} \tag{1.109}$$

where T is in kelvins and σ is in angstroms (1 angstrom = 1 Å = 10^{-10} m). The quantity Ω_μ is the *collision integral* and is tabulated in Table A.28; it is a weak function of temperature, becoming very nearly constant at high temperatures.

The Chapman–Enskog theory shows that, for a monatomic gas, the relation between thermal conductivity and viscosity is [17]

$$k = \frac{5}{2}c_v\mu, \quad \left(c_v = \frac{3}{2}\frac{\mathcal{R}}{M}\right)$$

Thus

$$k_{\text{translational}} = 8.32 \times 10^{-2}\frac{\sqrt{T/M}}{\sigma^2\Omega_k} \text{ W/m K} \tag{1.110}$$

The collision integral for thermal conductivity is identical to that for viscosity, $\Omega_k = \Omega_\mu$. Equation (1.110) is valid for monatomic gases; for polyatomic gases we add the *modified Eucken correction* to account for the contribution of the internal degrees of freedom:

$$k = k_{\text{translational}} + 1.32\left(c_p - \frac{5}{2}\frac{\mathcal{R}}{M}\right)\mu \tag{1.111}$$

We see that data for specific heats are required in order to calculate the thermal conductivity of polyatomic gases.

The mass diffusivity, in the form of the binary diffusion coefficient, is

$$\mathcal{D}_{12} = 1.86 \times 10^{-7}\frac{\sqrt{T^3\left(\dfrac{1}{M_1} + \dfrac{1}{M_2}\right)}}{\sigma_{12}^2\Omega_{\mathcal{D}}P} \text{ m}^2/\text{s} \tag{1.112}$$

where P is in atmospheres. The intermolecular potential field for a pair of unlike molecules, species 1 and 2, is approximated as

$$\phi_{12}(r) = 4\epsilon_{12}\left[\left(\frac{\sigma_{12}}{r}\right)^{12} - \left(\frac{\sigma_{12}}{r}\right)^6\right] \tag{1.113}$$

The collision integral for mass diffusivity $\Omega_{\mathcal{D}}$ differs from Ω_μ; values of $\Omega_{\mathcal{D}}$ are tabulated in Table A.28. The Lennard–Jones parameters σ_{12} and ϵ_{12} must be obtained from the empirical relations

$$\sigma_{12} = \frac{1}{2}(\sigma_1 + \sigma_2) \tag{1.114}$$

$$\epsilon_{12} = \sqrt{\epsilon_1\epsilon_2} \tag{1.115}$$

Mixture Rules

Next follows a prescription of how the viscosity and thermal conductivity of a gas mixture can be estimated from the values for the pure species. C. R. Wilke [19] simplified the rigorous kinetic theory prediction by introducing elements of the rigid-sphere model into the final result. The formulas are simple and have proven to be quite adequate.

$$\mu_{\mathrm{mix}} = \sum_{i=1}^{n} \frac{x_i \mu_i}{\sum_{j=1}^{n} x_j \Phi_{ij}} \tag{1.116}$$

$$k_{\mathrm{mix}} = \sum_{i=1}^{n} \frac{x_i k_i}{\sum_{j=1}^{n} x_j \Phi_{ij}} \tag{1.117}$$

where

$$\Phi_{ij} = \frac{\left[1 + \left(\mu_i/\mu_j\right)^{1/2} \left(M_j/M_i\right)^{1/4}\right]^2}{\sqrt{8}[1 + (M_i/M_j)]^{1/2}} \tag{1.118}$$

The important feature of these formulas is that the weighting is essentially with mole (number) fraction, as we would expect from simple kinetic theory.

Effective Binary Diffusion Coefficient

In a binary system, ordinary diffusion takes place whenever there is a concentration gradient, and is in the direction of decreasing concentration. It is tempting to intuitively generalize this result to a multicomponent system. However, as shown in advanced texts, in a multicomponent system it is possible for a species to move by ordinary diffusion up its concentration gradient. In a gas mixture, a light species such as H_2 or He can be pushed up its concentration gradient by collisions with heavier molecules in the mixture diffusing in that direction. Unfortunately, the exact theory of multicomponent diffusion is algebraically complicated and difficult to use. Thus, whenever possible, we try to solve engineering problems using the effective binary diffusion approach.

Fick's law, as we have defined it, is strictly valid only for binary mixtures—hence the use of a binary diffusion coefficient $\mathcal{D}_{12}$. If species 1 is in small concentration, we may safely define an *effective binary diffusion coefficient* for species 1 diffusing through the mixture, which we have denoted $\mathcal{D}_{1m}$ (see, for example, Section 1.3.3). This diffusion coefficient is an appropriate average value for interaction with the various species in the mixture or solution. For gas mixtures, a simple kinetic theory argument suggests

$$\mathcal{D}_{1m} \simeq \frac{(1 - x_1)}{\sum_{i=2}^{n} (x_i/\mathcal{D}_{1i})}, \qquad x_1 \ll 1 \tag{1.119}$$

This formula has a number of features:

1. For a true binary mixture of species 1 and 2, it gives the correct result of $\mathcal{D}_{1m} = \mathcal{D}_{12}$.
2. Similarly, in the limit of all the diffusion coefficients being equal, it recovers $\mathcal{D}_{1m} = \mathcal{D}_{12} = \mathcal{D}_{13} = \cdots$.
3. The addition of a small amount of a light gas (say species 3) to the mixture, small enough to leave m_1 essentially unaffected, does not have a significant effect on $\rho \mathcal{D}_{1m}$, even though $\mathcal{D}_{13}$ is much larger than the other $\mathcal{D}_{1i}$ values.

The Computer Programs GASMIX and AIRSTE

The computer program GASMIX calculates thermodynamic and transport properties of ideal gas mixtures. There is a menu of 36 species. The required inputs are temperature, pressure, and the mass fractions of the component species. Molecular weights and density are calculated using Table A.20. Specific heat is calculated from curve fits of the data listed in Table A.29. Transport properties are calculated using the Lennard–Jones potential model parameters listed in Tables A.26 and A.28.

The calculation of properties of mixtures containing water vapor poses a special problem, owing to the peculiar nature of the H_2O molecule. GASMIX is adequate for calculating properties of many mixtures containing water vapor, such as the combustion products of hydrocarbon fuels. However, for the widely encountered water vapor–air system, more accurate results are usually desirable. The computer program AIRSTE has been prepared for this purpose and is based on empirical data.

Example 1.14   Thermal Conductivity of Methane

Calculate the thermal conductivity of methane (CH_4) at a temperature of 350 K and 1 atm pressure.

Solution

Given: Methane gas at 350 K, 1 atm.
Required: Thermal conductivity.
Assumptions: The Lennard–Jones potential model is adequate.
As a first step, we calculate the thermal conductivity of methane as if it were a monatomic molecule, and its viscosity. From Table A.20, $M = 16.0$. From Table A.26, $\sigma = 3.758$ Å and $\epsilon/k = 149$ K; hence $kT/\epsilon = 350/149 = 2.35$. From Table A.28, $\Omega_\mu = \Omega_k = 1.114$. Using Eqs. (1.110) and (1.109),

$$k_{\text{translational}} = \frac{8.32 \times 10^{-2}\sqrt{350/16.0}}{(3.758)^2(1.114)} = 2.47 \times 10^{-2} \text{ W/m K}$$

$$\mu = \frac{2.67 \times 10^{-6}\sqrt{(16.0)(350)}}{(3.758)^2(1.114)} = 1.27 \times 10^{-5} \text{ kg/m s}$$

The actual thermal conductivity of CH_4 is then obtained from Eq. (1.111) and includes the contributions of both the translational and internal degrees of freedom:

$$k = k_{\text{translational}} + 1.32\left(c_p - \frac{5}{2}\frac{\mathcal{R}}{M}\right)\mu$$

The specific heat c_p can be obtained from thermodynamic tables; alternatively, since many polyatomic molecules do not have their vibrational degrees of freedom excited at normal temperatures, a simple formula for c_p is

$$c_p = (5 + N_r)\left(\frac{1}{2}\frac{\mathcal{R}}{M}\right)$$

where N_r is the number of rotational degrees of freedom (zero for a monatomic gas, two for a linear molecule, and three for a nonlinear one). Thus $N_r = 3$ for

CH$_4$ and

$$c_p - \frac{5}{2}\frac{\mathcal{R}}{M} = \frac{3}{2}\frac{\mathcal{R}}{M} = \frac{(3)(8314)}{(2)(16.0)} = 779.4 \text{ J/kg K}$$

Hence $k = 2.47 \times 10^{-2} + 1.32(779.4)(1.27 \times 10^{-5}) = 3.78 \times 10^{-2}$ W/m K.

Comment Compare this result with the value given by GASMIX. ▲

Example 1.15 Diffusivity of Sulfur Dioxide in Air

Calculate the binary diffusion coefficient of sulfur dioxide (SO$_2$) in air at a temperature of 500 K and 1.2×10^5 Pa pressure.

Solution

Given: An SO$_2$-air mixture at 500 K and 1.2×10^5 Pa.
Required: Binary diffusion coefficient.
Assumptions: The Lennard–Jones potential model is satisfactory.
Equation (1.112) gives the binary diffusion coefficient as

$$\mathcal{D}_{12} = 1.86 \times 10^{-7}\frac{\sqrt{T^3\,(1/M_1 + 1/M_2)}}{\sigma_{12}^2\Omega_{\mathcal{D}}P} \text{ m}^2/\text{s}$$

Using Tables A.20 and A.26,

Species	M	$\sigma\,[\text{Å}]$	$\frac{\epsilon}{k}\,[\text{K}]$
SO$_2$	64.1	4.112	335
Air	29.0	3.711	79

$$\sigma_{12} = \frac{1}{2}(\sigma_1 + \sigma_2) = \frac{1}{2}(4.112 + 3.711) = 3.912$$

$$\frac{\epsilon_{12}}{k} = \sqrt{\left(\frac{\epsilon_1}{k}\right)\left(\frac{\epsilon_2}{k}\right)} = \sqrt{(335)(79)} = 163$$

$$\frac{kT}{\epsilon} = \frac{500}{163} = 3.07$$

From Table A.28, $\Omega_{\mathcal{D}} = 0.943$. Thus,

$$\mathcal{D}_{12} = 1.86 \times 10^{-7}\frac{\sqrt{500^3\,(1/64.1 + 1/29.0)}}{(3.912)^2(0.943)(1.2 \times 10^5/1.0133 \times 10^5)}$$

$$= 27.2 \times 10^{-6} \text{ m}^2/\text{s}$$

Comment GASMIX gives $\mathcal{D}_{12} = 27.25 \times 10^{-6}$ m^2/s. ▲

Example 1.16 Viscosity of a Gaseous Fuel

A gaseous fuel has a composition of 70% (by volume) propane (C_3H_8) and 30% isobutane (iso-C_4H_{10}). Calculate the viscosity of the mixture at 300 K and 2 atm pressure.

Solution

Given: Mixture of propane and isobutane at 300 K and 2 atm.
Required: Viscosity.
Assumptions:

1. The Lennard–Jones potential model can be used for the pure species.

2. Wilke's mixture rule is adequate.

First we calculate the viscosity of the pure species C_3H_8 and iso-C_4H_{10} at 300 K from Eq. (1.109):

$$\mu = 2.67 \times 10^{-6} \frac{\sqrt{MT}}{\sigma^2 \Omega_\mu} \text{ kg/m s}$$

Using Tables A.20, A.26, and A.28:

Species	M	σ [Å]	$\frac{\epsilon}{k}$ [K]	$\frac{kT}{\epsilon}$	Ω_μ
C_3H_8	44	5.118	237	1.266	1.416
Iso-C_4H_{10}	58	5.278	330	0.909	1.667

$$\mu_1 = \frac{2.67 \times 10^{-6}\sqrt{(44)(300)}}{(5.118)^2(1.416)} = 8.27 \times 10^{-6} \text{ kg/m s}$$

$$\mu_2 = \frac{2.67 \times 10^{-6}\sqrt{(58)(300)}}{(5.278)^2(1.667)} = 7.58 \times 10^{-6} \text{ kg/m s}$$

The mixture rule is given by Eq. (1.116):

$$\mu_{\text{mix}} = \sum_{i=1}^{n} \frac{x_i \mu_i}{\sum_{j=1}^{n} x_j \Phi_{ij}} = \frac{x_1 \mu_1}{x_1 \Phi_{11} + x_2 \Phi_{12}} + \frac{x_2 \mu_2}{x_1 \Phi_{21} + x_2 \Phi_{22}}$$

where

$$
\Phi_{ij} = \frac{\left[1 + \left(\dfrac{\mu_i}{\mu_j}\right)^{1/2} \left(\dfrac{M_j}{M_i}\right)^{1/4}\right]^2}{\sqrt{8}[1 + (M_i/M_j)]^{1/2}}; \quad (\Phi_{11} = \Phi_{22} = 1)
$$

$$
\Phi_{12} = \frac{\left[1 + \left(\dfrac{8.27 \times 10^{-6}}{7.58 \times 10^{-6}}\right)^{1/2} \left(\dfrac{58}{44}\right)^{1/4}\right]^2}{\sqrt{8}[1 + (44/58)]^{1/2}} = 1.20
$$

$$
\Phi_{21} = \frac{\left[1 + \left(\dfrac{7.58 \times 10^{-6}}{8.27 \times 10^{-6}}\right)^{1/2} \left(\dfrac{44}{58}\right)^{1/4}\right]^2}{\sqrt{8}[1 + (58/44)]^{1/2}} = 0.83
$$

Then with the mole fractions $x_1 = 0.7$ and $x_2 = 0.3$ the mixture viscosity is

$$
\mu_{\text{mix}} = \frac{(0.7)(8.27 \times 10^{-6})}{(0.7)(1) + (0.3)(1.20)} + \frac{(0.3)(7.58 \times 10^{-6})}{(0.7)(0.83) + (0.3)(1)}
$$

$$
= 8.04 \times 10^{-6} \text{ kg/m s}
$$

Comments

1. Notice that viscosity is independent of pressure at normal pressures.
2. More accurate values of the molecular weights are 44.1 and 58.1 for C_3H_8 and iso-C_4H_{10}, respectively. However, the mixture rule is not of high accuracy, so the values used in the calculations are quite adequate. ▲

1.7.2 Dilute Liquid Solutions

Table A.18 gives Schmidt numbers for various species in dilute solution in water, from which diffusion coefficients can be calculated in a simple manner. We now present formulas for diffusion coefficients in liquids of more general applicability. Unfortunately, there is no theory of diffusion in liquids of rigor comparable to the Chapman–Enskog kinetic theory of gases. An approximate theory that has been reasonably successful is based on a hydrodynamic interpretation of the *Nernst–Einstein equation*, which was originally used to describe Brownian motion of particles in liquids. The extrapolation to diffusion of molecules is surprisingly good. In its most fundamental form, the Nernst–Einstein equation is

$$
\mathcal{D}_{12} = \frac{kT}{f_{12}} \tag{1.120}
$$

where f_{12} [N s/m] is the *friction coefficient* of solute 1 in solvent 2, and is defined as the force required to give a molecule of species 1 unit velocity. If f_{12} is evaluated assuming creeping flow (Re $\ll$ 1) and no slip, Stokes' law, Eq. (4.73) in *Heat Transfer*, applies and

$$\mathcal{D}_{12} = \frac{kT}{6\pi\,\mu_2 R_1} \tag{1.121}$$

where μ_2 is the viscosity of the pure solvent and R_1 is the radius of the diffusing molecule. Equation (1.121) shows that $\mathcal{D}_{12}$ should be inversely proportional to solvent viscosity and molecule size, and directly proportional to absolute temperature (in addition to the temperature dependence of μ_2). Equation (1.121) can be refined by allowing for slip between the molecule and solvent, but the result must still be empirically modified to give good agreement with measured values. Equation (1.121) is usually called the *Stokes–Einstein relation*. Wilke and Chang [20] proposed a useful formula, valid for dilute solutions of nondissociating solute 1 in solvent 2,

$$\mathcal{D}_{12} = 1.17 \times 10^{-16}\frac{(\phi_2 M_2)^{1/2}T}{\mu\tilde{V}_{1b}^{0.6}} \text{ m}^2/\text{s} \tag{1.122}$$

where $\tilde{V}_{1b}$ is the molar volume of the solute [m^3/kmol] as liquid at its normal boiling point, μ is the viscosity of the solution [kg/m s], and ϕ_2 is an "association parameter" for the solvent. Recommended values of ϕ_2 are 2.6 for water, 1.9 for methanol, 1.5 for ethanol, and 1.0 for heptane, ether, benzene, and other unassociated solvents. Equation (1.122) is usually accurate to ±10%. For dilute aqueous solutions, the Othmer–Thakar [21] correlation is accurate to about 11% and is particularly simple:

$$\mathcal{D}_{12} = \frac{1.11 \times 10^{-13}}{\mu_2^{1.1}\tilde{V}_{1b}^{0.6}} \text{ m}^2/\text{s} \tag{1.123}$$

Table A.27 lists values of molar volumes at the normal boiling point for various liquids. Alternatively, Benson's formula can be used,

$$\tilde{V}_c/\tilde{V}_b = \rho_b/\rho_c = 0.422 P_c \text{ [atm]} + 1.981 \tag{1.124}$$

where the subscript c refers to the critical state. Tables of critical-state properties are readily available or they can be estimated. Another popular method for determining $\tilde{V}_b$ is to use the additive method of Le Bas; see, for example, Reid et al. [22].

The preceding recommendations are for *dilute* solutions only. The theory of diffusion in concentrated liquid solutions is beyond the scope of this text: it is difficult because (1) multicomponent diffusion must be considered, and (2) concentrated solutions are generally nonideal solutions, and the appropriate driving potential for diffusion is not concentration but the *chemical potential*. Thus, the subject of diffusion coefficients in concentrated solutions will not be discussed either. As a final comment on liquid diffusion coefficients, we note that all liquid diffusion coefficients are of the order 10^{-9}m^2/s at room temperature, irrespective of the nature of the solute or solvent.

There are no simple mixture rules similar to those for gases that allow the viscosity and thermal conductivity of a liquid solution to be obtained. Some complex methods of limited accuracy are available [22], but generally the engineer must rely on measured values. Table A.13 gives some examples.

Example 1.17 Diffusivity of Oxygen and Chlorine in Water

Calculate the binary diffusion coefficients for oxygen and chlorine in dilute solution in water at 300 K using (i) the Wilke–Chang formula, (ii) the Othmer–Thakar formula, and (iii) Table A.18.

Solution

Given: O_2 and Cl_2 in dilute solution in water at 300 K.
Required: Binary diffusion coefficients using (i) the Wilke–Chang formula, (ii) the Othmer–Thakar formula, and (iii) Table A.18.

i. Let species 2 be water. From Eq. (1.122), the Wilke–Chang formula is

$$\mathcal{D}_{12} = 1.17 \times 10^{-16} \frac{(\phi_2 M_2)^{1/2} T}{\mu \tilde{V}_{1b}^{0.6}} \ \text{m}^2/\text{s}$$

Since the solution is dilute, μ can be taken as the value of viscosity for pure water; from Table A.8, $\mu = 8.67 \times 10^{-4} \text{kg/m s}$. From Table A.27, $\tilde{V}_{1b} = 25.6 \times 10^{-3}$ and $48.4 \times 10^{-3} \text{m}^3/\text{kmol}$ for oxygen and chlorine, respectively; also, for water $\phi_2 = 2.6$, $M_2 = 18$.

$$\text{Oxygen: } \mathcal{D}_{12} = 1.17 \times 10^{-16} \frac{(2.6 \times 18)^{1/2}(300)}{(8.67 \times 10^{-4})(25.6 \times 10^{-3})^{0.6}}$$

$$= 2.50 \times 10^{-9} \ \text{m}^2/\text{s}$$

$$\text{Chlorine: } \mathcal{D}_{12} = 1.17 \times 10^{-16} \frac{(2.6 \times 18)^{1/2}(300)}{(8.67 \times 10^{-4})(48.4 \times 10^{-3})^{0.6}}$$

$$= 1.70 \times 10^{-9} \ \text{m}^2/\text{s}$$

ii. From Eq. (1.123), the Othmer-Thakar equation is

$$\mathcal{D}_{12} = \frac{1.11 \times 10^{-13}}{\mu_2^{1.1} \tilde{V}_{1b}^{0.6}}$$

$$\text{Oxygen: } \mathcal{D}_{12} = \frac{1.11 \times 10^{-13}}{(8.67 \times 10^{-4})^{1.1}(25.6 \times 10^{-3})^{0.6}} = 2.34 \times 10^{-9} \ \text{m}^2/\text{s}$$

$$\text{Chlorine: } \mathcal{D}_{12} = \frac{1.11 \times 10^{-13}}{(8.67 \times 10^{-4})^{1.1}(48.4 \times 10^{-3})^{0.6}} = 1.59 \times 10^{-9} \ \text{m}^2/\text{s}$$

iii. Table A.18 gives Schmidt numbers for dilute solution in water at 300 K, and from Table A.8, $\nu = 0.87 \times 10^{-6} \ \text{m}^2/\text{s}$ for water at 300 K.

$$\text{Oxygen: } \mathcal{D}_{12} = \frac{\nu}{Sc_{12}} = \frac{0.87 \times 10^{-6}}{400} = 2.18 \times 10^{-9} \ \text{m}^2/\text{s}$$

$$\text{Chlorine: } \mathcal{D}_{12} = \frac{\nu}{Sc_{12}} = \frac{0.87 \times 10^{-6}}{670} = 1.30 \times 10^{-9} \ \text{m}^2/\text{s}$$

These results are summarized in the table below.

	Source	$\mathcal{D}_{12}$, $m^2/s \times 10^9$	
		O_2-water	Cl_2-water
(i)	Wilke–Chang	2.50	1.70
(ii)	Othmer–Thakar	2.34	1.59
(iii)	Table A.18	2.18	1.30

Comment The discrepancies shown in the table reflect the expected uncertainty in liquid diffusion coefficient data.   ▲

1.7.3 Brownian Diffusion of Particles

Small particles play an important role in many areas of technology. When small particles are suspended in a gas, they are usually called *aerosols*. Two types of aerosols are usually identified. *Dispersion* aerosols are formed by the fracture or atomization of solids or liquids. *Condensation* aerosols result from the condensation of a vapor, or from reactions between gaseous species to form a nonvolatile product. Liquid aerosols are in the form of spherical droplets, but a wide variety of shapes are encountered for solid aerosols. Liquid droplet aerosols are usually called *mists*, solid dispersion aerosols are *dusts*, and solid condensation aerosols are *smokes*. Suspensions of small particles in liquids are also of concern; for example, scale formation in heat exchangers used in geothermal wells can result from deposition of silica particles. *Soot* particles play an important role in furnaces burning hydrocarbon fuels because of their large effect on radiation absorption and emission.

The size of spherical droplets can be characterized by the diameter d_p, but for irregular solid particles the size is more difficult to characterize: often the measure of particle size relates to the particular instrument used to determine the particle size distribution. Aerosol particles vary in size from molecular clusters of 10^{-3} μm size (diameter) to droplets and solid particles as large as 100 μm. Since the mean free path in air at 1 atm and 25°C is $\ell = 0.065$ μm, the particle *Knudsen number*, Kn $= \ell/d_p$, varies from about 10^2 to 10^{-3}. As a result, the gas-particle interactions vary from being in the free molecule regime, through the transition regime, into the continuum regime. Inertial effects are very dependent on particle size: for a particle of unit specific gravity, inertial effects are unimportant for $d_p \leq 0.2$ μm and dominant for $d_p \geq 1.0$ μm. The larger particles settle out rapidly due to gravity, or larger particles in a flowing fluid cannot follow streamlines and thus impact on obstacles. For smaller particles, transfer from a flowing fluid to a surface is by diffusive mechanisms, whereas for larger particles it is primarily by sedimentation and impaction. Since the motion of large aerosol particles is governed by Newton's laws of motion, rather than by diffusion or convection, we will restrict our attention to smaller particles so as to be able to exploit the analogy to chemical species transfer and heat transfer.

The transport of small aerosol particles occurs in a similar manner to that of a molecular species. Small aerosols are convected by the mean flow, and in a turbulent flow they are transported by turbulent eddies in much the same way as a molecular species (of course, due to inertial effects, larger aerosols will not respond to the smallest turbulent fluctuations and may not follow streamlines in large turbulent eddies). Ordinary diffusion of aerosols is usually called *Brownian motion*, and, as might be expected, the Brownian diffusion coefficient decreases with increasing particle mass. Diffusion of aerosols due to a temperature gradient is called *thermophoresis* and often plays an important role. Concentration gradients of molecular species cause transport of aerosols by *diffusiophoresis*. *Forced diffusion* of a charged particle in an electrical field is similar to that for an ionized molecular species. Indeed, aerosol particles often carry electrostatic charges that affect both their transport and deposition. In this chapter attention is restricted to ordinary diffusion and convection. However, in Chapter 3 forced diffusion will be considered in connection with electrostatic precipitators for the removal of particulate matter from power plant and incinerator stack gases.

Brownian motion is governed by a Fick's-type law,

$$J = -\mathcal{D}_p \nabla \mathcal{N} \tag{1.125}$$

where J [particles/m^2 s] is the particle flux, $\mathcal{N}$ [particles/m^3] is the particle number density, and $\mathcal{D}_p$ [m^2/s] is the Brownian diffusion coefficient. The Nernst–Einstein relation, introduced in Section 1.7.2 for diffusion of a solute in a liquid solution, relates the Brownian diffusion coefficient to the friction coefficient f as [23]

$$\mathcal{D}_p = \frac{kT}{f} \tag{1.126}$$

For particles much larger than the mean free path of the gas, the friction coefficient is obtained from the Stokes' drag law, Eq. (4.73) in *Heat Transfer*, as

$$f = 3\pi \mu d_p; \quad d_p \gg \ell \tag{1.127}$$

to give the Stokes–Einstein relation. However, Brownian diffusion is usually important for very small particles for which $d_p \simeq \ell$ or $d_p \ll \ell$. Cunningham introduced a *slip correction factor C* into Stokes' law to account for the slip that occurs between the rarefied gas and particle surface, to give

$$f = \frac{3\pi \mu d_p}{C} \tag{1.128}$$

Rarefied gas effects are characterized by the Knudsen number, which is the ratio of the mean free path to a characteristic length. For a particle of diameter d_p,

$$\mathrm{Kn} = \frac{\ell}{d_p} \tag{1.129}$$

An interpolation formula for the slip correction factor that has the proper form in the limits of free molecule and Stokes flow is

$$C = 1 + \text{Kn}\left(C_1 + C_2 e^{-C_3/\text{Kn}}\right) \tag{1.130}$$

Based on experiment, Davies [24] recommends values of the constants for general use. If the mean free path for this purpose is defined[9] as

$$\ell = \frac{v}{0.499\bar{v}} \simeq \frac{2v}{\bar{v}} = v\left(\frac{\pi M}{2\mathcal{R}T}\right)^{1/2} \tag{1.131}$$

the values are $C_1 = 2.514$, $C_2 = 0.8$, and $C_3 = 0.55$. The precise values of these constants does depend on the nature of the molecule-particle collisions, and hence to some extent on physical properties of the particle material. Table A.30 gives values of C, $\mathcal{D}_p$, and $\text{Sc}_p = v/\mathcal{D}_p$ for spherical particles in air at 1 atm and 300 K. For nonspherical particles, it is common practice to introduce a *dynamic shape factor* χ into Eq. (1.128),

$$f = \frac{3\pi \mu d_{pe} \chi}{C}; \quad d_{pe} = (6m/\pi\rho_p)^{1/3} \tag{1.132}$$

where d_{pe} is the mass equivalent diameter. Table 1.5 gives some data for χ.

Brownian diffusion (or *motion*) also occurs in liquids, and indeed was discovered by Robert Brown in the early nineteenth century when he observed pollen grains in water [25]. It is usually in this context that the phenomenon is introduced in elementary physics texts. In a liquid, random molecular motion is on a length scale of the order of molecular size, and is thus much smaller than the particle size. Hence, for most purposes, Stokes' drag law can be used to obtain the friction coefficient f, as given by Eq. (1.127). *Macromolecules* (molecules with molecular weights of the order of 10,000 or greater) are important in many biological systems: for example, diffusion of proteins in aqueous solutions plays an important role in the life processes of animals and plants. Whether or not these macromolecules are modeled as solutes or as particles, the Stokes–Einstein relation applies. A useful approximation to this relation for macromolecules in dilute aqueous solutions is the Polson equation [26]

$$\mathcal{D}_{12} = \frac{9.40 \times 10^{-15} T}{\mu_2 M_1^{1/3}} \text{ m}^2/\text{s} \tag{1.133}$$

where T is in kelvins, μ_2 is the viscosity of water [kg/m s], and M_1 is the molecular weight of the macromolecule. Analysis of diffusion of smaller molecules in solutions containing macromolecules is made difficult by the presence of macromolecules. The problem is similar to that of multicomponent diffusion in gas mixtures and is beyond the scope of this text.

---

[9] In the use of the kinetic theory literature, a source of confusion is the different definitions of mean free path encountered. Care must be taken, since the relevant definition is often not prominently displayed.

Section 1.7 Transport Properties **97**

Table 1.5 Dynamic shape factor χ defined by Eq. (1.132).

Aerosol	$d_{pe} = (6m/\pi\rho_p)^{1/3}$ μm	χ
Sphere		1.0
Tetrahedron		1.17
Oblate spheroid		
Aspect ratio = 2		1.05
Aspect ratio = 5		1.24
Aspect ratio = 10		1.50
Prolate spheroid		
Aspect ratio = 2		1.05
Aspect ratio = 5		1.27
Aspect ratio = 10		1.60
Coal	0.56–4.27	1.88 ± 0.16
Quartz	0.65–1.85	1.84 ± 0.22
	> 4	1.23
Copper oxide	0.32	6.0
	0.70	8.5
Platinum oxide	0.038	1.1
	0.071	3.6
Uranium dioxide	0.19	2.6
	0.33	3.3
	0.55	1.2
U_3O_8	0.3–1	4–9
$(Pu_{0.14}U_{0.86})O_{2.22}$ aggregates (LFMBR fuel)		2–3
Sodium oxide aerosols formed in humid air		1.3
Iron oxide (Fe_2O_3) cluster aggregates	~ 1	4–9

Example 1.18   Brownian Diffusion Coefficient of an Uranium Oxide Particle

Calculate the Brownian diffusion coefficient of a UO_2 aerosol particle of mass equivalent diameter $10^{-2}\mu$m in steam at 400 K and 1 atm. The dynamic shape factor χ can be taken as 4. Also determine the particle Schmidt number.

Solution

Given: UO_2 particle in steam.

Required: Brownian diffusion coefficient and particle Schmidt number.

Assumptions: The nonsphericity effect can be accounted for by using a dynamic shape factor $\chi = 4$.

For steam at 400 K and 1 atm, Table A.7 gives $\mu = 1.40 \times 10^{-5}$ kg/m s, $\nu = 25.2 \times 10^{-6}$ m^2/s. From Eq. (1.131), the mean free path is

$$\ell \simeq \nu \left(\frac{\pi M}{2\mathcal{R}T}\right)^{1/2} = 25.2 \times 10^{-6}\left(\frac{18\pi}{2(8314)(400)}\right)^{1/2}$$

$$= 7.35 \times 10^{-8} \text{ m} = 0.0735 \ \mu\text{m}$$

The particle Knudsen number is

$$\text{Kn} = \frac{\ell}{d_{pe}} = \frac{0.0735}{0.01} = 7.35$$

Since this is the transitional flow regime, the friction coefficient is obtained from Eqs. (1.132) and (1.130) as

$$f = \frac{3\pi \mu d_{pe} \chi}{C}; \quad C = 1 + \text{Kn}\left(2.514 + 0.8e^{-0.55/\text{Kn}}\right)$$

$$C = 1 + 7.35\left(2.514 + 0.8e^{-0.55/7.35}\right) = 24.9$$

$$f = \frac{(3\pi)(1.40 \times 10^{-5})(10^{-8})(4)}{24.9} = 2.12 \times 10^{-13} \text{ kg/s}$$

Equation (1.126) gives the Brownian diffusion coefficient,

$$\mathcal{D}_p = \frac{kT}{f}$$

From Table A.16, the Boltzmann constant $k = 1.38054 \times 10^{-23}$ J/K and 1 J/K $= 1$ N m/K $= 1$ kg m^2/s^2 K; thus

$$\mathcal{D}_p = \frac{(1.38054 \times 10^{-23} \text{ kg m}^2/\text{s}^2 \text{ K})(400 \text{ K})}{2.12 \times 10^{-13} \text{ kg/s}} = 2.61 \times 10^{-8} \text{ m}^2/\text{s}$$

The particle Schmidt number is

$$\text{Sc}_p = \frac{\nu}{\mathcal{D}_p} = \frac{25.2 \times 10^{-6}}{2.61 \times 10^{-8}} = 966$$

Comments

1. This Schmidt number is of a comparable magnitude to that of glycerol dissolved in water (see Table A.18).
2. Transport of UO_2 aerosols is of concern to nuclear engineers in analyzing a hypothetical loss-of-coolant accident (LOCA). The aerosols are formed when there is a core meltdown and are subsequently transported in steam (or steam–hydrogen–air mixtures) through the containment vessel.   ▲

1.8 CLOSURE

This chapter provided an introduction to the subject of mass transfer, with special emphasis on topics of concern to mechanical engineers. Simple analyses of mass transfer by diffusion were presented, with the governing equation derived by application of the principle of conservation of chemical species to an appropriate elemental control volume. Where possible, the analogy between mass diffusion and heat conduction was exploited. Consideration of mass convection was initially limited to situations for which low mass transfer rate theory is of acceptable accuracy. The analogy between heat convection and mass convection allowed convective heat transfer correlations from Chapter 4 of *Heat Transfer* to be adapted for convective mass transfer problems in the present chapter. Particular attention was given to a variety of problems in which there was simultaneous heat and mass transfer. Mass transfer in porous catalysts was analyzed, with applications to catalysts used for automotive emissions control. To complete the chapter, methods for calculating transport properties of gas mixtures, liquid solutions, and small particles were presented.

Four computer programs were introduced in Chapter 1. The program MCONV is a useful tool for solving convection problems according to the theory presented in Sections 1.4 and 1.5. The program PSYCHRO calculates psychrometric and thermodynamic properties of water vapor and air mixtures. GASMIX calculates transport properties of gas mixtures and includes a menu of 36 chemical species. AIRSTE calculates transport properties of water vapor and air mixtures. Mass transfer calculations tend to require more effort than heat transfer calculations. Mass or mole fractions must be calculated, transport and thermodynamic properties can involve the use of mixture rules, and vapor pressure tables are frequently used. Thus, the use of computer programs to perform such calculations is particularly appropriate and can save the engineer considerable time and expense.

Chapter 2 is concerned with simultaneous diffusion and convection, which leads to high mass transfer rate theory for convection. However, in Chapter 3 all the analyses of mass transfer equipment are based on low mass transfer rate theory. Thus, if desired, the student can proceed directly to Chapter 3 and defer Chapter 2.

REFERENCES

1. Plumb, O. A., Spolek, G. A., and Olmstead, B. A., "Heat and mass transfer in wood during drying," *Int. J. Heat Mass Transfer*, 28, 1669–1678 (1985).

2. Perre, P., Moser, M., and Martin, M., "Advances in transport phenomena during convective drying with superheated steam and moist air," *Int. J. Heat Mass Transfer*, 36, 2725–2746 (1993).

3. Pesaran, A. A., and Mills, A. F., "Moisture transport in silica gel packed beds: I. Theoretical study," *Int. J. Heat Mass Transfer*, 30, 1037–1050 (1987).

4. Kafesjian, R., Plank, C. A., and Gerhard, E. R., "Liquid flow and gas phase mass transfer in wetted wall towers," *AIChE Journal*, 7, 463–466 (1961).

5. Saboya, F. E. M., and Sparrow, E. M., "Local and average transfer coefficients for one-row plate fin and tube heat exchanger configurations," *J. Heat Transfer*, 96, 265–272 (1974).

6. Goldstein, R. J., and Karni, J., "The effect of a wall boundary layer on local mass transfer from a cylinder in crossflow," *J. Heat Transfer*, 106, 260–267 (1984).

7. Cengel, Y. A., and Boles, M. A., *Thermodynamics: An Engineering Approach*, Chapter 13, McGraw-Hill, New York (1989).

8. Threlkeld, J. L., *Thermal Environmental Engineering*, 2nd ed., Prentice-Hall, Englewood Cliffs, N.J. (1970).

9. *Handbook of Air Conditioning, Heating and Ventilating*, eds. E. Stamper and R. L. Koral, 3rd ed., Industrial Press, New York (1979).

10. *ASHRAE Handbook: 1989 Fundamentals*, American Society of Heating, Refrigerating and Air Conditioning Engineers, Atlanta (1989).

11. Kennard, E. H., *Kinetic Theory of Gases*, McGraw-Hill, New York (1938).

12. Tien, C. L., and Lienhard, J., *Statistical Thermodynamics*, Holt, Rinehart & Winston, New York (1971).

13. Knuth, E. L., *Introduction to Statistical Thermodynamics*, McGraw-Hill, New York (1966).

14. Guggenheim, E. A., *Elements of the Kinetic Theory of Gases*, Pergamon Press, Oxford (1960).

15. Jeans, J., *An Introduction to the Kinetic Theory of Gases*, Cambridge University Press, New York (1940).

16. Edwards, D. K., Denny, V. E., and Mills, A. F., *Transfer Processes*, 2nd ed., Hemisphere/McGraw-Hill, Washington, D.C. (1979).

17. Hirschfelder, J. O., Curtiss, C. P., and Bird, R. B., *Molecular Theory of Gases and Liquids*, Wiley, New York (1954).

18. Chapman, S., and Cowling, T. G., *The Mathematical Theory of Nonuniform Gases*, 2nd ed., Cambridge University Press, New York (1964).

19. Wilke, C. R., "A viscosity equation for gas mixtures," *J. Chem. Phys.*, 18, 517–519 (1950).

20. Wilke, C. R., and Chang, P., "Correlation of diffusion coefficients in dilute solutions," *AIChE Journal*, 1, 264–270 (1955).

21. Othmer, D. F., and Thakar, M. S., "Correlating diffusion coefficients in liquids," *Ind. Eng. Chem.*, 45, 589–593 (1953).

22. Reid, R. C., Prausnitz, J. M., and Sherwood, T. K., *The Properties of Gases and Liquids*, 3rd ed., McGraw-Hill, New York (1977).

23. Einstein, A., *Investigations of the Theory of Brownian Movement*, Dover, New York (1956).

24. Davies, C. N., "Definitive equations for the fluid resistance of spheres," *Proc. Phys. Soc. Part A*, 57, 259–269 (1945).

25. Lavenda, B. H., "Brownian motion," *Scientific American*, 252, no. 2, 70–85 (1985).

26. Polson, A., "Some aspects of diffusion in solution and definition of colloidal particles," *J. Phys. Colloid Chem.*, 54, 649–652 (1950).

27. Won, Y. S., and Mills, A. F., "Correlation of the effects of viscosity and surface tension on gas absorption rates into freely falling turbulent liquid films," *Int. J. Heat Mass Transfer*, 25, 223–229 (1982).

28. Bartlett, E. P., Kendall, R. M., and Rindal, R. A., "An analysis of the coupled chemically reacting boundary layer and charring ablator, Part IV: A unified approximation for mixture transport properties for multicomponent boundary layer applications," NASA CR–1063 (1968).

EXERCISES

1-1 Derive Eqs. (1.10*a*) and (1.10*b*), namely,

$$m_i = \frac{x_i M_i}{\sum x_j M_j}; \quad x_i = \frac{m_i/M_i}{\sum m_j/M_j}$$

1-2

i. A mixture of noble gases contains equal mole fractions of helium, argon, and xenon. What is the composition in terms of mass fractions?

ii. If the mixture contains equal mass fractions of He, Ar, and Xe, what are the corresponding mole fractions?

1-3 Methane is burned with 20% excess air. At 1250 K the equilibrium composition of the product is:

Species i:	CO_2	H_2O	O_2	N_2	NO
x_i:	0.0803	0.160	0.0325	0.727	0.000118

Determine the mean molecular weight M and gas constant R of the mixture, and the mass fraction of the pollutant nitric oxide in parts per million.

1-4 A closed cylindrical vessel containing stagnant air stands with its axis vertical. Each end is maintained at a uniform temperature with the base colder than the upper end; the cylindrical wall is insulated. Neglecting the variation in hydrostatic pressure, ascertain whether there are vertical gradients of

i. partial density of oxygen.

ii. partial pressure of oxygen.

iii. mass fraction of oxygen.

1-5 Combustion products from a hydrogen burner contain 34.0% H_2O, 64.6% N_2, and 1.4% H_2 by volume at 600 K and 2 atm. Determine

i. the mole fractions, partial pressures, and mass fractions of the components.

ii. the mean molecular weight and gas constant of the mixture.

iii. the density and molar concentration of the mixture.

iv. the partial densities and molar concentrations of the components.

1-6 A 1-liter vessel contains 80% He and 20% N_2 by volume at 100 kPa and 300 K.

i. Find the mass fraction and partial density of the helium.

ii. If the cylinder is heated to 500 K, find the new values of P, m_{He}, and ρ_{He}.

iii. If 2×10^{-5} kmol of helium are now added to the vessel while it is maintained at 500 K, find the final values of P, m_{He}, and ρ_{He}.

1-7 Water at 310 K is sprayed as a very fine mist into a vessel containing gas samples at 100 kPa. Determine the mass and mole fractions of dissolved gas in water collected at the bottom of the vessel if the gas is

i. oxygen.

ii. carbon dioxide.

iii. hydrogen sulfide.

iv. ammonia.

1-8 Plot a graph of the mass fraction of water vapor in saturated water vapor–air mixtures at 1 atm pressure for 273.15 K $< T <$ 373.15 K.

1-9 A 1-liter soda bottle at 290 K is at a pressure of 1.2 bar. If it was charged with CO_2 at 290 K and 5 bar pressure, what volume of gas was added?

1-10 A recent determination of the diffusivity of carbon in BCC iron gave values of 1.25×10^{-8} cm^2/s at 500°C, and 2.17×10^{-5} cm^2/s at 1000°C. If $\mathcal{D}$ can be represented by an expression of the form $\mathcal{D} = \mathcal{D}_0 \exp(-E_a/\mathcal{R}T)$, determine E_a in kJ/kmol, and $\mathcal{D}_0$. Also determine the diffusivity at 1000 K.

1-11 Hydrogen, at 500°C and 5 bar, flows at 10 m/s in a 2 cm–O.D., 0.5 mm–wall thickness steel tube located in an evacuated enclosure. Estimate the rate of hydrogen loss for a 10-m length of tube. Solubility and diffusion coefficient data for the hydrogen-steel system are given in Example 1.1.

1-12 From the data given in Example 1.2, tabulate the permeability of helium in the laser shell glass for 300 K $< T <$ 500 K. Using these data, recalculate the rate of helium loss through the shell. Why is the permeability approach inappropriate for hydrogen diffusion through steel as considered in Example 1.1?

1-13 A pharmaceutical product is to be protected from exposure to oxygen by vacuum wrapping with a 0.2 mm–thick polyethylene film of 400 cm^2 surface area. Oxygen that penetrates the packing is rapidly consumed by reaction with the product. If storage is at 30°C, calculate the rate at which oxygen is available for reaction with the product. The permeability of O_2 in polyethylene at 30°C can be taken as 4.17×10^{-12} m^3 (STP)/m^2 s (atm/m).

1-14 A 4 cm–diameter composite membrane, consisting of a 0.2 mm–thick film of vulcanized rubber and a 1 mm–thick layer of polyethylene, separates pure hydrogen at 1.085×10^5 Pa and 25°C from atmospheric air. Calculate the rate at which hydrogen leaks through the membrane. Permeabilities of H_2 in the rubber and polyethylene, respectively, are 3.42×10^{-11} and 6.53×10^{-12} m^3 (STP)/m^2 s (atm/m).

1-15 A worm gear cut from 0.1% C steel is to be carburized (case-hardened) at 930°C until the carbon content is raised to 0.45% C at a depth of 0.5 mm. The carburizing gas holds the u-surface concentration at 1% C by weight.

   i. Calculate the time required.

   ii. What time is required at the same temperature to double this depth of penetration?

   iii. What temperature is required to obtain 0.45% C at 1 mm, in the same time as it was attained at 0.5 mm for a temperature of 930°C?

Take the diffusion coefficient of carbon in the steel as

$$\mathcal{D} = 2.67 \times 10^{-5} e^{-17,400/T} \text{ m}^2/\text{s}$$

where T is in kelvins.

1-16 The Loschmidt apparatus for measuring the diffusion coefficient in a binary gas mixture consists of a long tube, $0 < z < 2L$, with a diaphragm at $z = L$. The pure gases are loaded into each half of the tube, and at time $t = 0$ the diaphragm is removed. The composition of the mixture at $z = 0$ is monitored as a function of time and compared with the theoretically expected value. Derive an expression for x_1 at $z = 0$, valid for $\mathcal{D}_{12}t/L^2 > 0.2$, suitable for this purpose.

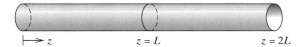

$\longmapsto z$ $z = L$ $z = 2L$

1-17 A 1 m–square, 1 m–deep pool of stagnant deaerated water at 290 K is suddenly exposed to pure oxygen at 2 atm pressure. Calculate the amount of O_2 absorbed after 12 hours has elapsed.

1-18 Water droplets, of 0.3 mm diameter, fall through a tower containing H_2S gas at 300 K and 10 atm.

    i. Use Henry's law to determine the u-surface H_2S mole fraction.

    ii. Determine the absorption rate at short times.

    iii. Determine the fractional approach to equilibrium as a function of time.

Assume no circulation inside the droplet, and refer to analogous heat conduction problems in Chapter 3 of *Heat Transfer*.

1-19 Upon inspection at the warehouse, a consignment of fir planks was found to have been insufficiently dried. Tests showed a residual moisture content of 13.3% by weight. The planks were returned to the mill for further drying. If the drying oven is able to maintain a u-surface concentration of moisture of 6.6%, determine the time required for the center of 5 cm–thick planks to dry to 8.0% moisture content. Take the effective diffusion coefficient for moisture in fir under these conditions as 0.86×10^{-9} m^2/s.

1-20 Experimental oxidation data for a titanium alloy, used to sheath hypersonic vehicle nose cones, indicates that the total weight gain can be described by a parabolic law of the form $w^2 = Ct$, where $C = 480 \exp(-E_a/\mathcal{R}T)$ g^2/cm^4 s, and $E_a = 61,800$ cal/mol. Examination of oxidized samples shows that about 20% of the oxygen goes to form a rutile (TiO_2) surface scale. Assuming that the diffusion coefficient for oxygen in alpha titanium can be expressed as $\mathcal{D}_{12} = \mathcal{D}_0 \exp(-E_a/\mathcal{R}T)$, with E_a again 61,800 cal/mol, estimate $\mathcal{D}_0$. Show also that the parabolic oxidation law implies diffusion across the rutile scale characterized by a linear concentration gradient. Use an alloy density of 4400 kg/m^3.

1-21 Naphthalene is used as a moth repellent in the form of spherical mothballs. If a 2 cm– diameter mothball is hung in a closet where the average temperature is 21°C, obtain a rough estimate of the lifetime of the mothball. Take $P = 1$ atm, and use data for naphthalene from Example 1.5.

1-22 At a particular location on a surface, the heat transfer coefficient to a laminar forced flow of 300 K water is 1000 W/m^2 K. Estimate the equivalent stagnant film thicknesses for

    i. heat transfer.

    ii. mass transfer for dissolution of sucrose.

1-23 A 0.3 mm–diameter carbon particle at 1650 K oxidizes in air at 1350 K and 1 atm pressure. At these temperatures, both the kinetic and diffusive resistances to mass transfer may be important, and the heterogeneous reaction is $C + O_2 \rightarrow CO_2$. The rate of combustion of carbon per unit surface area is $\Phi P_{O_2,s}^{1/2}$ where $\Phi = 1170 \exp(-30{,}200/T_s)$ kmol/m^2 s Pa$^{-1/2}$ for T_s in kelvins and P in pascals (the s-surface is located in the gas phase adjacent to the solid carbon surface).

    i. Model the diffusion process using the equivalent stagnant film concept on a *molar* basis.

    ii. Obtain an expression for the molar rate of consumption of carbon in terms of the film thickness δ_f, the free-stream oxygen concentration $x_{O_2,e}$, and Φ.

    iii. If the particle is entrained in the air flow so that there is a negligible relative velocity, estimate δ_f.

    iv. Determine the initial rate of decrease of particle diameter.

    v. Determine the particle lifetime.

1-24 A common procedure for surface micromachining involves chemical etching of sacrificial layers. In one procedure, phosphosilicate glass is etched by hydrofluoric acid, with an overall reaction of

$$6HF + SiO_2 \rightarrow H_2SiF_6 + 2H_2O$$

When the etching depth is small, the reaction rate is rate-controlled; but for deep etching, as required to machine channels, diffusion of acid through the solution also plays a role. Assuming quasi-steady diffusion and a one-dimensional model, obtain an expression for the etching depth as a function of time. The reaction is approximately first-order in HF molar concentration with rate constant k''. Hence determine the time required to etch a channel 100 μm deep if $k'' = 2 \times 10^{-6}$ m/s, $\mathcal{D}_{1m} = 2 \times 10^{-9}$ m²/s, and the bulk concentration of HF in an aqueous solution is 7.0 kmol/m³.

1-25 A proposed correlation for heat transfer from a jet to a disk normal to the flow has the form

$$Nu_L = CRe_L^{1/2}Pr^{1/3}$$

where L is the distance from the jet exit to the disk. An experiment to determine the constant C measured the sublimation rate of a naphthalene disk exposed to an air jet. Pertinent data include pressure $= 1$ atm, temperature $= 310$ K, jet velocity $= 10$ m/s, $L = 3$ mm, for which the measured recession rate was 430 μm/h. Determine C.

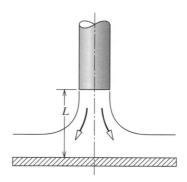

1-26 A new shape of particle for a packed-particle-bed catalytic reactor is under consideration. By means of a heat transfer experiment, it is found that, for superheated steam at 500 K and 1 atm, the heat transfer coefficient is 110 W/m² K at the design Reynolds number. If a stream of nitrogen containing 2% hydrogen by volume at 500 K flows through a bed at the same Reynolds number, estimate the maximum possible rate of formation of ammonia. The catalyst is iron promoted with ferric oxide. Property values at 500 K include Pr $= 0.94$, $k = 0.0365$ W/m K for the steam; and for the N_2–H_2–NH_3 mixture, Pr $= 0.70$, $Sc_{H_2m} = 0.21$, and $\mu = 25.1 \times 10^{-6}$ kg/m s.

1-27 A new catalyst for the oxidation of ammonia is being investigated. The catalyst is in the form of a 0.5 mm–diameter wire. The reactant gas is at 500 K and 1 atm pressure, and contains 12% ammonia by mass. If the flow across the wire is at a mass velocity of 0.3 kg/m^2 s, estimate the maximum rate at which ammonia can be oxidized per meter length of wire. Take the reactant gases to have properties similar to air.

1-28 A materials processing experiment on the space shuttle requires design data for convective heat transfer coefficients in an acoustically driven forced flow around a 1 cm–diameter sphere. A laboratory simulation is planned in Los Angeles, and to ensure acceptably small natural convection effects, Re^2/Gr should be greater than 10 (there is negligible natural convection in the near zero-g environment on the shuttle). If the characteristic velocity of the forced flow is V, what is the minimum value of V that should be used in the simulation if

    i. a heat transfer experiment is performed with a 1 cm–diameter sphere at a temperature 4 K above the ambient air at 298 K, 1 atm?

    ii. a mass transfer experiment is performed with a 1 cm–diameter naphthalene sphere in air at 300 K, 1 atm?

1-29 In a field experiment to measure deposition rates of sulfur dioxide gas, the test surface is a 10 cm–square paper plate impregnated with potassium carbonate. The reaction between SO_2 and KCO_3 is rapid enough to ensure that the concentration of SO_2 at the s-surface is zero. Expected concentrations of SO_2 in the atmosphere at the test site are expected to be of the order of 5 parts per million by volume, and the average wind speed over the plate is 0.3 m/s. The ambient air is at 295 K and 1 atm, and the plate temperature is also 295 K. If at least 1 mg of SO_2 must be collected for a reliable result, what is the minimum time of exposure required?

1-30 In absorption towers, gases are often absorbed into a thin film of liquid flowing over a packing. A simple model of this process is a laminar film falling down a vertical wall. From Section 7.2.1 of *Heat Transfer*, the surface velocity of the film is $u_\delta \simeq g\delta^2/2\nu_l$, where δ is the

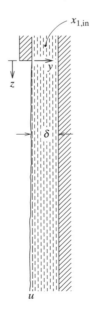

film thickness, and the characteristic small diffusion coefficient in liquids implies that the gas penetration is confined to a thin concentration boundary layer where the liquid velocity can be assumed constant at the surface value.

i. Using a volume element fixed in space, show that the equation governing the steady-state concentration of absorbed gas in the concentration boundary layer is

$$u_\delta \frac{\partial x_1}{\partial z} = \mathcal{D}_{12} \frac{\partial^2 x_1}{\partial y^2}$$

ii. By transforming to a coordinate system fixed on the film surface, show that the concentration distribution is

$$\frac{x_1 - x_{1,in}}{x_{1,u} - x_{1,in}} = \text{erfc}\frac{y}{(4\mathcal{D}_{12}z/u_\delta)^{1/2}}$$

where $x_{1,in}$ is the mole fraction of the gas in the liquid at $z = 0$.

iii. Show that the local Sherwood number is given by

$$\text{Sh}_z = \frac{1}{\pi^{1/2}} \text{Pe}_{mz}^{1/2}$$

where Pe_{mz} is the local mass transfer Peclet number, $\text{Pe}_{mz} = u_\delta z/\mathcal{D}_{12}$.

1-31 Water at 300 K and initially pure, falls as a 0.36 mm–thick film on the inside wall of a vertical tube of 2 cm I.D. and 1 m length. If pure CO_2 at 1 atm pressure flows through the tube, calculate the rate at which CO_2 is absorbed. Also estimate the thickness of the concentration boundary layer at the bottom of the tube to check the assumption of a small penetration distance.

1-32 In an experiment, water at 300 K and initially pure, falls as a film on the inside wall of a vertical tube of 2 cm I.D. and 1 m length. The water flow rate is 0.04767 kg/s. Pure CO_2 at 1 atm flows through the tube. At this water flow rate, the film is in the turbulent regime, for which the following correlation can be used to obtain the mole transfer conductance for gas absorption into the film [27]:

$$\text{St}_m = 6.79 \times 10^{-9} \text{Re}^n \text{Sc}^{-\alpha} \text{Ka}^{-1/2}; \quad \text{Re} > 1100;$$

$$n = 3.49 \text{Ka}^{0.068}; \quad \alpha = 0.36 + 2.43(\sigma \text{N/m})$$

where $\text{Ka} = \nu^4 \rho^3 g/\sigma^3$ is the *Kapitza number*, the Stanton number is defined as $\text{St}_m = \mathcal{G}_m/c(\nu g)^{1/3}$, and the Reynolds number is the film Reynolds number $\text{Re} = 4\Gamma/\mu$ (defined in Chapter 7 of *Heat Transfer*). Notice that, in addition to the usual Reynolds- and Schmidt-number dependence, the Stanton number also depends on surface tension (St_m increases with increasing σ).

i. Calculate $\mathcal{G}_m$ for the conditions of the experiment.

ii. Since the film is long, $\mathcal{G}_m$ can be assumed constant. Model the system as a single-stream exchanger, and hence obtain the outlet water bulk mole fraction of CO_2.

iii. Determine the rate at which CO_2 is absorbed in the tube.

1-33

    i. Reconsider Exercise 1–18 for the situation where the tower contains air with a bulk H_2S mole fraction of 0.03. Estimate the mass transfer Biot number, and hence ascertain if the gas absorption process remains liquid-side controlled.

    ii. Reconsider Exercise 1–17 for water exposed to air at 2 atm pressure. Will the gas absorption process remain liquid-side controlled? (*Hint:* Estimate a reasonable gas-side conductance and a characteristic length for diffusion into the pool.)

1-34 In an experiment, three moist air samples at 101 kPa and 25.0°C have respective wet-bulb temperatures of 15.0, 20.0, and 23.0°C. Prepare a table giving the water-vapor mass fraction and mole fraction, humidity ratio, and relative humidity of the three samples.

1-35 In a cooler–condenser performance test, the following data were obtained for the outlet steam–air mixture: $P = 22{,}300$ Pa, $T_{DB} = 299.7$ K, $T_{WB} = 301.0$ K. Determine the mass fraction of water vapor and the relative humidity.

1-36 Air at 290 K, 40% relative humidity blows at 5 m/s over a 10 m–square swimming pool near Albuquerque, New Mexico. If the ambient pressure is 870 mbar and the water surface temperature is measured to be 300 K, estimate the heat loss due to

    i. convective heat transfer.

    ii. evaporation.

1-37 A thick horizontal steel plate 1 m square has a temperature of 340 K. Water is spilled on the plate and forms a pool 0.2 mm deep. If the ambient air is still and is at a temperature of 310 K and a pressure of 1 atm, and it has a relative humidity of 30%, estimate the time required for the pool to evaporate.

1-38 A 1 m–square water bath is maintained at 320 K in surrounding still air at 290 K, 1 atm pressure, and 20% relative humidity. Estimate the convective, evaporative, and radiative heat losses.

1-39 A wet- and dry-bulb psychrometer is used to measure the composition of a mixture of methanol vapor and nitrogen. A sample at 100 kPa pressure gives readings of $T_{DB} = 9.2$°C and $T_{WB} = 5.0$°C. Determine the mass fraction of methanol. The chemical formula of methanol is CH_3OH, and its vapor pressure is tabulated below. At 5.0°C the enthalpy of vaporization of methanol is 1.18×10^6 J/kg.

P, mm Hg:	1	10	40	100	400	760
P, Pa:	133	1330	5330	13,330	53,300	101,330
T, °C:	−44.0	−16.2	5.0	21.2	49.9	64.7

1-40 A wet- and dry-bulb psychrometer is used to measure the composition of a mixture of ammonia vapor and nitrogen. A sample at 15 atm pressure gives readings of $T_{DB} = 250.5$ K and $T_{WB} = 240.0$ K. Determine the mass fraction of ammonia.

1-41 Air at 990 mbar flows over a wet- and dry-bulb psychrometer: the dry thermocouple measures 310.1 K, and the wet thermocouple measures 305.3 K. Determine the humidity ratio and relative humidity,

i. ignoring thermal radiation.

ii. correcting for thermal radiation.

The instrument is located in a duct with walls maintained at 290 K, and the air velocity is 1 m/s. The thermocouples can be approximated as cylinders of diameters 1.0 and 3.0 mm for the dry and wet thermocouples, respectively. Use an emittance of 0.3 for the dry thermocouple and 0.9 for the wet wick.

1-42 A wet- and dry-bulb psychrometer is used to measure the concentration of ethanol vapor in an air stream at 10 atm total pressure. The dry bulb reads 350 K, and the wet bulb reads 340 K. What is the ethanol mass fraction? Use the data in Table A.17b. The spherical bulb diameters are 2 mm, and the gas velocity is 4 m/s. The surroundings are at 320 K.

1-43 Determine the rate of sublimation of a south-facing snowbank on Mammoth Mountain in the Sierra Nevada, when the ambient air is at $-2°$C and 30% RH. Take the solar irradiation as 1000 W/m^2 and use a sky emittance of 0.75. The convective heat transfer coefficient for an equivalent surface has been estimated to be 6.0 W/m^2 K. Assume that heat transfer into the snowbank is negligible. Mammoth Mountain is at 2000 m altitude. Vapor pressure data for ice follows:

T, °C:	0	−2	−4	−6	−8	−10	−12	−14	−16	−18	−20
P_{sat}, Pa:	610	517	437	369	310	260	218	182	151	125	104

The enthalpy of sublimation is $h_{sg} = 2.827 \times 10^6$ J/kg at $0°$C, and can be taken to be a constant over the temperature range of concern.

1-44 A 50 m $\times$ 20 m swimming pool in Los Angeles is maintained at $28°$C during the winter. Typical nighttime conditions are a clear sky with still ambient air at $10°$C, 1000 mbar, and 40% relative humidity. Calculate the convective, evaporative, and radiative heat losses when

i. the pool is well stirred by swimmers.

ii. the pool is unoccupied and the estimated water-side heat transfer coefficient is 100 W/m^2 K.

1-45 One method of retarding evaporation from water reservoirs in hot arid regions is to add a small amount of a large-molecular-weight alcohol, such as cetyl alcohol, to the water. The alcohol spreads to form a surface monolayer, which presents a significant resistance to evaporating water molecules. In a test, the water surface is at $23°$C, and the ambient air is still at 1 atm, $20°$C, and 20% RH. The measured rate of evaporation from a 1 m–square pool is 1.21×10^{-5}kg/m^2 s. What is the reduction in evaporation rate due to the cetyl-alcohol monolayer?

1-46 A strip of wet textile passes through a dryer. Hot air at 1 atm, 360 K and 1% RH is blown on the top surface of the textile through a perforated-plate distributor. When the mass velocity of mixture is 0.8 kg/m^2 s, design charts for the distributor indicate a heat transfer Stanton number of 0.08. Calculate the rate at which water is removed from the textile in the constant-drying-rate period of the drying process (see Fig. 1.10). The back surface of the textile in contact with a conveyer belt can be assumed adiabatic.

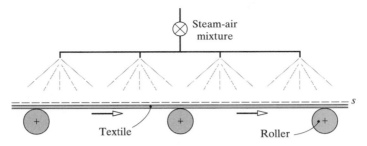

1-47 In a paper-drying unit the wet paper is on the outside of a rotating drum, the inside of which is maintained at 330 K by a hot water supply. Air at 305 K, 1 atm, and 40% RH is blown on the paper at a mass velocity of 0.3 kg/m^2 s. Design data for the configuration indicate a mass transfer Stanton number of 0.060. If the overall heat transfer coefficient from the hot water to the paper surface is estimated to be 50 W/m^2 K, calculate the paper drying rate in the constant-drying-rate period of the drying process (see Fig. 1.10).

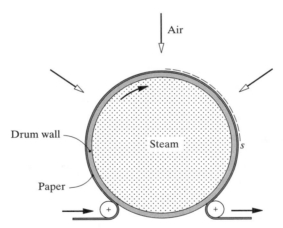

1-48 Air at 1 atm pressure, 360 K, and 1% RH flows at 10 m/s over a porous flat plate. Water flows through the plate at a rate just sufficient to keep the surface wet, and is supplied from a reservoir maintained at 280 K. Calculate the surface temperature, and the average evaporation rate for a 10 cm–long plate for

   i. a laminar boundary layer.

   ii. a turbulent boundary layer starting at the leading edge.

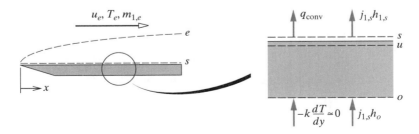

(*Hint:* First perform an energy balance on a control volume between the s-surface and an o-surface located far into the reservoir, where temperature gradients are negligible, to show that $h_c(T_e - T_s) = j_{1,s}[h_{fg} + c_{pl}(T_s - T_o)]$. Then show that the plate temperature is independent of location.) Neglect radiation heat transfer.

1-49 An operating hot tub loses heat continuously by a number of mechanisms, the most important being evaporative heat losses from the surface of the water, and into the air bubbling through the water. For steady operation, the heat loss must be balanced by adding heat to the circulating water. The design operating temperature for a small tub used indoors is 110°F (316.7 K) with the ambient air at 69°F (293.9 K), 1 atm, and 80% RH. The water circulation pump is rated at 1 kW, and its motor has an efficiency of 85%. Ambient air is blown into the pool at a rate of 7.3×10^{-3} kg/s. If the surface area of the water is 0.93 m^2 and the sides and bottom of the tub are well insulated, recommend a rating for the water heater.

1-50 The heat loss from a surface can be increased if it is kept wet. Consider air at 300 K, 1 atm, and 70% RH, flowing over a 3 cm–diameter sphere with a velocity of 2 m/s. Calculate the increase in heat transfer (wet versus dry) for sphere surface temperatures of 300, 305, 310, 320, and 330 K.

1-51 A 1 m–square wet towel is hung on a washline to dry on a day when there is a low overcast and no wind. The ambient air is at 21°C, 1 atm, and 50% RH. If the towel contains 0.5 kg of water, do you think the towel will dry in a reasonable time? Take $\epsilon = 0.9$ for the wet towel.

1-52 Develop an appropriate model for raindrops falling from a cumulus cloud at the beginning of a summer thunderstorm. Write down equations governing both the motion and size of a droplet, and describe a scheme for solving the problem numerically. Discuss any assumptions made and list the data you might require.

1-53 A platinum wire is suspended in a steady stream of air. It is heated electrically, and its temperature is deduced from its electrical resistance and measured current and voltage drop. At a standard power input, air stream velocity, temperature, and pressure, the wire temperature is found to be 700 K.

  i. This apparatus is used for measuring small concentrations of combustible gas that may be present after emptying oil tanker tanks. In a particular test, in which the power input and air temperature, pressure, and velocity have the standard values, the wire temperature is found to be 805 K. Estimate the mass fraction of combustible gas in the air if the platinum is a powerful catalyst for the combustion reaction. Neglect thermal radiation effects.

  ii. Make a new estimate allowing for radiative effects using the following data: wire diameter = 0.2 mm, wire emittance = 0.91, velocity of air perpendicular to the wire = 3 m/s, air temperature = 310 K, and pressure = 1 atm.

(*Note:* This exercise is a D. B. Spalding original.)

1-54 Porous catalysts are sometimes manufactured in the form of a thin plate or a thin-walled matrix. Show that the effectiveness of such a plate for a first-order reaction, when only one side of the plate is exposed to the reactant flow, is $\eta_p = (1/\Lambda) \tanh \Lambda$, where Λ is the Thiele modulus.

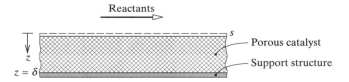

1-55 In designing a catalytic reactor, a plain wall was found to be inadequate, and thus it is proposed to bond a porous catalyst layer to the wall. If the layer thickness is 0.05 cm, the rate constant for a first-order reaction is $k'' = 4 \times 10^{-5}$ m/s, and $\mathcal{D}_{1,\text{eff}} = 1.4 \times 10^{-5}$ m^2/s, determine the area-to-volume ratio a_p m^2/m^3 required to achieve an effective rate constant based on s-surface area, of 0.07 m/s.

1-56 The catalyst bed of an automobile exhaust reactor is packed with CuO-on-alumina cylindrical pellets of 0.3 cm diameter and 0.6 cm long. The pellets have a volume void fraction of 0.60, tortuosity factor 3.5, average pore radius 0.7 μm, and catalytic surface area per unit volume of 6.81×10^5 cm^2/cm^3. Calculate the effectiveness of the pellet in promoting CO oxidation at 1 atm pressure and

    i. 700 K (idle).

    ii. 1000 K (full power).

The activation energy of the reaction is 18.6 kcal/mol, and the preexponential factor in the Arrhenius relation is 50.6 m/s.

1-57 Ammonia can be catalytically decomposed on a tungsten surface. When the ammonia is at a sufficiently high partial pressure, the catalyst surface is largely covered by adsorbed ammonia, and the reaction is consequently of zero order; that is, the reaction rate is independent of ammonia concentration and is given by $Ae^{-E_a/\mathcal{R}T}$ where A has units (moles reactant/unit area–unit time). Determine the concentration distribution in a spherical catalyst pellet promoting this reaction, assuming an isothermal pellet. Hence show if

$$\frac{V_p}{S_p}\left[\frac{k''a_p}{\mathcal{D}_{1,\text{eff}}c_{1,s}}\right]^{1/2}$$

is less than $\sqrt{2/3}$, the pellet effectiveness is equal to 1.0. (*Warning:* Of course, the reaction does not occur past the radius at which all the reactant has been consumed.)

1-58 A porous catalyst is used to burn fumes from a paint spray booth to environmentally safe CO_2 and H_2O. The reaction is first-order in mass fraction of fumes, with activation energy 18 kcal/mol and preexponential factor 10 m/s. The catalyst has specific surface area $a_p = 5 \times 10^5$ cm^2/cm^3 and average pore radius 0.5 μm, and is in the form of a matrix with 2 mm–thick walls and square cross-section passages with sides of 3 mm. The cross-sectional area of the reactor is large enough to ensure laminar flow through the passages. The catalyst operates at 500 K and 1 atm. At a location where the bulk mass fraction of fumes is 0.005, estimate the rate at which the fumes are oxidized per unit volume of reactor. For the fumes, take a molecular weight of 110 and an ordinary diffusion coefficient of 20×10^{-6} m^2/s. For the porous catalyst, take $\epsilon_v = 0.7$, $\tau = 4.0$.

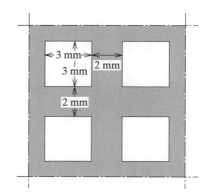

1-59 An automobile catalytic converter has cylindrical porous catalyst pellets of diameter 2.5 mm and length 5 mm. The pellets have a volume void fraction of 0.65, a tortuosity factor of 4.0, average pore radius 0.6 μm, and catalytic surface per unit volume 7.71×10^5 cm^2/cm^3. The catalyst promotes CO oxidation in a first-order reaction, with an activation energy of 19.1 kcal/mol and a preexponential factor in the Arrhenius relation equal to 43.3 m/s. Calculate the effectiveness of the pellet when the converter operates at 900 K and 1 atm.

1-60 To increase the rate of absorption of gas A into water, a reactant B is added to the water such that the homogeneous reaction A + B $\rightarrow$ C occurs. Analyze quasi-steady absorption into a pool of water of depth L for

i. a first-order reaction.

ii. a zero-order reaction.

The bottom of the container is impervious to species A.

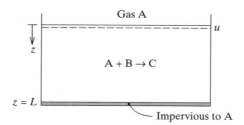

1-61 An important processing step in the production of semiconductors involves the formation of SiO$_2$ layers on silicon. These layers provide electrical insulation and dielectric properties necessary for semiconductor devices. Above 700° C, the layers form rapidly and follow exactly the topography of the underlying Si substrate. Develop a model for this *thermal oxidation* process with the following features:

The solubility of O$_2$ in SiO$_2$ is given by the linear relation $c_{O_2,u} = \text{K}P_{O_2,s}$.

Quasi-steady diffusion of O$_2$ through the SiO$_2$ layer.

The heterogeneous reaction Si + O$_2$ $\rightarrow$ SiO$_2$ occurs at the Si-SiO$_2$ interface and is first-order in molar concentration of O$_2$, with rate constant k''.

Hence show that the thickness-time relation is of the form

$$\frac{\delta}{k_L} + \frac{\delta^2}{k_P} = t$$

where k_L and k_P are so-called linear and parabolic rate constants for the overall oxidation process. Also solve for $\delta(t)$.

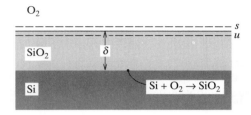

1-62

i. Referring to Exercise 1–30, consider the situation where the solute gas is consumed in a first-order homogeneous reaction with rate constant k'''. When $x_{1,\text{in}} = 0$, show that the absorption rate is given by

$$J_{1,u} = c x_{1,u} \left(\mathcal{D}_{12} k''' \right)^{1/2} \left[\operatorname{erf} \left(k''' z/u_\delta \right) + \frac{e^{-k''' z/u_\delta}}{\left(\pi k''' z/u_\delta \right)^{1/2}} \right]$$

and that for a very fast reaction, the mole transfer conductance is

$$\mathcal{G}_{m1} = c \left(\mathcal{D}_{12} k''' \right)^{1/2}$$

ii. A cross-flow falling-film reactor is to be designed for a chemical laser. Chlorine gas flows between walls wetted with basic hydrogen peroxide to produce singlet-delta oxygen according to the reaction

$$2\text{KOH} + \text{H}_2\text{O}_2 + \text{Cl}_2 \rightarrow 2\text{H}_2\text{O} + 2\text{KCl} + \text{O}_2 \left({}^1\Delta \right)$$

The reaction is first-order, with a rate constant $k''' \simeq 10^8 \text{ s}^{-1}$. If the exchanger height is 1 m and the film surface velocity is 0.03 m/s, estimate the liquid-side mole transfer conductance and penetration distance. Take $\mathcal{D}_{12} \simeq 4 \times 10^{-10} \text{ m}^2/\text{s}$ and $c = 36.0$ kmol/m^3.

(*Hint:* Attempt this exercise after you have studied Laplace transforms in a mathematics or controls course.)

1-63 A sample of air at 300 K and 1 atm has an approximate composition of 23.2% O_2 and 76.8% N_2 by mass. Estimate the viscosity and thermal conductivity of the mixture using the Chapman–Enskog kinetic theory formulas of Section 1.7.1.

1-64 Estimate the diffusion coefficient of water vapor in air at 1 atm in the temperature range $0°\text{C} - 100°\text{C}$ using

i. the formula given in Table A.17a.

ii. the Chapman–Enskog kinetic theory formula given in Section 1.7.1.

1-65 Estimate the diffusion coefficient in air at $25°\text{C}$ and 1 atm for

i. helium.

ii. methane.

iii. carbon tetrachloride.

1-66 A useful approximate method for treating multicomponent diffusion is based on a bifurcation approximation for the binary diffusion coefficients. The approach is to correlate the binary diffusion coefficients in a particular system as

$$\mathcal{D}_{ij} = \frac{\overline{\mathcal{D}}}{F_i F_j}$$

where $\overline{\mathcal{D}}$ is a reference diffusion coefficient that contains the pressure and temperature dependence, and the F_i are *diffusion factors* that depend on species i properties only. If $\overline{\mathcal{D}}$ is taken to be the self-diffusion coefficient of O_2, then a simple correlation

$$F_i = \left(\frac{M_i}{26}\right)^{0.481}$$

has been shown to work well for mixtures of species formed from the elements C, H, O, and N, that is, for products of combustion of hydrocarbon fuels, and graphite or carbon-phenolic heat shields [28]. Check the validity of this correlation for a mixture of air, H_2, and H_2O at 400 K and 1.5 atm.

1-67 Estimate the diffusion coefficients for dilute aqueous solutions of methanol and n-propyl alcohol at 300 K using the Wilke–Chang and Othmer–Thakar formulas. Compare your results with values calculated from the Schmidt numbers given in Table A.18.

1-68 Estimate the diffusion coefficients for dilute aqueous solutions of ammonia in the temperature range 280 K to 320 K using the Wilke–Chang and Othmer–Thakar formulas. Compare your results with values calculated from the Schmidt number and temperature correction factor in Table A.18.

1-69 Aluminum sparks caused by arc welding or clashing electrical power cables are small droplets ($\sim$1 mm diameter) of molten aluminum surrounded by a flame in which aluminum vapor and atmospheric oxygen react to form aluminum oxide Al_2O_3 as a smoke of very small aerosol particles. Estimate the Brownian diffusion coefficient for Al_2O_3 particles in air at 3300 K, 1 atm. Assume a dynamic shape factor $\chi = 2.0$, and obtain results for mass equivalent diameters $d_{pe} = 10^{-3}, 10^{-2}$, and 10^{-1} μm.

1-70 For particle deposition from a flow onto a surface by Brownian diffusion, we can define a particle transfer conductance G_p that is analogous to the mole transfer conductance G_m. Then the diffusive flux of particles across the s-surface is

$$J_p = G_p(\mathcal{N}_s - \mathcal{N}_e) \text{ particles/m}^2\text{s}$$

Since number density $\mathcal{N}$ has units [particles/m^3], G_p has units [m/s], that is, the same as a velocity. Usually we can assume that all particles reaching the s-surface are captured; thus, $\mathcal{N}_s = 0$ and the particle deposition rate is $-J_p = G_p\mathcal{N}_e$. The Sherwood number is $Sh_p = G_p L/\mathcal{D}_p$. Air at 300 K and 1 atm flows over a 10 cm–long flat plate. Prepare a graph of the particle transfer conductance $\overline{G}_p$ as a function of particle size, for air velocities of 0.3, 1, and 3 m/s. Use the data in Table A.30.

1-71 In a field experiment to measure "dry" deposition rates of atmospheric pollutants in the form of aerosols, the test surface is a 10 cm–diameter Teflon disk in the center of a Frisbee-shaped airfoil. A control surface is located nearby and is coated with naphthalene. Both surfaces are maintained at the same temperature. After exposure for 12 hours at a temperature of 300 K, the Teflon disk is washed in a small quantity of pure water to extract the pollutants, and the water is analyzed. Also, the weight loss of the naphthalene disk is determined. In a particular test, a deposition of 3560 ng of aerosol X is found, and the mass loss of naphthalene is 1.46 g. If aerosol X has a density of 2100 kg/m^3 and an average size of 0.05 μm, estimate the number density of aerosol X in the air at the test site.

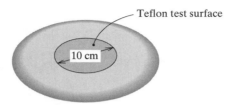

1-72 Air at 1 atm pressure and 400 K flows at 20 m/s over a 10 cm–long flat plate. The air contains 10^{15} particles/m^3 of spherical aerosol particles of diameter 3×10^{-3} μm. Estimate the rate of deposition at the trailing edge for the following conditions:

  i. A laminar boundary layer.

  ii. A turbulent boundary layer starting at the leading edge.

  iii. A turbulent boundary layer starting at the leading edge of a plate roughened with 1 mm–diameter close-packed sand grains.

1-73 Experimental data give the diffusion coefficient of soybean protein in dilute aqueous solution at 20°C as 2.91×10^{-11} m^2/s. If the molecular weight of this protein is 361,800, use the Polson equation to estimate the diffusion coefficient, and compare your result to the experimental value.

1-74 In an experiment to investigate silica scaling in a geothermal energy power system, brine at 350 K flows at 0.366 kg/s in a 15.2 mm–I.D. pipe, 76 m long. At the inlet, the concentration of amorphous silica particles is 10^{19} particles/m^3, and their average diameter is 6 nm. Estimate

  i. the rate of scaling near the pipe inlet.

  ii. the outlet concentration of silica particles.
The silica density can be taken as 2200 kg/m^3.

High Mass Transfer Rate Theory

2.1 INTRODUCTION

In Chapter 1 we considered some simple mass diffusion and mass convection problems. In Section 1.3, transport of a chemical species was by diffusion only; that is, we restricted our attention to a stationary medium. Actually, we did not even bother to define precisely what we mean by *stationary*. Since we have seen that in a mixture the various species can move relative to each other, a suitable definition of *stationary* is not immediately obvious. What we had in mind was an imprecise idea based on physical experience. For example, diffusion of a trace species in a solid would qualify as diffusive transport. Similarly, transport of a chemical species in the porous catalysts considered in Section 1.6 was by diffusion only. Convective mass transport was considered in Sections 1.4 and 1.5; however, we used a simple physics-based analogy to convective heat transfer. We did not formulate the governing equations in a rigorous manner. We considered only *low mass transfer rate theory,* which assumes that mass transport across the s-surface is by diffusion only. At higher mass transfer rates, there is also significant mass transport across the s-surface by convection, because there is a velocity component normal to the surface associated with net mass transfer. Unfortunately, the concepts involved in the analysis of diffusion in a moving medium, and in **high mass transfer rate** convection, are not simple. It was for this reason that the limit cases of diffusion in a stationary medium and low mass transfer rate convection were treated first, as a stepping-stone to the complete picture, with such details as the precise definition of a stationary medium ignored.

In this chapter we go back to square one and develop a rigorous and more complete mass transfer theory. In Section 2.2 velocities and fluxes in a mixture are carefully defined, a general species conservation equation is derived, and Fick's law is more precisely stated. In Section 2.3 some one-dimensional problems involving high mass transfer rate diffusion are analyzed. In Section 2.4 high mass transfer rate convection is introduced with the analysis of a simple Couette–flow model of real boundary layers. The results form the basis of an engineering problem-solving procedure that uses **blowing factors** to account for the effects of high mass transfer rates on mass transfer, skin friction, and heat transfer. Heat transfer in forced laminar boundary layers was analyzed in Section 5.4 of *Heat Transfer*; in Section 2.5 this analysis is extended to include mass transfer, and exact self-similar solutions are obtained. The chapter continues with a general problem-solving procedure that can be used to solve a wide variety of convective heat and mass transfer problems. The chapter closes with a rigorous treatment of transport in multicomponent gas mixtures and liquid solutions, and conservation equations for model multicomponent gas mixtures.

2.2 VELOCITIES, FLUXES, AND THE SPECIES CONSERVATION EQUATION

When diffusion occurs in a moving medium, a chemical species is transported both by diffusion and by motion of the medium as a whole (i.e., convection). Thus, the flux of the species relative to stationary coordinate axes consists of two components, one due to diffusion and one due to convection. We now examine this *simultaneous diffusion and convection*. In Sections 2.2.1 and 2.2.2 velocities and fluxes are defined, and these allow the general species conservation equation to be derived in Section 2.2.3. We define Fick's law precisely in Section 2.2.4 after examining what we mean by a stationary medium.

2.2.1 Definitions of Velocities

In a multicomponent system the various species may move at different velocities. Let v_i denote the absolute velocity of species i, that is, the velocity relative to stationary coordinate axes. In this sense, the velocity is not that of an individual molecule of species i. Rather, it is the local average of the species, that is, the sum of the velocities of all molecules of species i within an elemental volume divided by the number of such molecules. Then the local **mass-average velocity, v,** is defined as

$$\mathbf{v} = \frac{\sum \rho_i \mathbf{v}_i}{\sum \rho_i} = \frac{\sum \rho_i \mathbf{v}_i}{\rho} = \sum m_i \mathbf{v}_i \tag{2.1}$$

The quantity $\rho \mathbf{v}$ is the local mass flux, that is, the rate at which mass passes through a unit area placed normal to the velocity vector $\mathbf{v}$. From Eq. (2.1), $\rho \mathbf{v} = \sum \rho_i \mathbf{v}_i$; that is, the local mass flux is the sum of the local species mass fluxes. The velocity $\mathbf{v}$ is

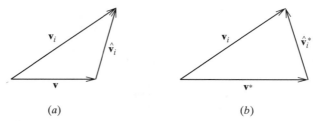

Figure 2.1   Velocity vectors: (a) mass basis,
(b) molar basis.

the velocity that would be measured by a Pitot tube and corresponds to the velocity $\mathbf{v}$ used in considering pure fluids. Of particular importance is that this velocity $\mathbf{v}$ is the velocity field described by the Navier–Stokes equations and hence by Newton's second law of motion.

The local **molar-average velocity**, $\mathbf{v}^*$, is defined in an analogous manner:

$$\mathbf{v}^* = \frac{\sum c_i \mathbf{v}_i}{\sum c_i} = \frac{\sum c_i \mathbf{v}_i}{c} = \sum x_i \mathbf{v}_i \tag{2.2}$$

The quantity $c\mathbf{v}^*$ is the local molar flux, that is, the rate at which moles pass through a unit area placed normal to the velocity $\mathbf{v}^*$.

The velocity of a particular species relative to the mass- or molar-average velocity is termed a **diffusion velocity**, because a diffusion mechanism is required for a species to move relative to the average velocity. We define two such velocities:

$$\hat{\mathbf{v}}_i = \mathbf{v}_i - \mathbf{v} \equiv \text{Diffusion velocity of species } i \text{ relative to } \mathbf{v} \tag{2.3a}$$

$$\hat{\mathbf{v}}_i^* = \mathbf{v}_i - \mathbf{v}^* \equiv \text{Diffusion velocity of species } i \text{ relative to } \mathbf{v}^* \tag{2.3b}$$

Figure 2.1 illustrates these velocities.

Example 2.1   Molar and Mass Average Velocities

A gas mixture contains 50% He and 50% O_2 by weight. The absolute velocities of each species are $\mathbf{v}_{He} = -9\mathbf{i}$ m/s, $\mathbf{v}_{O_2} = +9\mathbf{i}$ m/s, where $\mathbf{i}$ denotes the unit vector in the x direction. Determine the mass and molar average velocities.

Solution

Given: Mixture of He and O_2.
Required: Mass- and molar-average velocities, $\mathbf{v}$ and $\mathbf{v}^*$.
From Eq. (2.1) we find the mass-average velocity $\mathbf{v}$ as

$$\mathbf{v} = \sum_{i=1}^{2} m_i \mathbf{v}_i = (0.5)(-9\mathbf{i}) + (0.5)(+9\mathbf{i}) = 0$$

The mole fractions are found from Eq. (1.10b):

$$x_{He} = \frac{m_{He}/M_{He}}{m_{He}/M_{He} + m_{O_2}/M_{O_2}} = \frac{0.5/4}{(0.5/4) + (0.5/32)} = 8/9$$

$$x_{O_2} = 1 - x_{He} = 1/9$$

Then, from Eq. (2.2), the mole-average velocity is

$$\mathbf{v}^* = \sum_{i=1}^{2} x_i \mathbf{v}_i = (8/9)(-9\mathbf{i}) + (1/9)(+9\mathbf{i}) = -7\mathbf{i} \text{ m/s}$$

Comment It can be seen that although the mass-average velocity is zero, the molar-average velocity is large. Is the mixture stationary? We will soon be able to answer this question. ▲

2.2.2 Definitions of Fluxes

The mass (or molar) flux of species i is a vector quantity giving the mass (or moles) of species i that pass per unit time through a unit area perpendicular to the vector. Such fluxes may be defined relative to stationary coordinate axes or to either of the two local average velocities. We define the absolute mass and molar fluxes of species i, that is, relative to stationary coordinate axes, as

$$\text{Mass flux: } \mathbf{n}_i = \rho_i \mathbf{v}_i \qquad (2.4a)$$

$$\text{Molar flux: } \mathbf{N}_i = c_i \mathbf{v}_i \qquad (2.4b)$$

The mass diffusion flux relative to the mass-average velocity $\mathbf{v}$ of species i is

$$\mathbf{j}_i = \rho_i \hat{\mathbf{v}}_i = \rho_i (\mathbf{v}_i - \mathbf{v}) \qquad (2.5a)$$

The molar diffusion flux relative to the molar-average velocity $\mathbf{v}^*$ of species i is

$$\mathbf{J}_i^* = c_i \hat{\mathbf{v}}_i^* = c_i (\mathbf{v}_i - \mathbf{v}^*) \qquad (2.5b)$$

From a mathematical viewpoint, any one of these flux definitions is adequate for all diffusion situations; however, in a given situation there is usually one definition that leads to minimum algebraic complexity. An important example is when the convective transport present requires a solution of the conservation-of-momentum equation; the solution yields the mass-average velocity field, and it is then most convenient to use the mass flux relative to the mass-average velocity, that is, $\mathbf{j}_i$. Conditions of constant pressure and temperature, often encountered by chemical engineers, have often led to the choice of the absolute molar flux $\mathbf{N}_i$ because of simplifications that result from the molar density c being constant.

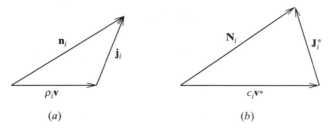

(a) (b)

Figure 2.2 Flux vectors: (a) mass basis, (b) molar basis.

The definitions of fluxes lead directly to a number of important relations that can be easily derived. The absolute mass flux of the mixture (mass velocity) is

$$\mathbf{n} = \sum \mathbf{n}_i = \rho \mathbf{v} \qquad (2.6a)$$

and on a molar basis,

$$\mathbf{N} = \sum \mathbf{N}_i = c\mathbf{v}^* \qquad (2.6b)$$

Also,

$$\sum \mathbf{j}_i = \sum \mathbf{J}_i^* = 0 \qquad (2.7)$$

$$\mathbf{N}_i = \frac{\mathbf{n}_i}{M_i} \qquad (2.8)$$

$$\mathbf{n}_i = \rho_i \mathbf{v} + \mathbf{j}_i = m_i \sum \mathbf{n}_i + \mathbf{j}_i \qquad (2.9a)$$

$$\mathbf{N}_i = c_i \mathbf{v}^* + \mathbf{J}_i^* = x_i \sum \mathbf{N}_i + \mathbf{J}_i^* \qquad (2.9b)$$

Notice that Eq. (2.9a) states that the absolute mass flux can be expressed as a sum of the convective flux $\rho_i \mathbf{v}$ and the diffusive flux $\mathbf{j}_i$. Equation (2.9b) can be interpreted similarly. Figure 2.2 illustrates the flux definitions.

Example 2.2   Evaporation of Water

A water surface is at 350 K and evaporates into air at 1 atm at a rate of 0.01 kg/m^2 s. Determine the convective and diffusive components of the water vapor flux across the s-surface.

Solution

Given: Water at 350 K evaporating at 0.01kg/m^2 s into air at 1 atm.
Required: Convective and diffusive components of the water vapor flux across the s-surface.
Assumptions: Air is negligibly soluble in water.
Let water be species 1 and air species 2. By mass conservation the absolute mass fluxes across the u- and s-surfaces are equal,

$$n_u = n_s = 0.01 \text{kg/m}^2 \text{s}$$

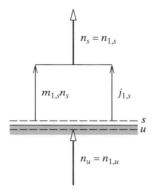

Using Eq. (2.9a), the absolute flux of species 1 across the s-surface can be written as a sum of convective and diffusive components,

$$n_{1,s} = m_{1,s}n_s + j_{1,s}$$

From PSYCHRO, for $T_s = 350$ K, $P = 1.013 \times 10^5$ Pa, and RH $= 100\%$, $m_{1,s} = 0.303$.

Hence, the convective component is

$$m_{1,s}n_s = (0.303)(0.01) = 0.00303 \text{ kg/m}^2 \text{ s}$$

Also, $n_{2,s} = 0$ since air is negligibly soluble in water. Hence,

$$n_s = n_{1,s} + n_{2,s} = n_{1,s} = 0.01 \text{ kg/m}^2 \text{ s}$$

$$j_{1,s} = n_s(1 - m_{1,s}) = (0.01)(1 - 0.303) = 0.00697 \text{ kg/m}^2 \text{ s}$$

Comment   The ratio of diffusive to convective components is equal to $(1-m_{1,s})/m_{1,s}$. As $T \to T_{\text{BP}}$, $m_{1,s} \to 1$ and the diffusive component goes to zero.   ▲

2.2.3 The General Species Conservation Equation

We take an Eulerian viewpoint and consider a Cartesian coordinate volume element Δx by Δy by Δz, fixed in space, as shown in Fig. 2.3. Conservation of a chemical species i requires that the time rate of storage of species i within the volume equals the net rate of inflow of species i across the boundary plus the production rate of species i within the volume due to chemical reactions.

The time rate of storage of species i within the control volume is simply

$$\frac{\partial \rho_i}{\partial t} \Delta x \Delta y \Delta z$$

The gross rate of inflow of species i is

$$n_{ix}|_x \Delta y \Delta z + n_{iy}|_y \Delta x \Delta z + n_{iz}|_z \Delta x \Delta y$$

Similarly, the gross rate of outflow is

$$n_{ix}|_{x+\Delta x} \Delta y \Delta z + n_{iy}|_{y+\Delta y} \Delta x \Delta z + n_{iz}|_{z+\Delta z} \Delta x \Delta y$$

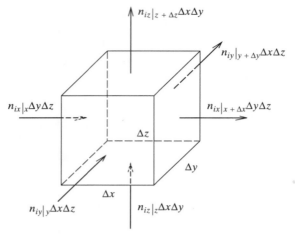

Figure 2.3 Cartesian elemental control volume for
derivation of the species conservation equation.

The net rate of inflow is found by subtracting the outflow from the inflow, expanding
quantities evaluated at $x + \Delta x$, $y + \Delta y$, and $z + \Delta z$ in Taylor series, and retaining
only first-order terms to obtain

$$-\left(\frac{\partial n_{ix}}{\partial x} + \frac{\partial n_{iy}}{\partial y} + \frac{\partial n_{iz}}{\partial z}\right)\Delta x \Delta y \Delta z$$

Species i may be produced at the rate $\dot{r}_i'''$ mass per unit volume per unit time
$[\text{kg/m}^3\,\text{s}]$ due to homogeneous chemical reactions (reactions that take place *within*
the medium). The rate of production of i within the volume is then

$$\dot{r}_i''' \Delta x \Delta y \Delta z$$

Substituting in our conservation statement and dividing throughout by the volume
$\Delta x \Delta y \Delta z$,

$$\frac{\partial \rho_i}{\partial t} = -\left(\frac{\partial n_{ix}}{\partial x} + \frac{\partial n_{iy}}{\partial y} + \frac{\partial n_{iz}}{\partial z}\right) + \dot{r}_i''' \tag{2.10}$$

which is one form of the species conservation equation. Summation of Eq. (2.10)
over all species i gives

$$\frac{\partial}{\partial t}\sum \rho_i = -\left(\frac{\partial}{\partial x}\sum n_{ix} + \frac{\partial}{\partial y}\sum n_{iy} + \frac{\partial}{\partial z}\sum n_{iz}\right) + \sum \dot{r}_i'''$$

Since $\sum \rho_i = \rho$, $\sum n_i = n$, and $\sum \dot{r}_i''' = 0$ (mass is conserved in chemical reactions),
there results

$$\frac{\partial \rho}{\partial t} + \left(\frac{\partial n_x}{\partial x} + \frac{\partial n_y}{\partial y} + \frac{\partial n_z}{\partial z}\right) = 0 \tag{2.11}$$

For velocity components u, v, w in the x-, y-, and z-coordinate directions, we have $n_x = \rho u$, $n_y = \rho v$, and $n_z = \rho w$; hence,

$$\frac{\partial \rho}{\partial t} + \left(\frac{\partial \rho u}{\partial x} + \frac{\partial \rho v}{\partial y} + \frac{\partial \rho w}{\partial z} \right) = 0 \tag{2.12a}$$

or, in vector form,

$$\frac{\partial \rho}{\partial t} + \nabla \cdot \rho \mathbf{v} = 0; \quad \mathbf{v} = \mathbf{i}u + \mathbf{j}v + \mathbf{k}w \tag{2.12b}$$

where $\nabla \cdot$ is the divergence operator. Equation (2.12) is the mass conservation (or continuity) equation. Equation (2.10) can also be written in compact form using the divergence operator,

$$\frac{\partial \rho_i}{\partial t} + \nabla \cdot \mathbf{n}_i = \dot{r}_i''' \tag{2.13}$$

In addition, if we wish to write Eq. (2.13) in coordinate systems other than the Cartesian one, we need only to express $\nabla \cdot$ appropriately. For example, in spherical coordinates with spherical symmetry, we obtain

$$\frac{\partial \rho_i}{\partial t} + \frac{1}{r^2} \frac{\partial}{\partial r} \left(r^2 n_{ir} \right) = \dot{r}_i''' \tag{2.14}$$

The species equation can also be derived on a molar basis as

$$\frac{\partial c_i}{\partial t} + \nabla \cdot \mathbf{N}_i = \dot{R}_i''' \tag{2.15}$$

where $\dot{R}_i'''$ is the molar rate of production of species i [kmol/m^3 s]. Summing Eq. (2.15) over all species gives

$$\frac{\partial c}{\partial t} + \nabla \cdot (c\mathbf{v}^*) = \sum_i \dot{R}_i''' \tag{2.16}$$

where, in general, $\sum \dot{R}_i''' \neq 0$, since moles need not be conserved in chemical reactions.

The final step to obtain a differential equation governing the concentration distribution is to write the absolute flux as the sum of convective and diffusive fluxes, and introduce appropriate physical laws for the diffusive flux $\mathbf{j}_i$ (or $\mathbf{J}_i^*$). Also, if there are rate-controlled homogeneous chemical reactions, chemical kinetics expressions are required for the production term $\dot{r}_i'''$ (or $\dot{R}_i'''$).

2.2.4 A More Precise Statement of Fick's Law

We have noted how a mass species may be transported by convection and diffusion. Convection is of its nature a *bulk* motion and thus transports the mixture as a whole. A given species can be transported relative to this bulk motion if diffusion occurs. Thus, it is clear that a precise definition of Fick's first law must describe diffusion relative to an average velocity of the mixture. Our definitions of velocities and fluxes, Eqs. (2.1) through (2.5), were of a mathematical nature and depend on no physical laws. However, they were chosen so that the introduction of physics via Fick's first law is straightforward. We now propose that the law should be written

$$\mathbf{j}_1 = -\rho \mathcal{D}_{12} \nabla m_1 \tag{2.17}$$

that is, exactly as Eq. (1.15), but now $\mathbf{j}_1$ has been precisely defined as the mass flux of species 1 *relative to the mass-average velocity*. The corresponding law written for species 2 is

$$\mathbf{j}_2 = -\rho \mathcal{D}_{21} \nabla m_2$$

Since $\nabla m_1 = -\nabla m_2$ in a binary system, and from Eq. (2.7), $\mathbf{j}_1 + \mathbf{j}_2 = 0$, it then follows immediately that

$$\mathcal{D}_{12} = \mathcal{D}_{21} \tag{2.18}$$

The molar form of Fick's first law equivalent to Eq. (2.17) is

$$\mathbf{J}_1^* = -c \mathcal{D}_{12} \nabla x_1 \tag{2.19}$$

Algebraic manipulation will show that Eqs. (2.19) and (2.17) are mathematically equivalent. However, we see that the diffusive molar flux $\mathbf{J}_1^*$ is the molar flux *relative to the molar-average velocity*.

It is now possible to give a correct interpretation of what we mean by a stationary medium. If we are working in mass units, we require that the mass-average velocity be zero; if we are working in molar units, we require that the molar-average velocity be zero. Then, in a stationary medium, mass transport is by diffusion only. Since a Pitot tube or anemometer measures the mass-average velocity, the interpretation of stationary as zero mass-average velocity is more in accord with our physical intuition. A zero molar-average velocity corresponds to a zero molecule average velocity, and is a concept more appropriate to interpretation at a molecular level.

Finally, the substitution of Fick's law into Eqs. (2.9a) and (2.9b), written for a binary system, yields two important relations:

$$\mathbf{n}_1 = \rho_1 \mathbf{v} - \rho \mathcal{D}_{12} \nabla m_1 = m_1(\mathbf{n}_1 + \mathbf{n}_2) - \rho \mathcal{D}_{12} \nabla m_1 \tag{2.20a}$$

$$\mathbf{N}_1 = c_1 \mathbf{v}^* - c \mathcal{D}_{12} \nabla x_1 = x_1(\mathbf{N}_1 + \mathbf{N}_2) - c \mathcal{D}_{12} \nabla x_1 \tag{2.20b}$$

We see that the absolute flux of a species can always be conveniently expressed as the sum of two components, one due to convection and the other due to diffusion.

2.3 HIGH MASS TRANSFER RATE DIFFUSION

Net mass transfer across a surface results in a velocity component normal to the surface, and an associated convective flux in the direction of mass transfer. If the mass transfer rate is high, this convection cannot be ignored, as was done in Chapter 1 in using low mass transfer rate theory. In the problems analyzed in this section, the only convection is that induced by the mass transfer process itself, and it is always in the direction of the mass transfer. The induced convective flow is called a *Stefan flow*. In this section we are not concerned with situations where there is also a forced- or natural-convection flow along the transfer surface: such situations are dealt with in Sections 2.4 and 2.5. The analyses that follow consider a variety of evaporation and combustion problems.

2.3.1 Diffusion with One Component Stationary

Diffusion in a binary mixture where one component is stationary occurs in many situations involving evaporation, condensation, or transpiration. To develop the concepts involved, a particular situation will be considered, that of a heatpipe into which has leaked a small amount of gas. Fluid flow and heat transfer in heatpipes were analyzed in Chapter 7 of *Heat Transfer*. We now consider a feature of heatpipe operation that requires a mass transfer analysis.

One-Dimensional Model of a Heatpipe

We wish to determine the effect of a small amount of noncondensable gas added to the vapor in a fixed-conductance heatpipe. Corrosion may have generated gas within the heatpipe, or a construction defect may have allowed gas to leak into the heatpipe. A simple one-dimensional analysis gives insight into this phenomenon. We will consider the case of a heatpipe with the evaporator and condenser located at the *ends* of the heatpipe pipe only. (We shall see later that this is a poor design.) Figure 2.4 depicts an idealized model of such a heatpipe.

The vapor is designated as species 1 and the gas as species 2. Heat is applied at the surface $z = 0$, and heat is removed at the surface $z = L$. The sides of the heatpipe are assumed insulated. The temperatures at $z = 0$ and $z = L$ are T_s and T_e, respectively. The corresponding mole fractions of vapor adjacent to the liquid surfaces are $x_{1,s}$ and $x_{1,e}$. These values are the equilibrium values obtained from the saturation vapor pressures corresponding to the liquid surface temperatures T_s and T_e, respectively. The reason for choosing subscripts s and e (rather than, say, E and C for evaporator and condenser) will become apparent later. The solubility of gas in the liquid is assumed zero, and thus species 1 is the only substance transferred across the interfaces. At steady state there is net transfer of vapor along the heatpipe, whereas conservation of species requires that the gas, species 2, remains stationary. The variations of pressure and absolute temperature along the heatpipe are small; thus, if an ideal gas mixture is assumed, the molar concentration c is virtually independent of position z. Advantage of this fact may be taken by performing the analysis on a molar basis. For this problem Eq. (2.15) written for the vapor reduces to

$$\frac{d}{dz}(N_1) = 0 \qquad (2.21)$$

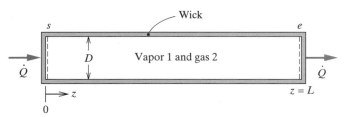

Figure 2.4   A heatpipe with the evaporator and condenser located at its ends.

Likewise, Eq. (2.15) written for the gas reduces to

$$\frac{d}{dz}(N_2) = 0 \tag{2.22}$$

The subscript z has been dropped from N_1 and N_2 since it is not needed for this one-dimensional situation. Integration yields $N_2 = $ constant. Since the gas is insoluble in the liquid, $N_2|_{z=0} = 0$, and it follows from Eq. (2.22) that N_2 is identically zero; that is, the gas is stationary.

The molar flux N_1 can be expressed as the sum of a convective and a diffusive component; from Eq. (2.20b),

$$N_1 = x_1(N_1 + N_2) - c\mathcal{D}_{12}\frac{dx_1}{dz}$$

But $N_2 = 0$; thus, solving for N_1 yields

$$N_1 = -\frac{c\mathcal{D}_{12}}{1 - x_1}\frac{dx_1}{dz} \tag{2.23}$$

Substitution of this expression in the differential equation Eq. (2.21) gives

$$\frac{d}{dz}\left(-\frac{c\mathcal{D}_{12}}{1 - x_1}\frac{dx_1}{dz}\right) = 0 \tag{2.24}$$

The product of the molar concentration c and the binary diffusion coefficient is assumed constant ($\mathcal{D}_{12}$ is also independent of composition). Therefore, the differential equation governing the concentration distribution may be written as

$$\frac{d}{dz}\left(\frac{1}{1 - x_1}\frac{dx_1}{dz}\right) = 0 \tag{2.25}$$

which is to be solved subject to the boundary conditions

$$z = 0, \quad x_1 = x_{1,s} \tag{2.26a}$$

$$z = L, \quad x_1 = x_{1,e} \tag{2.26b}$$

Integrating twice and evaluating the constants of integration from the boundary conditions yields the concentration distribution of the vapor,

$$\left(\frac{1 - x_1}{1 - x_{1,s}}\right) = \left(\frac{1 - x_{1,e}}{1 - x_{1,s}}\right)^{z/L} \tag{2.27}$$

Alternatively, the concentration distribution of the gas is given by

$$\left(\frac{x_2}{x_{2,s}}\right) = \left(\frac{x_{2,e}}{x_{2,s}}\right)^{z/L} \tag{2.28}$$

The concentration distributions are illustrated in Fig. 2.5. The rate of vapor transport through the pipe is the rate at which the liquid evaporates and may be evaluated as follows:

$$N_1|_{z=0} = N_{1,s} = -\frac{c\mathcal{D}_{12}}{1 - x_{1,s}}\frac{dx_1}{dz}\bigg|_{z=0}$$

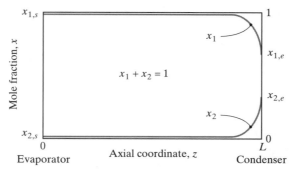

Figure 2.5 Vapor (species 1) and gas (species 2) concentration profiles along a heatpipe.

Differentiation of Eq. (2.27) and evaluation at $z = 0$ gives

$$N_1 \equiv N_{1,s} = \frac{c\mathcal{D}_{12}}{L} \ln \frac{1 - x_{1,e}}{1 - x_{1,s}} \tag{2.29}$$

The rate of heat transfer along the heatpipe is simply $M_1 N_1 h_{fg}(\pi/4)D^2$, where h_{fg} is the enthalpy of vaporization:

$$\frac{\dot{Q}}{(\pi/4)D^2} = \frac{M_1 c \mathcal{D}_{12} h_{fg}}{L} \ln \frac{1 - x_{1,e}}{1 - x_{1,s}} = \frac{M_1 c \mathcal{D}_{12} h_{fg}}{L} \ln \frac{x_{2,e}}{x_{2,s}} \tag{2.30}$$

An Alternative Form for the Vapor Transport Rate

An alternative form of Eq. (2.29) may be obtained by introducing the *logarithmic mean value* of the mole fraction x_2, defined as

$$(x_2)_{lm} = \frac{x_{2,e} - x_{2,s}}{\ln(x_{2,e}/x_{2,s})} \tag{2.31}$$

Algebraic manipulation of Eq. (2.29) and substitution of Eq. (2.31) then yields

$$N_1 = \frac{c\mathcal{D}_{12}}{L(x_2)_{lm}}(x_{1,s} - x_{1,e}) \tag{2.32}$$

For low transfer rates ($N_1 \to 0$), the value of $(x_2)_{lm}$ approaches its limiting value of unity, and there is a linear Ohm's law type relation between N_1 and $(x_{1,s} - x_{1,e})$. As the concentration difference Δx_1 and transfer rate increase, $(x_2)_{lm}$ decreases, and N_1 is no longer a linear function of Δx_1; this nonlinearity at high mass transfer rates is an important characteristic of high mass transfer rate theory.

Even though we have said that one component is stationary, there is, of course, diffusion of both species. The concentration gradients of components 1 and 2 are equal and opposite, and hence gas 2 diffuses toward the evaporating surface. Although the gas is insoluble, its concentration does not build up at the liquid surface. There is a convective motion in the positive z direction such that at steady state, the net flux N_2, which is the sum of convective component $x_2 N$ and diffusive component

$-c\mathcal{D}_{12}(dx_2/dz)$, is exactly zero. A small total pressure gradient is required to overcome wall friction in driving this convective flow along the heatpipe. Convection is, by nature, a bulk movement of the mixture, and hence vapor 1 is also convected in the positive z-direction; this was, of course, accounted for in Eq. (2.20b) by writing the total flux of vapor 1, N_1, as a sum of convective and diffusive components. This induced convective flow that augments the diffusive flux of vapor 1 is called the *Stefan flow*.

Location of the evaporator and condenser at the ends of the heatpipe allows the use of a simple one-dimensional model of diffusion and convection. The practical consequences of having the condenser at the end of the heatpipe are discussed at the end of Example 2.3. A different concern is whether the one-dimensional analysis is appropriate for the model problem. We have seen that there is a flow along the tube induced by the diffusion process, and we assumed a uniform profile of the velocity $v_z = N_1/c$. In reality, the no-slip condition for a viscous fluid must apply at the walls of the pipe, and the flow pattern in the heatpipe is rather complex. Notwithstanding this complication, our simple one-dimensional model is quite adequate to demonstrate the effect of a noncondensable gas on heatpipe performance.

Example 2.3   Effect of Air on Heatpipe Performance

A water heatpipe 70 cm long and of 2 cm inside diameter is taken from a storehouse and its performance checked prior to use. For a heat load of 100 W and an evaporator temperature of $100°C$, the condenser temperature is measured to be $95.0°C$. An air leak or gas generation is suspected; calculate the amount of air that might have leaked into the heatpipe.

Solution

Given: Test data for a water heatpipe.
Required: Amount of air that has leaked into the heatpipe.
Assumptions: A one-dimensional model is appropriate.
First we determine the expected temperature drop if no air were present. The vapor velocity V is calculated from the heat load, evaluating the steam properties for saturated steam at 1 atm.

$$\rho V \left(\frac{\pi}{4}\right) D^2 h_{fg} = \dot{Q} = 100 \text{ W}$$

$$V = \frac{\dot{Q}}{\rho(\pi/4)D^2 h_{fg}} = \frac{100}{(0.5977)(\pi/4)(0.02)^2(2.257 \times 10^6)}$$

$$= 0.236 \text{ m/s}$$

The Reynolds number is

$$\text{Re}_D = \frac{VD}{\nu} = \frac{(0.236)(0.02)}{20.1 \times 10^{-6}} = 235$$

The friction factor for fully developed laminar flow is

$$f = \frac{64}{\text{Re}_D} = \frac{64}{235} = 0.272$$

and hence the pressure drop is

$$\Delta P \simeq f \frac{L}{D} \left(\frac{1}{2} \rho V^2 \right) = (0.272) \left(\frac{70}{2} \right) (1/2)(0.5977)(0.236)^2 = 0.158 \text{ Pa}$$

Steam tables show that the corresponding temperature drop is quite negligible.

Now consider the situation with air present. We will assume (and check later) that the air concentration at the evaporator end is very small; then, since the evaporator is at $100°$C, the total pressure there is 101,330 Pa and varies negligibly along the pipe. From steam tables, the saturation pressure corresponding to $95°$C is 84,520 Pa, and the mole fraction of air is

$$x_{2,e} = \frac{P_{2,e}}{P_{\text{total}}} = \frac{101,330 - 84,520}{101,330} = 0.166$$

Equation (2.30) may be rearranged to read

$$\ln \frac{x_{2,e}}{x_{2,s}} = \frac{\dot{Q}L}{M_1 c \mathcal{D}_{12} h_{\text{fg}} (\pi/4) D^2}$$

The binary diffusion coefficient for water vapor in air is conveniently given by the formula in Table A.17a for $273 < T < 373$ K,

$$\mathcal{D}_{12} = 1.97 \times 10^{-5} \left(\frac{P_0}{P} \right) \left(\frac{T}{T_0} \right)^{1.685} \text{ m}^2/\text{s}; \quad P_0 = 1 \text{ atm}, T_0 = 256 \text{ K}$$

$$\mathcal{D}_{12} = 1.97 \times 10^{-5} \left(\frac{370.5}{256} \right)^{1.685} = 36.7 \times 10^{-6} \text{ m}^2/\text{s}$$

and the total molar concentration c is

$$c = P/\mathcal{R}T = (101,330)/(8314)(370.5) = 3.29 \times 10^{-2} \text{ kmol/m}^3$$

where both $\mathcal{D}_{12}$ and c have been evaluated at the average temperature along the heatpipe, $97.5°$C $\simeq 370.5$ K. Let the ratio $x_{2,e}/x_{2,s}$ be denoted by r; then

$$\ln r = \frac{(100)(0.70)}{(18)(3.29 \times 10^{-2})(36.7 \times 10^{-6})(2.257 \times 10^6)(\pi/4)(0.02)^2}$$

$$= 4.54 \times 10^3$$

which is very large; our assumption of a negligibly small value of $x_{2,s}$ is thus fully justified.

The total moles of air in the heatpipe is the volume integral of the air concentration,

$$W = (\pi/4)D^2 \int_0^L x_2 c \, dz$$

Using Eq. (2.28) we obtain

$$W = (\pi/4)D^2 cx_{2,s} L \int_0^1 r^{z/L} \, d(z/L)$$

$$= (\pi/4)D^2 cx_{2,s} L \frac{r-1}{\ln r} = (\pi/4)D^2 cx_{2,e} \frac{L}{r} \frac{r-1}{\ln r} = \frac{(\pi/4)D^2 cx_{2,e} L}{\ln r}$$

since r is large. Hence,

$$W = \frac{(\pi/4)(0.02)^2(3.29 \times 10^{-2})(0.166)(0.70)}{4.54 \times 10^3} = 2.65 \times 10^{-10} \text{ kmol}$$

The mass of air in the pipe w is equal to $W M_2$; thus,

$$w = (2.65 \times 10^{-10})(29) = 7.69 \times 10^{-9} \text{ kg} = 7.69 \ \mu\text{g}$$

Comments

1. The fact that such a small amount of air as 7.69 μg would cause a 5°C temperature drop is due to the geometrical arrangement considered. Location of the condenser on the end of the heatpipe results in direct confrontation between the gas and the vapor attempting to condense. Since a minute amount of gas can completely shut down the system, precise control would be almost impossible. A superior design would be one where the condenser is located along the side wall of the heatpipe; the working fluid can then simply push the gas aside out of the main stream to an unused portion of the condenser. Indeed, this is the principle of the variable-conductance heatpipe described in Section 7.7.3 of *Heat Transfer*, for which a relatively large amount of gas is present in the heatpipe.

2. Notice the formula used for the diffusion coefficient of water vapor in air. We *cannot* use a Schmidt number $Sc = \nu/\mathcal{D}_{12} = 0.61$ to obtain $\mathcal{D}_{12}$ as we did in Sections 1.4 and 1.5. For the water vapor–air system, $Sc = 0.61$ is appropriate only for small concentrations of water vapor and low temperatures (see Exercise 2–7). ▲

2.3.2 Heterogeneous Combustion

In Section 1.3.3 diffusion with a heterogeneous reaction was considered for the special case of a catalytic reaction. The analysis was particularly simple because no net mass was consumed in the reaction, and hence there was no velocity component or associated convective transport normal to the catalyst surface. Similarly, the low-temperature carbon combustion problem of Section 1.5.3 was analyzed on a molar basis to take advantage of the fact that the molar-average velocity normal to the carbon surface was zero: one mole of CO_2 is produced for each mole of O_2 consumed. We now consider a situation where net mass is consumed and there is convective transport normal to the reacting surface, irrespective of whether the analysis is performed on a mass or molar basis. The example problem is that of a small carbon particle entrained and burning in a high-temperature air flow, as shown in Fig. 2.6. At temperatures above about 2300 K, the reaction is

$$2C + O_2 \rightarrow 2CO$$

and is in equilibrium (diffusion controlled) with two moles of CO produced for one mole of O_2 consumed: thermodynamic data indicate $m_{O_2,s} \simeq 0$ and that no carbon dioxide forms in the gas phase. Oxygen is therefore inert in the gas phase. Denoting O_2 as species 1 and assuming quasi-steady, spherically symmetric diffusion, Eq. (2.14) reduces to

$$\frac{d}{dr}(r^2 n_1) = 0 \tag{2.33}$$

where the subscript r on n_1 has been dropped for convenience. The mass conservation equation Eq. (2.12b) reduces to

$$\frac{d}{dr}(r^2 \rho v) = \frac{d}{dr}(r^2 n) = 0 \tag{2.34}$$

Integrating Eq. (2.34),

$$r^2 n = \text{Constant} = r^2 n|_{r=R} = R^2 n_s \tag{2.35}$$

where n_s is the mass flux across the s-surface. Equation (2.35) is equivalent to the statement that the mass flow $\dot{m} = 4\pi r^2 n$ is a constant, independent of r. A mass balance on a control volume located between the s- and u-surfaces requires that $n_s = n_u$ since mass is conserved. We denote

$$n_s = n_u = \dot{m}'' \tag{2.36}$$

and term $\dot{m}'' = \dot{m}/A$ the *mass transfer rate* [kg/m^2 s]. This distinctive notation is useful since our major objective is usually to determine the mass transfer rate. Thus, Eq. (2.35) becomes

$$r^2 n = R^2 \dot{m}'' \tag{2.37}$$

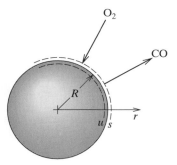

Figure 2.6 Schematic of a carbon particle burning in high-temperature air. The surface reaction is $2C + O_2 \rightarrow 2CO$.

Integrating Eq. (2.33),

$$r^2 n_1 = \text{Constant} = R^2 n_{1,s}$$

or, expressing the absolute flux n_1 as a sum of its convective and diffusive components,

$$r^2 \left(m_1 n - \rho \mathcal{D}_{1m} \frac{dm_1}{dr} \right) = R^2 n_{1,s} \tag{2.38}$$

Substitution of $r^2 n = R^2 \dot{m}''$ from Eq. (2.37), rearranging, and integrating with $m_1 = m_{1,s}$ at $r = R$ and $m_1 \rightarrow m_{1,e}$ as $r \rightarrow \infty$ gives

$$\int_{m_{1,s}}^{m_{1,e}} \frac{dm_1}{m_1 - (n_{1,s}/\dot{m}'')} = \frac{\dot{m}'' R^2}{\rho \mathcal{D}_{1m}} \int_R^\infty \frac{dr}{r^2}$$

where $\rho \mathcal{D}_{1m}$ has been assumed constant. Then,

$$\ln \frac{m_{1,e} - (n_{1,s}/\dot{m}'')}{m_{1,s} - (n_{1,s}/\dot{m}'')} = \frac{\dot{m}'' R}{\rho \mathcal{D}_{1m}}$$

Solving for $\dot{m}''$ and rearranging,

$$\dot{m}'' = \frac{\rho \mathcal{D}_{1m}}{R} \ln \left[1 + \frac{m_{1,e} - m_{1,s}}{m_{1,s} - (n_{1,s}/\dot{m}'')} \right] \tag{2.39}$$

For air, $m_{1,e} = 0.231$, and we have already noted that $m_{1,s} \simeq 0$. The ratio $(n_{1,s}/\dot{m}'')$ is obtained from the stoichiometry of the reaction, as follows:

$$2C + O_2 \rightarrow 2\,CO$$
$$2 \times 12\ \text{kg} + 32\ \text{kg} \rightarrow 2 \times 28\ \text{kg} \tag{2.40}$$

For each kilogram of carbon consumed, 4/3 kg of oxygen is consumed: thus, for each kilogram of carbon transferred across the u-surface, 4/3 kg of oxygen crosses the s-surface in the negative direction; that is,

$$\frac{n_{1,s}}{n_{C,u}} = -\frac{4}{3}$$

But $n_{C,u} = \dot{m}''$; thus, $(n_{1,s}/\dot{m}'') = -4/3$. Substituting in Eq. (2.39),

$$\dot{m}'' = \frac{\rho \mathcal{D}_{1m}}{R} \ln \left[1 + \frac{0.231 - 0}{0 + 4/3} \right] = \frac{\rho \mathcal{D}_{1m}}{R} \ln[1 + 0.173] = 0.160 \frac{\rho \mathcal{D}_{1m}}{R} \tag{2.41}$$

which is a particularly simple result for the rate of mass transfer across the particle surface.

Particle Lifetime
The lifetime of the particle can be obtained by performing a mass balance on the particle of volume V and surface area A_s:

$$\frac{d}{dt} (V \rho_{\text{solid}}) = -A_s \dot{m}''$$

where ρ_{solid} is the density of solid carbon. Thus,

$$\frac{d}{dt}\left(\frac{4}{3}\pi R^3 \rho_{\text{solid}}\right) = -(4\pi R^2)\left(0.160\frac{\rho \mathcal{D}_{1m}}{R}\right)$$

$$\frac{dR}{dt} = -0.160\frac{\rho \mathcal{D}_{1m}}{\rho_{\text{solid}} R} \tag{2.42}$$

Integrating with $\rho \mathcal{D}_{1m}$ taken as constant, and $R = R_0$ at $t = 0$, $R = 0$ at $t = \tau$,

$$\int_{R_0}^{0} R\,dR = -\frac{0.160\rho \mathcal{D}_{1m}}{\rho_{\text{solid}}}\int_{0}^{\tau} dt$$

Hence

$$\tau = \frac{\rho_{\text{solid}} R_0^2}{0.32\rho \mathcal{D}_{1m}} = \frac{\rho_{\text{solid}} D_0^2}{1.28\rho \mathcal{D}_{1m}} \tag{2.43}$$

Property Evaluation

Calculation of τ requires an appropriate reference state for the $\rho \mathcal{D}_{1m}$ product. Since the molecular weight of CO is almost the same as that of air, it is sufficient to use air properties evaluated at the mean film temperature. In order to obtain the mean film temperature, a simultaneous heat and mass transfer problem must be solved to determine the particle temperature, as was done in Example 1.11 for low-temperature oxidation of a carbon particle. Exercise 2–10 requires an approximate heat transfer analysis; a more exact heat transfer analysis is given in Example 2.14.

CO Concentration at the s-Surface

Notice that it is the rate at which O_2 can diffuse to the carbon surface that determines the rate of reaction: hence, the reaction is termed *diffusion-controlled.* We did not use the species conservation equation for the reaction product, CO. If we were to do so, we would obtain an equation identical to Eq. (2.39), with the subscript 2, denoting CO, replacing the subscript 1:

$$\dot{m}'' = \frac{\rho \mathcal{D}_{2m}}{R}\ln\left[1 + \frac{m_{2,e} - m_{2,s}}{m_{2,s} - (n_{2,s}/\dot{m}'')}\right] \tag{2.44}$$

But since we cannot specify $m_{2,s}$, the concentration of CO at the s-surface, the equation cannot be used to determine $\dot{m}''$. In fact, with $\dot{m}''$ known, this equation allows the determination of $m_{2,s}$. In physical terms, the concentration of the product at the s-surface adjusts itself to the value required for transfer away from the particle at the required rate: diffusion of the product away from the particle does not control the rate of reaction in this situation. Now $m_{2,e} = 0$, since there is no CO far from the particle; also, $n_{2,s}/n_{C,u} = 28/12 = 7/3 = n_{2,s}/\dot{m}''$. Substituting in Eq. (2.44) and using Eq. (2.41) for $\dot{m}''$ gives

$$0.160\mathcal{D}_{1m} = \mathcal{D}_{2m}\ln\left[1 + \frac{0 - m_{2,s}}{m_{2,s} - 7/3}\right] \tag{2.45}$$

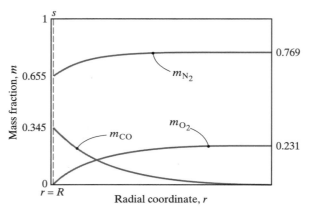

Figure 2.7 Concentration profiles for a carbon particle burning in high-temperature air.

The effective diffusion coefficients for O_2 and CO in the mixture are not too different since their molecular weights are close; assuming $\mathcal{D}_{1m} = \mathcal{D}_{2m}$ and solving gives $m_{2,s} = 0.345$, which is the mass fraction of the reaction product CO at the s-surface. Concentration profiles are shown in Fig. 2.7.

2.3.3 General Chemical Reactions

We now consider the situation where a number of reactions occur. These reactions can be either heterogeneous or homogeneous. As a specific example, we will analyze the combustion of a small carbon particle of radius R entrained in an air stream at moderate temperatures, as shown in Fig. 1.2d. If the carbon surface temperature is in the range 1800–2300 K and the ambient air is relatively cool, there is a surface (heterogeneous) reaction in which carbon dioxide is reduced to carbon monoxide,

$$C + CO_2 \rightarrow 2\,CO$$

and in the gas phase there is a homogeneous reaction in which carbon monoxide is oxidized to carbon dioxide,

$$2\,CO + O_2 \rightarrow 2\,CO_2$$

The resulting concentration profiles are shown in Fig. 2.8. The gas-phase reaction actually occurs in a very narrow region called a flame front—the flame one sees in a charcoal-fired barbecue. The gas-phase reaction is exothermic, whereas the surface reaction is endothermic: the resulting temperature profile is also shown in Fig. 2.8. Assuming a quasi-steady state and spherical symmetry, the species conservation equation, Eq. (2.14), written for each species in turn is

$$\frac{1}{r^2}\frac{d}{dr}(r^2 n_{O_2}) = \dot{r}_{O_2}^{'''} \tag{2.46a}$$

$$\frac{1}{r^2}\frac{d}{dr}(r^2 n_{CO_2}) = \dot{r}_{CO_2}^{'''} \tag{2.46b}$$

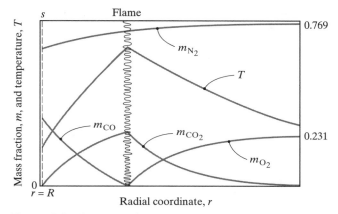

Figure 2.8 Concentration and temperature profiles for a
carbon particle burning in air at intermediate
temperatures. The surface reaction is $C + CO_2 \rightarrow 2CO$;
the reaction in the flame is $2CO + O_2 \rightarrow 2CO_2$.

$$\frac{1}{r^2}\frac{d}{dr}(r^2 n_{CO}) = \dot{r}'''_{CO} \tag{2.46c}$$

$$\frac{1}{r^2}\frac{d}{dr}(r^2 n_{N_2}) = \dot{r}'''_{N_2} = 0, \quad \text{since } N_2 \text{ is inert} \tag{2.46d}$$

where again the subscript r on n_i has been dropped for convenience. The production
rates $\dot{r}'''_i$ are given by complex expressions, possibly involving chemical kinetics or
equilibrium relations. However, the problem can be solved to obtain the carbon
oxidation rate without knowing the $\dot{r}'''_i$ expressions, provided we are prepared to
assume that all the species diffusion coefficients are equal. For this chemical system,
such an assumption is quite realistic. In order to execute the analysis, three definitions
relating to *chemical elements* need to be introduced.

1. The mass fraction of chemical element α in species i is denoted $m_{(\alpha)i}$ and
   is obtained directly from the chemical formula for the species. For example,
   $m_{(O)H_2O} = 16/18 = 8/9$, and $m_{(C)CO_2} = 12/44 = 3/11$.
2. The mass fraction of chemical element α at any location in a mixture is given
   by

$$m_{(\alpha)} = \sum_i m_{(\alpha)i} m_i \tag{2.47}$$

   That is, $m_{(\alpha)}$ is the local mass fraction of element α, irrespective of the chemical
   species in which it is present.
3. The absolute flux of chemical element α is given by

$$\mathbf{n}_{(\alpha)} = \sum_i m_{(\alpha)i} \mathbf{n}_i \tag{2.48}$$

We now multiply Eq. (2.46a) by the mass fraction of element oxygen in species O_2, $m_{(O)O_2}$, Eq. (2.46b) by $m_{(O)CO_2}$, Eq. (2.46c) by $m_{(O)CO}$, and then add:

$$\frac{1}{r^2}\frac{d}{dr}\left[r^2\left(m_{(O)O_2}n_{O_2} + m_{(O)CO_2}n_{CO_2} + m_{(O)CO}n_{CO}\right)\right]$$
$$= m_{(O)O_2}\dot{r}'''_{O_2} + m_{(O)CO_2}\dot{r}'''_{CO_2} + m_{(O)CO}\dot{r}'''_{CO} \tag{2.49}$$

or

$$\frac{1}{r^2}\frac{d}{dr}\left(r^2\sum_i m_{(O)i}n_i\right) = \sum_i m_{(O)i}\dot{r}'''_i \tag{2.50}$$

The first term on the right-hand side of Eq. (2.49) is the rate at which element O appears (or disappears) in the form of O_2, as O_2 is produced (or consumed) by the chemical reactions; the second term is the rate at which O appears in the form of CO_2; and so on. Since there cannot be a net creation or destruction of a chemical element, these terms must sum to zero, $\sum_i m_{(O)i}\dot{r}'''_i = 0$. Or, more generally, for any chemical element α,

$$\sum_i m_{(\alpha)i}\dot{r}'''_i = 0 \tag{2.51}$$

Using Eqs. (2.48) and (2.51), Eq. (2.49) becomes

$$\frac{d}{dr}(r^2 n_{(O)}) = 0$$

Integrating once,

$$r^2 n_{(O)} = \text{Constant} = R^2 n_{(O),s}$$

Substituting back for $n_{(O)}$ from Eq. (2.48) and expressing each absolute species flux as a sum of convective and diffusive components gives

$$r^2\sum_i m_{(O)i}\left(m_i n - \rho\mathcal{D}_{im}\frac{dm_i}{dr}\right) = R^2 n_{(O),s}$$

If we now assume all diffusion coefficients are equal, that is $\mathcal{D}_{O_2 m} = \mathcal{D}_{CO_2 m} = \mathcal{D}_{COm} = \mathcal{D}$,

$$r^2\left[n\sum_i m_{(O)i}m_i - \rho\mathcal{D}\frac{d}{dr}\sum_i m_{(O)i}m_i\right] = R^2 n_{(O),s}$$

or

$$r^2\left[nm_{(O)} - \rho\mathcal{D}\frac{dm_{(O)}}{dr}\right] = R^2 n_{(O),s} \tag{2.52}$$

At this point, the mathematical problem has been reduced to a form identical to that in Section 2.3.2, since Eq. (2.52) is of the same form as Eq. (2.38). We need not repeat the algebra of that analysis: from Eq. (2.39) the result is

$$\dot{m}'' = \frac{\rho\mathcal{D}}{R}\ln\left[1 + \frac{m_{(O),e} - m_{(O),s}}{m_{(O),s} - (n_{(O),s}/\dot{m}'')}\right] \tag{2.53}$$

In order to use this result to calculate the carbon oxidation rate $\dot{m}''$, we must be able to evaluate $m_{(O),e}$, $m_{(O),s}$ and $n_{(O),s}/\dot{m}''$ from known data. For ambient air,

$$m_{(O),e} = m_{(O)O_2}m_{O_2,e} = m_{O_2,e} = 0.231$$

At the s-surface,

$$m_{(O),s} = \sum_i m_{(O)i}m_{i,s} = \frac{1}{1}m_{O_2,s} + \frac{32}{44}m_{CO_2,s} + \frac{16}{28}m_{CO,s}$$

In the surface temperature range under consideration, the forward reactions of carbon with oxidizing species are very rapid. Thus, the concentration CO_2 (and O_2) at the s-surface may be taken to be zero, and then

$$m_{(O),s} = \frac{16}{28}m_{CO,s}$$

Carbon monoxide is the product of the oxidation reactions: its concentration at the s-surface is unknown. Thus, $m_{(O),s}$ is unknown, and as a result we cannot use Eq. (2.53) directly to obtain $\dot{m}''$. Instead we proceed as follows. First, Eq. (2.53) is rewritten as

$$m_{(O),s} - (n_{(O),s}/\dot{m}'') = [m_{(O),e} - (n_{(O),s}/\dot{m}'')]\exp(-\dot{m}''R/\rho\mathcal{D}) \quad (2.54a)$$

and, recognizing that the analysis could be repeated for the element carbon, we can also write

$$m_{(C),s} - (n_{(C),s}/\dot{m}'') = [m_{(C),e} - (n_{(C),s}/\dot{m}'')]\exp(-\dot{m}''R/\rho\mathcal{D}) \quad (2.54b)$$

Multiplying Eq. (2.54a) by ζ and Eq. (2.54b) by ξ, adding, and rearranging gives

$$\dot{m}'' = \frac{\rho\mathcal{D}}{R}\ln\left[1 + \frac{(\zeta m_{(O)} + \xi m_{(C)})_e - (\zeta m_{(O)} + \xi m_{(C)})_s}{(\zeta m_{(O)} + \xi m_{(C)})_s - (\zeta n_{(O)} + \xi n_{(C)})_s/\dot{m}''}\right] \quad (2.55)$$

We are free to choose any values for the multipliers ζ and ξ. Let us see if values can be chosen such that the terms in Eq. (2.55) can be evaluated. The e-state presents no problem. For the s-state,

$$(\zeta m_{(O)} + \xi m_{(C)})_s = \zeta\left(m_{O_2,s} + \frac{32}{44}m_{CO_2,s} + \frac{16}{28}m_{CO,s}\right)$$

$$+ \xi\left(\frac{12}{44}m_{CO_2,s} + \frac{12}{28}m_{CO,s}\right)$$

$$= m_{CO,s}\left(\frac{16}{28}\zeta + \frac{12}{28}\xi\right), \quad \text{since } m_{O_2,s} = m_{CO_2,s} = 0$$

If we choose $\zeta = 12/16$ and $\xi = -1$, then $[(16/28)\zeta + (12/28)\xi] = 0$, and we need not know $m_{CO,s}$ to evaluate the s-state. For this choice of multipliers,

$$\left(\frac{12}{16}m_{(O)} - m_{(C)}\right)_e = \frac{12}{16}m_{O_2,e} = \frac{3}{4}m_{O_2,e}$$

To evaluate $n_{(O),s}$ and $n_{(C),s}$ we perform element balances on a control volume located between the u- and s-surfaces. Since only the element carbon crosses the u-surface,

$$n_{(O),s} = n_{(O),u} = 0; \quad n_{(C),s} = n_{(C),u} = n_s = \dot{m}''$$

Substituting in Eq. (2.55) with $m_{(O),e} = 0.231$, $m_{(C),e} = 0$, gives

$$\dot{m}'' = \frac{\rho \mathcal{D}}{R} \ln \left[1 + \frac{(3/4)(0.231) - 0}{0 - (-\dot{m}''/\dot{m}'')} \right] = 0.160 \frac{\rho \mathcal{D}}{R} \tag{2.56}$$

which is identical to Eq. (2.41), except that Eq. (2.41) requires the diffusion coefficient to be that of O_2 in the mixture, whereas Eq. (2.56) requires $\mathcal{D}$ to be some average of $\mathcal{D}_{O_2,m}$, $\mathcal{D}_{CO_2,m}$, and $\mathcal{D}_{CO,m}$. Apart from this difference, the fact that the detailed reactions and even the final product are not the same has no effect on the oxidation rate.

The general principle underlying the above analysis is that the assumption of equal diffusion coefficients for all species containing the reactants allows the elimination of the species production rate expressions $\dot{r}_i'''$, and leads to an analytical solution in terms of mass fractions of chemical elements. In the combustion literature, this method is often called the *Shvab–Zeldovich transformation*. Although it allows the mass transfer rate to be calculated, the analysis does not yield information as to where the gas-phase reaction actually occurs; that is, in the carbon particle problem, it does not give the radius of the flame front. Exercise 2–11 requires an analysis of this problem that does allow the flame-front radius to be determined.

Property Evaluation

Use of Eq. (2.56) requires evaluation of the $\rho \mathcal{D}$ product. At 1000 K, the binary diffusion coefficients of O_2, CO, and CO_2 in air are $1.52 \times 10^{-4} \, \mathrm{m^2/s}$, $1.54 \times 10^{-4} \, \mathrm{m^2/s}$, and $1.24 \times 10^{-4} \, \mathrm{m^2/s}$, respectively. These values are not too different, and using the value for O_2 will be adequate for most purposes. Evaluation of ρ poses a tricky problem because of the unusual concentration profiles. Since the concentration of CO_2 goes to zero at both the s-surface and at infinity, its presence does not affect evaluation of density at the mean film composition. However, N_2 is the dominant species, and use of the mean film composition should be quite adequate for most purposes. The temperature profile shown in Fig. 2.8 indicates that the choice of an appropriate reference temperature for property evaluation is also difficult. Use of the mean of the ambient temperature and the adiabatic flame temperature for combustion of CO is one possible choice.

Example 2.4   Combustion of a Carbon Particle in a Steam–Oxygen Mixture

In a coke gasification process, small carbon particles are entrained in a flow of an oxygen and steam mixture. Determine the effect of the steam content on particle oxidation rate at the inlet of the reactor.

Solution

Given: Carbon particles oxidizing in O_2–H_2O mixtures.

Required: Effect of H_2O mass fraction on particle oxidation rate.

Assumptions:

1. Concentrations of O_2, CO_2, and H_2O at the s-surface are zero.
2. The $\rho \mathcal{D}$ product of the mixture is independent of H_2O content.

Equation (2.55) applies and gives the oxidation rate as

$$\dot{m}'' = \frac{\rho \mathcal{D}}{R} \ln \left[1 + \frac{(\zeta m_{(O)} + \xi m_{(C)})_e - (\zeta m_{(O)} + \xi m_{(C)})_s}{(\zeta m_{(O)} + \xi m_{(C)})_s - (\zeta n_{(O)} + \xi n_{(C)})_s / \dot{m}''} \right]$$

The s-state is evaluated as

$$(\zeta m_{(O)} + \xi m_{(C)})_s = \zeta \left(m_{O_2,s} + \frac{32}{44} m_{CO_2,s} + \frac{16}{28} m_{CO,s} + \frac{16}{18} m_{H_2O,s} \right)$$
$$+ \xi \left(\frac{12}{44} m_{CO_2,s} + \frac{12}{28} m_{(CO),s} \right)$$

If temperatures are sufficiently high, we can assume diffusion-controlled reactions and $m_{O_2,s} = m_{H_2O,s} = m_{CO_2,s} = 0$. Then,

$$(\zeta m_{(O)} + \xi m_{(C)})_s = m_{CO,s} \left(\frac{16}{28} \zeta + \frac{12}{28} \xi \right)$$

Thus, as before, a choice of $\zeta = 12/16$, $\xi = -1$ will eliminate the unknown $m_{CO,s}$. Also as before, $n_{(O),s} = 0$, $n_{(C),s} = \dot{m}''$. Substituting in Eq. (2.55),

$$\dot{m}'' = \frac{\rho \mathcal{D}}{R} \ln \left[1 + \left(\frac{12}{16} m_{(O)} - m_{(C)} \right)_e \right]$$

At the inlet of the reactor, the free stream contains only O_2 and H_2O; hence $m_{CO_2,e} = m_{CO,e} = 0$, and

$$\dot{m}'' = \frac{\rho \mathcal{D}}{R} \ln \left[1 + \frac{12}{16} \left(m_{O_2,e} + \frac{16}{18} m_{H_2O,e} \right) \right]$$

The table shows the effect of steam content $m_{H_2O,e}$ on the dimensionless oxidation rate $\dot{m}'' R / \rho \mathcal{D}$.

$m_{H_2O,e}$	$\dot{m}'' R / \rho \mathcal{D}$
0	0.560
0.1	0.555
0.2	0.550
0.3	0.545
0.4	0.540
0.5	0.536

Comments

1. The calculated effect of steam content on oxidation rate is small.

2. Since the effect is small, secondary factors related to the effect of steam content on the $\rho \mathcal{D}$ product may play a more important role. The $\rho \mathcal{D}$ product depends on both gas-mixture composition and temperature. As the H_2O and H_2 content of the mixture increases, ρ decreases but the effective value of $\mathcal{D}$ increases. Note, however, we do not have to include $\mathcal{D}_{H_2 m}$ when choosing an appropriate value for the assumed equal $\mathcal{D}_{O_2,m}$, $\mathcal{D}_{CO_2,m}$, $\mathcal{D}_{CO,m}$, and $\mathcal{D}_{H_2O,m}$, since H_2 does not contain element (O) or (C).

3. The reactor is a mass exchanger in which the concentrations of the reaction products H_2 and CO increase along the reactor: the above result applies only at the inlet. In a 100% effective exchanger, the outlet mixture will contain H_2 and CO only.   ▲

2.3.4 Droplet Evaporation

Low mass transfer rate theory was applied to the evaporation of a water droplet in Example 1.10. We now develop a high mass transfer rate theory for evaporation of a small droplet entrained in a gas flow, as shown in Fig. 2.9a. The droplet contains a single component, species 1, and the gas is species 2. The temperature of the droplet surface is T_s (to be determined later), and the corresponding mass fraction of vapor 1 at the s-surface is $m_{1,s} = m_{1,s}(T_s, P)$ from saturation vapor-pressure tables. Assuming a quasi-steady state and spherical symmetry, the governing conservation equation is again Eq. (2.33),

$$\frac{d}{dr}(r^2 n_1) = 0 \tag{2.32}$$

The solution of Eq. (2.33) for $m_1 = m_{1,s}$ at $r = R$ and $m_1 = m_{1,e}$ as $r \to \infty$, is Eq. (2.39) with $\mathcal{D}_{1m}$ replaced by $\mathcal{D}_{12}$:

$$\dot{m}'' = \frac{\rho \mathcal{D}_{12}}{R} \ln\left[1 + \frac{m_{1,e} - m_{1,s}}{m_{1,s} - (n_{1,s}/\dot{m}'')}\right] \tag{2.57}$$

If the gas is negligibly soluble in the droplet, then $n_{2,s} = 0$; hence $\dot{m}'' = n_{1,s} + n_{2,s} = n_{1,s}$, and $n_{1,s}/\dot{m}'' = 1$. Equation (2.57) thus becomes

$$\dot{m}'' = \frac{\rho \mathcal{D}_{12}}{R} \ln\left[1 + \frac{m_{1,e} - m_{1,s}}{m_{1,s} - 1}\right] \tag{2.58}$$

which gives the droplet evaporation rate in terms of the unknown s-surface vapor mass fraction, $m_{1,s}$.

As we learned in Example 1.10, droplet evaporation is a simultaneous heat and mass transfer problem. For quasi-steady evaporation, the enthalpy of vaporization must be supplied by heat transfer from the gas, and the droplet temperature is approximately equal to the psychrometric wet-bulb temperature. To perform a heat

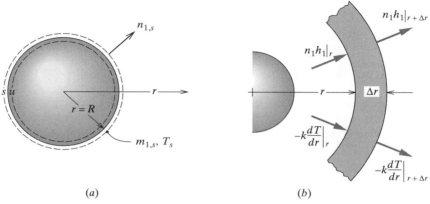

Figure 2.9 (*a*) Schematic of droplet evaporating into a gas. (*b*) Elemental control volume for application of the steady-flow energy equation.

transfer analysis, the steady-flow energy equation, Eq. (1.4) in *Heat Transfer*, is applied to the elemental control volume shown in Fig. 2.9*b* and requires that

$$4\pi r^2 \left(n_1 h_1 - k\frac{dT}{dr} \right)\Big|_r - 4\pi r^2 \left(n_1 h_1 - k\frac{dT}{dr} \right)\Big|_{r+\Delta r} = 0$$

since $n_2 = 0$: only species 1 transports enthalpy in and out of the control volume. Thus,

$$r^2 \left(n_1 h_1 - k\frac{dT}{dr} \right) = \text{Constant} = R^2 \left(\dot{m}'' h_{1,s} - k\frac{dT}{dr}\Big|_s \right) \qquad (2.59)$$

since $n_{1,s} = \dot{m}''$. Also, applying the steady-flow energy equation between the *s*- and *u*-surfaces shown in Fig. 2.9*a*,

$$\dot{m}'' h_{1,s} - k\frac{dT}{dr}\Big|_s = \dot{m}'' h_{1,u} - k\frac{dT}{dr}\Big|_u \qquad (2.60)$$

Based on Example 1.10, we assume that the initial transient in which the droplet temperature changes from its initial value to a steady value is short, and that the steady value is attained when $-k(dT/dr)|_u \simeq 0$. Then, using Eq. (2.60), Eq. (2.59) becomes

$$r^2 \left(n_1 h_1 - k\frac{dT}{dr} \right) = R^2 \dot{m}'' h_{1,u} \qquad (2.61)$$

Since there are no chemical reactions, we are free to choose enthalpy datum states as we please. A nice simplification of the analysis is achieved if we take the vapor enthalpy h_1 to be zero at temperature T_s. Also, for simplicity we will assume that the vapor specific heat is independent of temperature. Then, for the vapor,

$$h_1 = \int_{T_s}^{T} c_{p1} dT = c_{p1}(T - T_s) \qquad (2.62a)$$

and for the liquid at the *u*-surface,

$$h_{1,u} = -h_{\text{fg}} \qquad (2.62b)$$

Substituting in Eq. (2.61),

$$r^2 \left[n_1 c_{p1}(T - T_s) - k\frac{dT}{dr} \right] = -R^2 \dot{m}'' h_{\mathrm{fg}}$$

Rearranging using $n_1 r^2 = \dot{m}'' R^2$ from mass conservation,

$$R^2 \dot{m}'' [h_{\mathrm{fg}} + c_{p1}(T - T_s)] = r^2 k \frac{dT}{dr}$$

Separating variables and integrating from R to ∞,

$$\frac{\dot{m}'' R^2}{k} \int_R^\infty \frac{dr}{r^2} = \int_{T_s}^{T_e} \frac{dT}{h_{\mathrm{fg}} + c_{p1}(T - T_s)}$$

where the mixture thermal conductivity has been assumed constant. The result is

$$\dot{m}'' = \frac{k/c_{p1}}{R} \ln \left[1 + \frac{c_{p1}(T_e - T_s)}{h_{\mathrm{fg}}} \right] \tag{2.63}$$

where $c_{p1}(T_e - T_s)/h_{\mathrm{fg}}$ is recognized to be the vapor-phase Jakob number, Ja_v. The Jakob number was introduced in Chapters 3 and 7 of *Heat Transfer*. Like Eq. (2.58), Eq. (2.63) gives the evaporation rate $\dot{m}''$, this time in terms of the unknown surface temperature T_s. The *simultaneous* heat and mass transfer problem requires the simultaneous solution of Eqs. (2.58) and (2.63) together with the vapor pressure relation,

$$\dot{m}'' = \frac{\rho \mathcal{D}_{12}}{R} \ln \left(1 + \frac{m_{1,e} - m_{1,s}}{m_{1,s} - 1} \right)$$

$$= \frac{k/c_{p1}}{R} \ln \left(1 + \frac{c_{p1}(T_e - T_s)}{h_{\mathrm{fg}}} \right) \tag{2.64a, b}$$

$$m_{1,s} = m_{1,s}(T_s, P) \tag{2.64c}$$

which is a set of three equations in the three unknowns $\dot{m}''$, T_s, and $m_{1,s}$. Temperature T_s is the adiabatic vaporization temperature for a droplet evaporating at high mass transfer rates, and will be close to the psychrometric wet-bulb temperature (defined by Eq. (1.57) for the water–air system).

Property Evaluation
To use Eqs. (2.64), the mixture properties ρ, $\mathcal{D}_{12}$, and k must be evaluated at a suitable reference state, since these properties were assumed constant in the analysis (the weak temperature dependence of c_{p1} is of little consequence). Notice that we did not have to assume $c_{p1} = c_{p2} = c_p$ in the analysis. Thus, our result correctly accounts for the difference in specific heats of the two species, and often this difference is large. One possible scheme is to evaluate ρ, $\mathcal{D}_{12}$, and k at the mean film temperature and composition, and c_{p1} at the mean film temperature. However, the validity of this scheme has not been checked against exact numerical solutions or experimental data. An alternative scheme proposed by Hubbard et al. [1] was established by comparing numerical solutions in which all properties were calculated exactly, with a constant-property analysis in which all properties, *including* c_p, were assumed constant at

a reference state. A *1/3 rule* result was obtained for which ρ, $\mathcal{D}_{12}$, k, and c_p are evaluated at

$$m_{1,r} = m_{1,s} + (1/3)(m_{1,e} - m_{1,s}) \tag{2.65a}$$

$$T_r = T_s + (1/3)(T_e - T_s) \tag{2.65b}$$

In using this scheme, c_{p1} in Eq. (2.64a) is set equal to c_p of the mixture evaluated at the reference state. Alternative schemes for property evaluation are available [2, 3].

Example 2.5   Evaporation of a Water Droplet

A 50 μm–diameter droplet is entrained in a dry air stream at 1 atm, 1650 K. Estimate the droplet lifetime.

Solution

Given: A small water droplet entrained in a dry air stream.
Required: Droplet lifetime.
Assumptions:

1. The initial transient is short.
2. Radiation heat transfer is negligible.
3. Hubbard's 1/3 rule for property evaluation is appropriate.

For quasi-steady evaporation, T_s and $m_{1,s}$ are constant and can be obtained from Eqs. (2.64) by setting $m_{1,e} = 0$ for dry air, and simultaneously solving

$$\ln\left(\frac{1}{1 - m_{1,s}}\right) = \frac{k/c_{p1}}{\rho \mathcal{D}_{12}} \ln\left(1 + \frac{c_{p1}(T_e - T_s)}{h_{\text{fg}}}\right) \tag{1}$$

$$m_{1,s} = m_{1,s}(T_s, P) \tag{2}$$

We guess $T_s = 350\,\text{K}$, and from PSYCHRO we obtain $m_{1,s} = 0.303$. Using Hubbard's 1/3 rule to evaluate properties, we set $c_{p1} = c_p$ and use the reference state

$$m_{1,r} = 0.303 + (1/3)(0 - 0.303) = 0.202$$

$$T_r = 350 + (1/3)(1650 - 350) = 783\,\text{K}$$

From AIRSTE, mixture properties at the reference state are $\rho = 0.4017\,\text{kg/m}^3$, $c_p = 1296\,\text{J/kg K}$, $k = 0.0577\,\text{W/m K}$, $\mathcal{D}_{12} = 1.452 \times 10^{-4}\,\text{m}^2/\text{s}$, and from Table A.12a at $T = T_s = 350\,\text{K}$, $h_{\text{fg}} = 2.316 \times 10^6\,\text{J/kg}$. Then the left-hand side of Eq. (1) is

$$\ln\left(\frac{1}{1 - 0.303}\right) = 0.361$$

and the right-hand side of Eq. (1) is

$$\frac{0.0577/1296}{(0.4017)(1.452 \times 10^{-4})} \ln\left(1 + \frac{1296(1650 - 350)}{2.316 \times 10^6}\right) = 0.417$$

Since Eq. (1) is not satisfied, iteration is required and gives the result $T_s = 352.7\,\text{K}$, $m_{1,s} = 0.345$, for which reference state properties include $\rho = 0.3946\,\text{kg/m}^3$, $\mathcal{D}_{12} = 1.459 \times 10^{-4}\,\text{m}^2/\text{s}$. From Eq. (2.64a), the evaporation rate is

$$\dot{m}'' = \frac{(0.3946)(1.459 \times 10^{-4})}{R} \ln\left(\frac{1}{1 - 0.345}\right) = \frac{2.44 \times 10^{-5}}{R}\,\text{kg/m}^2\,\text{s}$$

A mass balance on the droplet gives

$$\frac{d}{dt}[(4/3)\pi R^3 \rho_l] = -\dot{m}''(4\pi R^2)$$

$$\frac{dR}{dt} = -\frac{(2.44 \times 10^{-5})}{R\rho_l}$$

Integrating with $R = R_0$ at $t = 0$, and $R = 0$ at $t = \tau$ gives the lifetime as

$$\tau = \frac{\rho_l R_0^2}{(2)(2.44 \times 10^{-5})}$$

At 353 K, Table A.8 gives $\rho_l = 971\,\text{kg/m}^3$. Thus,

$$\tau = \frac{(971)(50 \times 10^{-6})^2}{(2)(2.44 \times 10^{-5})} = 0.0497\,\text{s}$$

Comments

1. Rework this example evaluating ρ, $\mathcal{D}_{12}$, and k at the mean film temperature and composition, and c_{p1} as water vapor at the mean film temperature. The result is $T_s = 350.6\,\text{K}$, $m_{1,s} = 0.312$, and $\tau = 0.0466\,\text{s}$ (a 6% shorter lifetime).

2. Vary T_e to see when low mass transfer rate theory becomes invalid.   ▲

2.3.5 Droplet Combustion

In many furnaces and combustion chambers liquid fuels are atomized to form a spray of small droplets that burn rapidly and efficiently. Examples include fuel oil in a power plant furnace, kerosene in a gas turbine combustor, and hydrazine in a rocket motor. Although the practical problem involves a spray, it has proven useful to understand the behavior of a single isolated droplet. Analysis of single-droplet combustion assists interpretation of experimental data for single-droplet combustion and also yields results that can be extended to spray combustion. Of particular importance is the lifetime of the droplet, because the required size of the combustion chamber depends in part on the time required for burning.

Combustion of Liquid Hydrocarbon Fuel Droplets

During combustion of a volatile fuel droplet in air, the fuel vapor and oxygen react in a thin flame, as shown in Fig. 2.10a. For hydrocarbon fuels burning in air, the flame diameter is about four to six times the droplet diameter. For a stationary droplet under zero gravity conditions, the flame is spherical; at normal gravity, natural

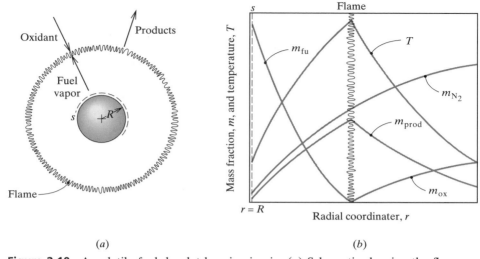

Figure 2.10 A volatile fuel droplet burning in air. (*a*) Schematic showing the flame surrounding the droplet. (*b*) Concentration and temperature profiles.

convection produces a plume above the droplet. Heat is transferred from the flame to the droplet and serves to vaporize the fuel. In the flame the vapor reacts with oxygen to form gaseous products, primarily CO_2 and H_2O. The characteristic temperature and concentration profiles are shown in Fig. 2.10*b*. Owing to the very rapid reaction kinetics, there is negligible penetration of the flame by the fuel vapor or oxidant: fuel is on one side of the flame, and oxidant is on the other. When a fuel droplet is injected into a combustion chamber, there is a short initial transient during which the droplet heats up, until further conduction into the droplet is negligible and the droplet attains a steady temperature (approximately the wet-bulb temperature). Due to the high temperature of the flame and the volatility of typical liquid hydrocarbon fuels, the wet-bulb temperature is very close to the boiling point of the liquid. The gas-phase reaction at the flame front can be modeled as a single-step reaction,

$$\text{Fuel} + \text{Oxidant} \rightarrow \text{Products} \qquad (2.66)$$

with a constant stoichiometric ratio r and a heat of combustion Δh_c J/kg of fuel. The stoichiometric ratio is typically about 3–4 for hydrocarbon fuels burning in air.[1]

Analysis

This problem involves simultaneous heat and mass transfer. In order to sustain steady combustion, heat must be transferred from the flame to the droplet and the oxidant must diffuse to the flame front. However, if we are prepared to make some simplifying assumptions, we can obtain the combustion rate from a heat transfer analysis only, as will now be shown. We will consider zero-g combustion of a stationary droplet in order to have a spherically symmetric problem: Fig. 2.11 shows the

---

[1] In this analysis the stoichiometric ratio is defined as the ratio (kg oxidant/kg fuel), and not (kg air/kg fuel) as is the practice in internal-combustion-engine analysis.

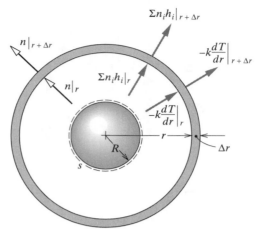

Figure 2.11 A volatile fuel droplet burning in air: elemental control volume for application of the mass conservation equation and the steady-flow energy equation.

coordinate system and an elemental control volume between radii r and $r + \Delta r$. A mass balance on the elemental volume requires that at steady state,

$$4\pi r^2 n|_r - 4\pi r^2 n|_{r+\Delta r} = 0$$

or

$$r^2 n = \text{Constant} = r^2 n|_{r=R} = R^2 \dot{m}'' \tag{2.67}$$

Of course, Eq. (2.67) can also be obtained from the mass conservation equation, Eq. (2.12b). Similarly, the steady-flow energy equation applied to the volume requires that[2]

$$4\pi r^2 \left(\sum n_i h_i - k\frac{dT}{dr} \right)_r - 4\pi r^2 \left(\sum n_i h_i - k\frac{dT}{dr} \right)_{r+\Delta r} = 0$$

We have recognized that a convection of enthalpy $n_i h_i$ is associated with the absolute flux of each species, n_i (h_i is the species enthalpy). The conductivity k is the conductivity of the gas mixture. Thermal radiation has been ignored: its effect will be discussed later. Hence,

$$r^2 \left(\sum n_i h_i - k\frac{dT}{dr} \right) = \text{Constant} = R^2 \left(\sum n_{i,s} h_{i,s} - k\frac{dT}{dr}\Big|_s \right) \tag{2.68}$$

---

[2] The steady-flow energy equation, Eq. (1.4) in *Heat Transfer*, has been extended to apply to a number of streams of distinct chemical species entering and leaving the control volume.

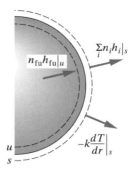

Figure 2.12 Interface energy
balance for combustion of a
volatile fuel droplet.

Referring to Fig. 2.12, the constant in Eq. (2.68) is conveniently evaluated by noting that the steady-flow energy equation applied between the s- and u-surfaces requires that

$$R^2 \left(\sum n_{i,s} h_{i,s} - k \frac{dT}{dr}\Big|_s \right) = R^2 n_{\text{fu},u} h_u \tag{2.69}$$

because only liquid fuel crosses the u-surface, and $dT/dr|_u = 0$ if the droplet temperature is steady. The enthalpy h_u is the enthalpy of liquid fuel at temperature $T_u = T_s$, and $n_{\text{fu},u} = \dot{m}''$. Equation (2.68) then becomes

$$r^2 \left(\sum_i n_i h_i - k \frac{dT}{dr} \right) = R^2 \dot{m}'' h_u \tag{2.70}$$

The absolute fluxes n_i are now expressed in terms of their convective and diffusive components as

$$n_i = m_i n + j_i = m_i n - \rho \mathcal{D}_{im} \frac{dm_i}{dr}$$

where the $\mathcal{D}_{im}$ are effective binary diffusion coefficients. Substituting in Eq. (2.70),

$$r^2 \left(n \sum_i m_i h_i - \sum_i \rho \mathcal{D}_{im} \frac{dm_i}{dr} h_i - k \frac{dT}{dr} \right) = R^2 \dot{m}'' h_u \tag{2.71}$$

This equation can be simplified as follows. The mixture enthalpy is $h = \sum m_i h_i$; differentiating,

$$dh = \sum m_i dh_i + \sum h_i dm_i$$

$$= \sum m_i c_{pi} dT + \sum h_i dm_i$$

$$= c_p dT + \sum h_i dm_i$$

where the mixture specific heat is $c_p = \sum m_i c_{pi}$. Hence,

$$\frac{k}{c_p}\frac{dh}{dr} = k\frac{dT}{dr} + \frac{k}{c_p}\sum_i h_i \frac{dm_i}{dr} \tag{2.72}$$

We now make a key assumption, namely, that the Lewis numbers for all the species are equal to 1. By definition, $Le_i = Pr/Sc_i$. Hence, $Le_i = 1$ requires that

$$\frac{Pr}{Sc_i} = \frac{c_p \mu/k}{\mu/\rho \mathcal{D}_{im}} = \frac{\rho \mathcal{D}_{im}}{k/c_p} = 1$$

That is, $\rho \mathcal{D}_{im} = k/c_p$ for all species i in the mixture. Note that ρ, k, and c_p are *mixture* properties. Then, substituting Eq. (2.72) in Eq. (2.71) and rearranging using Eq. (2.67) gives

$$R^2 \dot{m}'' h - r^2 \frac{k}{c_p}\frac{dh}{dr} = R^2 \dot{m}'' h_u \tag{2.73}$$

We have succeeded in obtaining a differential equation in terms of a single dependent variable, the mixture enthalpy h. Integrating with $h = h_s$ at $r = R$ and $h = h_e$ as $r \to \infty$,

$$\int_{h_s}^{h_e} \frac{dh}{h - h_u} = \frac{\dot{m}'' R^2}{k/c_p}\int_R^\infty \frac{dr}{r^2}$$

where we have assumed k/c_p to be constant, independent of r. Then

$$\ln \frac{h_e - h_u}{h_s - h_u} = \frac{\dot{m}'' R}{k/c_p}$$

which can be rearranged to give the mass transfer rate,

$$\dot{m}'' = \frac{k/c_p}{R}\ln\left[1 + \frac{h_e - h_s}{h_s - h_u}\right] \tag{2.74}$$

We now assume that, for the purpose of evaluating Eq. (2.74), we can take T_s to equal the boiling point of the fuel, which will be known. (As mentioned earlier, $T_s = T_{WB}$, but T_{WB} is close to T_{BP} in this situation.) Then $m_{fu,s} \simeq 1$, and Eq. (2.74) can be evaluated to obtain $\dot{m}''$ using standard enthalpy tables to calculate h_e, h_s, and h_u. However, we can obtain a more convenient form of this result as follows.

By definition, the enthalpy of species i is

$$h_i = h_i^0 + \int_{T_0}^T c_{pi}\, dT \tag{2.75}$$

where h_i^0 is the *heat of formation* of species i at the datum temperature T_0. Consistent with Eq. (2.66), we will consider a mixture of fuel vapor, oxidant, products, and inerts. We are free to arbitrarily choose the datum temperature T_0, and to assign the heat of combustion (Δh_c J/kg fuel) to the heat of formation of either the fuel, oxidant, or products. If we choose $T_0 = T_s$ and assign the heat of combustion to the oxidant, and if, in addition, we assume that all species specific heats are equal and constant, we obtain

$$h_{\text{fu}} = c_p(T - T_s) \quad \text{(vapor)} \tag{2.76a}$$

$$h_{\text{ox}} = \frac{\Delta h_c}{r} + c_p(T - T_s) \tag{2.76b}$$

$$h_{\text{prod}} = c_p(T - T_s) \tag{2.76c}$$

$$h_{\text{inerts}} = c_p(T - T_s) \tag{2.76d}$$

Then h_e, h_s, and h_u can be evaluated as

$$h_e = \sum m_{i,e} h_{i,e} = m_{\text{ox},e} \frac{\Delta h_c}{r} + m_{\text{ox},e} c_p(T_e - T_s) + m_{\text{inerts},e} c_p(T_e - T_s)$$

since $m_{\text{fu},e} = m_{\text{prod},e} = 0$. But $m_{\text{ox},e} + m_{\text{inerts},e} = 1$; thus,

$$h_e = m_{\text{ox},e} \frac{\Delta h_c}{r} + c_p(T_e - T_s) \tag{2.77a}$$

$$h_s = 0, \quad \text{since there is no oxidant at the } s\text{-surface} \tag{2.77b}$$

$$h_u = h_{\text{fu},u} = -h_{\text{fg}}, \quad \text{the enthalpy of vaporization of the fuel} \tag{2.77c}$$

Substituting into Eq. (2.74) gives the mass transfer rate as

$$\dot{m}'' = \frac{k/c_p}{R} \ln\left[1 + \frac{m_{\text{ox},e}\Delta h_c/r + c_p(T_e - T_s)}{h_{\text{fg}}}\right] \tag{2.78}$$

which is easily evaluated. The mass transfer rate is simply the rate at which the fuel is consumed; a mass balance on the droplet gives the droplet lifetime.

Droplet Lifetime

A mass balance on the droplet requires that

$$\frac{d}{dt}\left(\frac{4}{3}\pi R^3 \rho_l\right) = -4\pi R^2 \dot{m}''$$

Substituting from Eq. (2.78) and rearranging gives the rate of change of droplet diameter D as

$$\frac{dD}{dt} = -\frac{4k/c_p}{\rho_l D} \ln\left[1 + \frac{m_{\text{ox},e}\Delta h_c/r + c_p(T_e - T_s)}{h_{\text{fg}}}\right] \tag{2.79}$$

Integrating with $D = D_0$ at $t = 0$, and $D = 0$ at $t = \tau$ gives the droplet lifetime as

$$\tau = \frac{\rho_l D_0^2}{8\left(\dfrac{k}{c_p}\right) \ln\left[1 + \dfrac{m_{\text{ox},e}\Delta h_c/r + c_p(T_e - T_s)}{h_{\text{fg}}}\right]} \tag{2.80}$$

The droplet lifetime is proportional to droplet diameter squared (as was the case for droplet evaporation and spherical particle sublimation and combustion). Equation (2.80) was first derived by Godsave [4] and by Spalding [5].

Property Evaluation

In deriving Eq. (2.80), we have made a number of assumptions regarding properties:

1. The Lewis numbers of all species are unity, or $\rho \mathcal{D}_{im} = k/c_p$ for all species i (which requires all diffusion coefficients $\mathcal{D}_{im}$ to be equal).
2. All species specific heats are equal.
3. The k/c_p ratio is a constant.

In reality, the diffusion coefficients are not quite equal, the Lewis numbers are not quite unity, and the k/c_p ratio varies, due to both temperature and composition changes, from the droplet surface, across the flame, and into the surrounding air. Thus, Eq. (2.80) will be useful only if some simple and general reference property scheme can be recommended for evaluating k and c_p, one that will give good agreement between theory and experiment. A scheme that works quite well for alkane fuel droplets burning in air was proposed by Law and Williams [6]. The reference temperature is $T_r = (1/2)(T_{BP} + T_{flame})$, where T_{flame} is the adiabatic flame temperature. The reference specific heat is $c_{pr} = c_{pfu}$, and the reference thermal conductivity is $k_r = 0.4k_{fu} + 0.6k_{air}$.

The Burning Rate Constant and the Transfer Number

Combustion specialists often quote two parameters that characterize droplet combustion. The *burning rate constant*, β, is defined by the relation

$$\frac{d}{dt}(D^2) = -\beta; \quad \beta = \frac{8k}{\rho_l c_p} \ln \left[1 + \frac{m_{ox,e} \Delta h_c/r + c_p(T_e - T_s)}{h_{fg}} \right] \quad (2.81)$$

That is, β characterizes the linear decrease of droplet diameter squared for quasi-steady combustion. Figure 2.13 shows typical experimental data plotted as D^2 versus t. After an initial transient there is a linear decrease yielding a line of slope $-\beta$. The goal of experimental studies of single-droplet combustion is usually to determine β: values of β are on the order of 10^{-6} m²/s for hydrocarbon fuel droplets burning in air.

The *transfer number* $\mathcal{B}$ is defined as

$$\mathcal{B} = \frac{m_{ox,e} \Delta h_c/r + c_p(T_e - T_s)}{h_{fg}} \quad (2.82)$$

and Eq. (2.80) can be written more compactly as

$$\tau = \frac{\rho_l D_0^2}{8(k/c_p) \ln(1 + \mathcal{B})} \quad (2.83)$$

In mass transfer theory $\mathcal{B}$ is termed an enthalpy-based *mass transfer driving force*. For hydrocarbon fuel droplets burning in air, $\mathcal{B}$ ranges from 5 to 15.[3] Enthalpy-based mass transfer driving forces are further developed in Sections 2.5.2 and 2.6.2.

---

[3] The parameter $\mathcal{B}$ scales the mass transfer rate; in general use, we can say that the mass transfer rate is low when $\mathcal{B} < 0.1$. Thus, mass transfer rates for combustion of volatile hydrocarbon fuel droplets are very high.

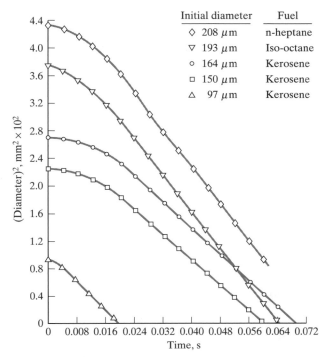

Initial diameter		Fuel
◇	208 μm	n-heptane
▽	193 μm	Iso-octane
○	164 μm	Kerosene
□	150 μm	Kerosene
△	97 μm	Kerosene

Figure 2.13   Experimental data for hydrocarbon fuel droplets burning in air confirming the "D^2 law" [7].

Effect of Radiation

Recent research has shown that the effect of radiation from the CO_2 and H_2O products is not negligible for single-droplet combustion [8]. Analysis including the effect of radiation predicts lower values of the burning rate constant β than when radiation is ignored. Thus, actual droplet lifetimes can be expected to be longer than those predicted by Eq. (2.83). However, when a property evaluation scheme is used to match Eq. (2.83) to experimental data, as was done by Law and Williams [6], the effect of radiation is accounted for. In combustion chambers burning sprays of droplets, radiation heat transfer is large, and our simple single droplet combustion theory does not apply.

Example 2.6   Combustion of a *n*-Hexane Droplet

Estimate the lifetime of a 1 mm–diameter *n*-hexane droplet burning at nearly zero gravity in air at 1 atm pressure and 300 K. For *n*-hexane (n-C_6H_{14}), take $\rho_l = 610 \text{ kg/m}^3$, $h_{fg} = 3.35 \times 10^5$ J/kg, $\Delta h_c = 4.48 \times 10^7$ J/kg, and $T_{BP} = 342$ K. The flame temperature is $T_f = 2320$ K. At the Law and Williams reference temperature of $T_r = (1/2)(T_f + T_{BP}) = 1330$ K, property values for *n*-hexane vapor include $k = 0.179$ W/m K and $c_p = 4303$ J/kg K.

Solution

Given: A 1 mm–diameter n-hexane droplet burning in air at 1 atm, 300 K.
Required: Droplet lifetime.
Assumptions:
1. Spherical symmetry.
2. Quasi-steady combustion with the droplet at its boiling point.
3. Properties can be evaluated using the reference state scheme of Law and Williams.

Equation (2.83) gives the droplet lifetime τ as

$$\tau = \frac{\rho_l D_0^2}{8(k/c_p)\ln(1+\mathcal{B})}; \quad \mathcal{B} = \frac{m_{ox,e}\Delta h_c/r + c_p(T_e - T_s)}{h_{fg}}$$

We first evaluate the transfer number $\mathcal{B}$. The reaction is

$$C_6H_{14} + 9.5\,O_2 \rightarrow 6\,CO_2 + 7\,H_2O$$

$$86.17 \text{ kg} + (9.5)(32) \text{ kg} \rightarrow$$

Hence, the stoichiometric ratio is $r = (9.5)(32)/86.17 = 3.53$. The mass fraction of oxygen in air $m_{ox,e} = 0.231$, and we take $T_s \simeq T_{BP} = 342$ K. The Law and Williams reference state scheme requires that c_p be evaluated for pure hexane vapor at $T_r = (1/2)(T_f + T_{BP}) = 1330$ K, which is given as 4303 J/kg. The transfer number is

$$\mathcal{B} = \frac{(0.231)(4.48 \times 10^7)/3.53 + 4303(300 - 342)}{3.35 \times 10^5}$$

$$= \frac{(29.3 - 1.8) \times 10^5}{3.35 \times 10^5} = 8.21$$

The reference state value of the thermal conductivity is

$$k_r = 0.4k_{fu} + 0.6k_{air}$$

evaluated at $T_r = 1330$ K. From Table A.7, $k_{air} = 0.084$ W/m K; hence

$$k_r = (0.4)(0.179) + (0.6)(0.084) = 0.122 \text{ W/m K}$$

The droplet lifetime is then

$$\tau = \frac{(610)(1 \times 10^{-3})^2}{8(0.122/4303)\ln(1+8.21)} = 1.21 \text{ s}$$

Comments

1. Notice how the sensible heat term $c_p(T_e - T_s)$ is almost negligible compared with the heat of combustion term $m_{ox,e}\Delta h_c/r$. Thus, the air temperature has little effect on droplet lifetime.
2. The Law and Williams reference state scheme is based on experimental data and thus should account for the various shortcomings of our model.
3. Law and Williams [6] also give correction factors for natural or forced convection flow around the droplet.   ▲

2.4 HIGH MASS TRANSFER RATE CONVECTION

The analyses in Section 2.3 were examples of *high mass transfer rate theory*, because convection of mass across the interface was properly accounted for. In the heatpipe problem of Example 2.3, the mass transfer rates are *very* high. We can talk in terms of strong blowing at the evaporator end and strong suction at the condenser end. A similar strong suction is characteristic of condensation of steam in power-plant condensers, where the steam always contains a small amount of noncondensable gas. The mass transfer rate for hydrocarbon fuel droplet combustion is also very high; there is a strong blowing away from the droplet surface. Strong blowing is also characteristic of ablation and transpiration cooling, for example, ablation of the heat shield on the Apollo reentry vehicle. Notice that both in the heatpipe evaporator and at the droplet surface, the mass fraction of the evaporating species is close to its limiting value of unity. To simplify the analyses, the only convection considered was that normal to the interface induced by the mass transfer process itself. Forced or natural convection flows along the interface, such as the relative motion between the carbon particle or fuel droplet and the gas flow, were not considered. We now analyze high mass transfer rate forced convection. For this purpose we will use a Couette–flow model and restrict our attention to binary inert mixtures with transfer of a single species. Our purpose is to develop problem-solving procedures for convective mass, momentum, and heat transfer applicable to situations where the low mass transfer theory of Section 1.4 is invalid.

2.4.1 The Couette–Flow Model

In Section 5.2 of *Heat Transfer* a Couette–flow model of the high-speed boundary layer was used to give insight into the effect of viscous dissipation on heat transfer. In Section 7.2.4 of *Heat Transfer* a Couette–flow model of a vapor boundary layer was used to obtain useful relations for the effects of vapor shear and superheat on film condensation. The Couette–flow model will now be used to examine the essential features of boundary layer flows with mass transfer, in particular, the effects of high mass transfer rates. We will first consider a low-speed Couette flow of an inert binary mixture with only one component (species 1) transferred, and examine mass, momentum, and heat transfer.

Mass Transfer

Figure 2.14 shows the model and coordinate system. For a steady state and no reactions, the species conservation equation, Eq. (2.10), reduces to

$$\frac{\partial n_{1x}}{\partial x} + \frac{\partial n_{1y}}{\partial y} = 0$$

But for a Couette flow, the dependent variables are independent of x; thus, $\partial n_{1x}/\partial x = 0$ and then

$$\frac{\partial n_{1y}}{\partial y} = 0$$

$$n_{1y} = \text{Constant} \tag{2.84}$$

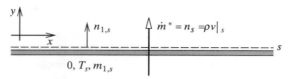

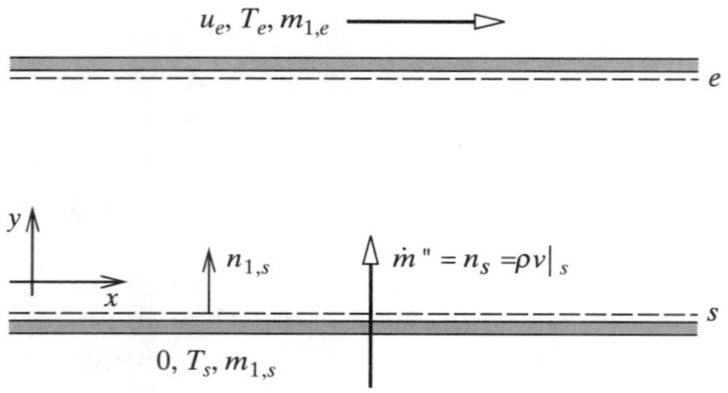

Figure 2.14 Schematic of a Couette–flow model for a boundary layer with mass transfer.

Similarly, the mass conservation equation, Eq. (2.11), reduces to

$$n_y = \text{Constant} \tag{2.85}$$

Since the problem is now one-dimensional, the subscript y will be dropped. Then Eq. (2.85) becomes

$$n = \text{Constant} = n|_{y=0} = n_s \equiv \dot{m}'' \tag{2.86}$$

where $\dot{m}''$ [kg/m^2 s] is the mass transfer rate that was introduced in Section 2.3.2. Substituting in Eq. (2.84),

$$n_1 = m_1 n + j_1 = m_1 \dot{m}'' - \rho \mathcal{D}_{12} \frac{dm_1}{dy} = \text{Constant} = n_{1,s} \tag{2.87}$$

Rearrangement and integration between the limits of $m_{1,s}$ at $y = 0$ and m_1 at y gives the concentration profile,

$$\int_{m_{1,s}}^{m_1} \frac{dm_1}{m_1 \dot{m}'' - n_{1,s}} = \int_0^y \frac{dy}{\rho \mathcal{D}_{12}}$$

and if $\rho \mathcal{D}_{12}$ is assumed constant,

$$\frac{m_1 - n_{1,s}/\dot{m}''}{m_{1,s} - n_{1,s}/\dot{m}''} = \exp\left(\frac{\dot{m}'' y}{\rho \mathcal{D}_{12}}\right) \tag{2.88}$$

In many problems of interest we can assume that only one species is transferred, for example, when water evaporates into air. If species 1 is the species transferred, $n_{2,s} = 0$ and $\dot{m}'' = n_s = n_{1,s}$; that is, $n_{1,s}/\dot{m}'' = 1$, and Eq. (2.88) becomes

$$\frac{m_1 - 1}{m_{1,s} - 1} = \exp\left(\frac{\dot{m}'' y}{\rho \mathcal{D}_{12}}\right)$$

which can be rewritten as

$$1 + \frac{m_1 - m_{1,s}}{m_{1,s} - 1} = \exp\left(\frac{\dot{m}'' y}{\rho \mathcal{D}_{12}}\right) \tag{2.89}$$

This expression for the concentration profile cannot be evaluated immediately, since both $m_{1,s}$ and $\dot{m}''$ are required. We proceed as follows: writing Eq. (2.89) for $y = \delta$ where $m_1 = m_{1,e}$ is specified, gives

$$1 + \frac{m_{1,e} - m_{1,s}}{m_{1,s} - 1} = \exp\left(\frac{\dot{m}''\delta}{\rho\mathcal{D}_{12}}\right) \tag{2.90}$$

Case 1: $m_{1,s}$ is specified; for example, the wall at $y = 0$ might be an evaporating liquid at temperature T_s, and $m_{1,s} = m_{1,s}(P, T_s)$ from vapor pressure data. Then Eq. (2.90) gives the evaporation rate $\dot{m}''$, and if it is substituted in Eq. (2.89), the concentration profile can be obtained. In this case, notice that $n_{2,s} = 0$ because the free-stream gas is assumed to be insoluble in the liquid.

Case 2: $\dot{m}''$ is specified; for example, the wall at $y = 0$ might be porous, and species 1 is blown through the wall as is done in transpiration cooling. Then Eq. (2.90) gives $m_{1,s}$, and if it is substituted in Eq. (2.89), the concentration profile can be obtained. In this case, notice that $n_{2,s} = 0$ because the flux of the free-stream component into the wall is zero at steady state.

The Mass Transfer Driving Force and Blowing Factor

Although the Couette–flow problem is solved at this point, it is useful to further rearrange Eq. (2.90). First, recall the definition of the mass transfer conductance $\mathcal{g}_{m1}$: from Eq. (1.36),

$$\mathcal{g}_{m1} = \frac{j_{1,s}}{m_{1,s} - m_{1,e}} \tag{2.91}$$

Notice that $\mathcal{g}_{m1}$ is defined in terms of the *diffusive flux* $j_{1,s}$, rather than the absolute flux $n_{1,s}$ across the surface. By definition,

$$n_{1,s} = m_{1,s}n_s + j_{1,s}$$

and since only species 1 is transferred, $n_{2,s} = 0$, $n_s = n_{1,s} + n_{2,s} = n_{1,s} = \dot{m}''$; hence

$$\dot{m}'' = m_{1,s}\dot{m}'' + j_{1,s} \tag{2.92}$$

Substituting from Eq. (2.91) and rearranging,

$$\dot{m}'' = \mathcal{g}_{m1}\frac{m_{1,e} - m_{1,s}}{m_{1,s} - 1}$$

or

$$\dot{m}'' = \mathcal{g}_{m1}\mathcal{B}_{m1}; \quad \mathcal{B}_{m1} = \frac{m_{1,e} - m_{1,s}}{m_{1,s} - 1} \tag{2.93}$$

where $\mathcal{B}_{m1}$ is the **mass transfer driving force**. Notice that Eq. (2.93) is not quite in the Ohm's law form, flux equals conductance times driving potential, since the mass transfer driving force $\mathcal{B}_{m1}$ is a nonlinear function of $m_{1,s}$. Also, Eq. (2.93) is quite general: it is based on definitions and applies to any convective situation, not just to a Couette flow. Introducing $\mathcal{B}_{m1}$ into Eq. (2.90) gives

$$1 + \mathcal{B}_{m1} = \exp\left(\frac{\dot{m}''\delta}{\rho\mathcal{D}_{12}}\right)$$

and, solving for $\dot{m}''$,

$$\dot{m}'' = \frac{\rho \mathcal{D}_{12}}{\delta} \ln(1 + \mathcal{B}_{m1}) \tag{2.94}$$

or

$$\dot{m}'' = \frac{\rho \mathcal{D}_{12}}{\delta} \frac{\ln(1 + \mathcal{B}_{m1})}{\mathcal{B}_{m1}} \mathcal{B}_{m1} \tag{2.95}$$

Comparing Eqs. (2.93) and (2.95), we can identify

$$g_{m1} = \frac{\rho \mathcal{D}_{12}}{\delta} \frac{\ln(1 + \mathcal{B}_{m1})}{\mathcal{B}_{m1}} \tag{2.96}$$

We will find it useful to normalize the mass transfer conductance with its value in the limit of zero mass transfer rate. Equation (2.95) shows that $\dot{m}'' \to 0$ as $\mathcal{B}_{m1} \to 0$, and

$$\lim_{\mathcal{B}_{m1} \to 0} \frac{\ln(1 + \mathcal{B}_{m1})}{\mathcal{B}_{m1}} = \lim_{\mathcal{B}_{m1} \to 0} \frac{\mathcal{B}_{m1} - (1/2)\mathcal{B}_{m1}^2 + \cdots}{\mathcal{B}_{m1}} = 1$$

which, upon substitution in Eq. (2.96), gives

$$\lim_{\dot{m}'' \to 0} g_{m1} = \frac{\rho \mathcal{D}_{12}}{\delta} \equiv g_{m1}^* \tag{2.97}$$

The * superscript is used here to denote the limit of zero mass transfer rate, and must not be confused with the * used to denote a molar flux relative to the mole average velocity. Since we very seldom analyze mass convection on a molar basis, there should be little chance of confusion.

Notice the definition of $\mathcal{B}_{m1} = (m_{1,e} - m_{1,s})/(m_{1,s} - 1)$ implies that $\mathcal{B}_{m1} \to 0$ as $m_{1,s} \to m_{1,e}$—that is, as the concentration gradients go to zero—and intuition tells us that this corresponds to the limit $\dot{m}'' \to 0$. It is also instructive to see that as $\dot{m}'' \to 0$, the concentration profile is linear. For $\dot{m}'' \to 0$, Eq. (2.87) becomes

$$-\rho \mathcal{D}_{12} \frac{dm_1}{dy} = \text{Constant}$$

Integrating and applying the boundary conditions gives

$$\frac{m_1 - m_{1,s}}{m_{1,e} - m_{1,s}} = \frac{y}{\delta}$$

which is linear, as shown in Fig. 2.15. Then

$$j_{1,s} = -\rho \mathcal{D}_{12} \frac{dm_1}{dy}\bigg|_{y=0} = \frac{\rho \mathcal{D}_{12}}{\delta}(m_{1,s} - m_{1,e}) \tag{2.98}$$

Hence, $g_{m1}^* = j_{1,s}/(m_{1,s} - m_{1,e}) = \rho \mathcal{D}_{12}/\delta$, as before. With the relation $g_{m1}^* = \rho \mathcal{D}_{12}/\delta$ established, we can rewrite Eq. (2.95) as

$$\dot{m}'' = \underbrace{g_{m1}^*}_{\substack{\text{zero mass} \\ \text{transfer} \\ \text{limit} \\ \text{conductance}}} \times \underbrace{\frac{\ln(1 + \mathcal{B}_{m1})}{\mathcal{B}_{m1}}}_{\substack{\text{blowing} \\ \text{factor}}} \times \underbrace{\mathcal{B}_{m1}}_{\substack{\text{driving} \\ \text{force}}} \tag{2.99}$$

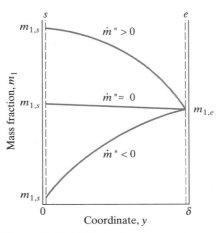

Figure 2.15 Effect of mass transfer
on concentration profiles for a
Couette flow. For evaporation or
transpiration of species 1, $\dot{m}'' > 0$; for
condensation of species 1, $\dot{m}'' < 0$.

and notice that details of the Couette–flow model do not appear in this result. Recall that our Couette–flow model is intended only as a *model* of a real boundary layer flow, which we have analyzed to examine the effects of mass transfer. We have isolated this effect in Eq. (2.99) as a **blowing factor**, $\ln(1 + \mathcal{B}_{m1})/\mathcal{B}_{m1}$.

In anticipation of the fact that our Couette–flow model blowing factor might be modified using the results of more exact analysis or experimental data, we will write our final result as

$$\dot{m}'' = g_{m1}\mathcal{B}_{m1}; \quad \mathcal{B}_{m1} = \frac{m_{1,e} - m_{1,s}}{m_{1,s} - 1} \tag{2.100}$$

$$\frac{g_{m1}}{g_{m1}^*} = \frac{\ln(1 + \mathcal{B}_{m1})}{\mathcal{B}_{m1}} \quad \text{(Couette–flow model)} \tag{2.101}$$

where Eq. (2.100) is quite general since it is simply a combination of definitions, and Eq. (2.101) is the Couette–flow result. It sometimes proves useful to work in terms of a parameter that is directly proportional to $\dot{m}''$; we define a **blowing parameter**

$$\mathcal{B}_{m1} = \frac{\dot{m}''}{g_{m1}^*} = \frac{\dot{m}''}{\rho u_e \mathrm{St}_m^*} \tag{2.102}$$

where St_m is the mass transfer Stanton number. Then, for the Couette–flow model,

$$\frac{\mathrm{St}_m}{\mathrm{St}_m^*} = \frac{g_{m1}}{g_{m1}^*} = \frac{\mathcal{B}_{m1}}{\exp \mathcal{B}_{m1} - 1} \tag{2.103}$$

This form is more convenient for problem solving when $\dot{m}''$ is prescribed, for example, in transpiration cooling.

The preceding Couette–flow analysis is essentially identical to the analysis of diffusion with one component stationary in Section 2.3.1, since the streamwise velocity component u did not enter into the problem. Since species 2 is stationary, Eq. (2.29) with δ replacing L can be rewritten as

$$N_{1,s} = \frac{c\mathcal{D}_{12}}{\delta} \ln\left[1 + \frac{x_{1,e} - x_{1,s}}{x_{1,s} - 1}\right] \tag{2.104}$$

which is the molar equivalent of Eq. (2.94). Equation (2.104) can be viewed as the result of a *stagnant-film* model. Thus, whether we talk of a Couette–flow analysis or a stagnant-film analysis to determine the effects of high mass transfer rates is immaterial: the analyses are essentially identical. Chemical engineers have traditionally preferred to think in terms of a stagnant-film model, while mechanical engineers have preferred a Couette–flow model.[4] Recall also that Eq. (2.104) was derived assuming a constant total molar concentration c, whereas Eq. (2.94) required the generally poorer assumption of a constant mass density ρ.

Momentum Transfer

Considering next momentum transfer, our starting point is the constant-property conservation-of-momentum equation in the form derived in Section 5.4 of *Heat Transfer*, which remains valid if second-order effects associated with the diffusion velocities are ignored. Equation (5.39) is

$$u\frac{\partial u}{\partial x} + v\frac{\partial u}{\partial y} = \nu\frac{\partial^2 u}{\partial y^2}$$

and, for Couette flow with $\partial u/\partial x = 0$, reduces to

$$v\frac{du}{dy} = \nu\frac{d^2 u}{dy^2}$$

Multiplying through by ρ and once again noting that $\rho v = n = n_s = \dot{m}''$,

$$\dot{m}''\frac{du}{dy} = \mu\frac{d^2 u}{dy^2} \tag{2.105}$$

Integrating with $u = 0$ at $y = 0$ and $u = u_e$ at $y = \delta$ gives the velocity profile as

$$\frac{u}{u_e} = \frac{\exp(\dot{m}''y/\mu) - 1}{\exp(\dot{m}''\delta/\mu) - 1} \tag{2.106}$$

and the shear stress on the wall is

$$\tau_s = \mu\frac{du}{dy}\bigg|_{y=0} = \frac{\dot{m}''u_e}{\exp(\dot{m}''\delta/\mu) - 1} \tag{2.107}$$

Notice that when $\dot{m}''$ is positive and large, $\tau_s \to 0$: we talk about the flow being blown off the wall, giving a zero shear stress. When $\dot{m}''$ is negative and large, $\tau_s \to -\dot{m}''u_e$: in this *strong suction* limit, the shear stress is independent of viscosity and equals

---

[4] Stagnant-film models in cylindrical and spherical coordinates also give results that can be put in the form of Eq. (2.99).

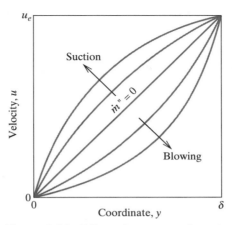

Figure 2.16   Effect of mass transfer on velocity profiles for a Couette flow.

the momentum lost by the transferred fluid as it is decelerated from a velocity u_e to zero. The strong suction limit case was used to account for vapor drag on film condensation in Section 7.2.4 in *Heat Transfer*. Also, the limit of τ_s for $\dot{m}'' \to 0$ is

$$\tau_s^* = \frac{\mu u_e}{\delta} \qquad (2.108)$$

which corresponds to a linear velocity profile, as shown in Fig. 2.16. Hence,

$$\frac{\tau_s}{\tau_s^*} = \frac{\dot{m}''\delta/\mu}{\exp(\dot{m}''\delta/\mu) - 1} \qquad (2.109)$$

A blowing parameter for momentum transfer is defined by analogy to Eq. (2.102) for mass transfer:

$$B_f = \frac{\dot{m}'' u_e}{\tau_s^*} \qquad (2.110)$$

Substituting for τ_s^* from Eq. (2.108) gives $B_f = \dot{m}''\delta/\mu$; then, introducing the skin friction coefficient $C_f = \tau_s/(1/2)\rho u_e^2$, we obtain

$$\frac{C_f}{C_f^*} = \frac{\tau_s}{\tau_s^*} = \frac{B_f}{\exp B_f - 1} \qquad (2.111)$$

which is the momentum transfer analogy to the mass transfer result Eq. (2.103).

Heat Transfer

Finally, we consider heat transfer in a low-speed flow, for which we derive the governing conservation equation from first principles. Figure 2.17 shows an elemental control volume of cross-sectional area A and thickness Δy. Energy can flow in and out of the control volume by conduction $-k(dT/dy)$, and species 1 can transport enthalpy $n_1 h_1$. Species 2 is stationary ($n_2 = n_{2,s} = 0$) and does not transport enthalpy. For a Couette flow, there is no x-direction temperature gradient, and hence

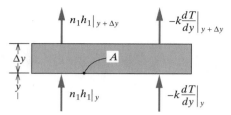

Figure 2.17 Elemental control volume for application of the steady-flow energy equation in a low-speed Couette flow, with transfer of a single chemical species.

no x-direction energy flow. At steady state, the steady-flow energy equation applied to the control volume requires that

$$A \left(n_1 h_1 - k \frac{dT}{dy} \right)_{y+\Delta y} - A \left(n_1 h_1 - k \frac{dT}{dy} \right)_y = 0$$

Dividing by $A \Delta y$ and letting $\Delta y \to 0$,

$$\frac{d}{dy} \left(n_1 h_1 - k \frac{dT}{dy} \right) = 0 \tag{2.112}$$

As before, species conservation requires

$$n_1 = \text{Constant} = n_{1,s} = \dot{m}'' \tag{2.113}$$

For convenience in the analysis that follows, we take the specific heat of species 1 to be independent of temperature, and choose an enthalpy datum state of $h = 0$ at $T = 0$; hence $h_1 = c_{p1}T$. Substituting in Eq. (2.112),

$$\frac{d}{dy} \left(\dot{m}'' c_{p1} T - k \frac{dT}{dy} \right) = 0 \tag{2.114}$$

Integrating twice with boundary conditions $y = 0$, $T = T_s$; $y = \delta$, $T = T_e$; and k constant, gives the temperature profile as

$$\frac{T - T_s}{T_e - T_s} = \frac{\exp(\dot{m}'' c_{p1}/k)y - 1}{\exp(\dot{m}'' c_{p1}/k)\delta - 1} \tag{2.115}$$

Differentiation gives the conduction heat flux across the s-surface as

$$-k \frac{dT}{dy} \bigg|_s = \frac{\dot{m}'' c_{p1}(T_s - T_e)}{\exp(\dot{m}'' c_{p1}/k)\delta - 1} \tag{2.116}$$

Even though there is mass transfer through the wall, we define the heat transfer coefficient by an equation identical to that for an impermeable surface, namely, Eq. (4.3) in *Heat Transfer*:

$$h_c = \frac{-k(dT/dy)|_s}{T_s - T_e}$$

Substituting Eq. (2.116) then gives

$$h_c = \frac{\dot{m}'' c_{p1}}{\exp(\dot{m}'' c_{p1}/k)\delta - 1} \tag{2.117}$$

The limit form as $\dot{m}'' \to 0$ is obtained by applying L'Hopital's rule,

$$\lim_{\dot{m}'' \to 0} h_c = h_c^* = \frac{k}{\delta} \tag{2.118}$$

which is simply the value for conduction across a slab δ thick, since in this limit the temperature profile is linear. Combining Eqs. (2.117) and (2.118) gives

$$\frac{h_c}{h_c^*} = \frac{\dot{m}'' c_{p1}/h_c^*}{\exp(\dot{m}'' c_{p1}/h_c^*) - 1} \tag{2.119}$$

A blowing parameter for heat transfer is now defined as

$$B_h = \frac{\dot{m}'' c_{p1}}{h_c^*} \tag{2.120}$$

and Eq. (2.119) becomes

$$\frac{h_c}{h_c^*} = \frac{B_h}{\exp B_h - 1} \tag{2.121}$$

which is the heat transfer analog to Eqs. (2.103) and (2.111). Equation (2.121) allows the heat transfer coefficient h_c for a finite mass transfer rate to be calculated from h_c^*, its value for zero mass transfer, and the mass transfer rate $\dot{m}''$.

To complete the picture for heat transfer, we look at an energy balance on the control volume located between the s- and u-surfaces, as shown in Fig. 2.18. If radiation heat transfer is negligible,

$$-k\frac{dT}{dy}\Big|_u + n_{1,u}h_{1,u} = -k\frac{dT}{dy}\Big|_s + n_{1,s}h_{1,s}$$

But $n_{1,s} = n_{1,u} = \dot{m}''$; thus

$$k\frac{dT}{dy}\Big|_u = k\frac{dT}{dy}\Big|_s - \dot{m}''(h_{1,s} - h_{1,u}) \tag{2.122}$$

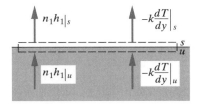

Figure 2.18 Interface energy balance for transfer of a single chemical species.

Case 1: If the surface is a liquid evaporating into a gas stream, then $h_{1,s} - h_{1,u} = h_{fg}$, the enthalpy of vaporization, and with $k(\partial T/\partial y)|_s = h_c(T_e - T_s)$:

$$k\frac{dT}{dy}\bigg|_u = h_c(T_e - T_s) - \dot{m}''h_{fg} \tag{2.123}$$

Equation (2.123) states that the heat transfer into the liquid (or porous wall filled with liquid) is equal to the convective heat transfer to the surface minus the evaporation rate times vaporization enthalpy. This result is the basis of sweat cooling: the heat transfer into the water is less than that to the surface by the amount $\dot{m}''h_{fg}$. In addition, h_c is less than h_c^* due to the blowing effect given by Eq. (2.121), and hence the convective heat transfer itself is reduced by the sweat cooling process.

Case 2: If the wall is porous and species 1 is a gas blown through the wall, then $h_{1,s} = h_{1,u}$ and

$$k\frac{dT}{dy}\bigg|_u = h_c(T_e - T_s) \tag{2.124}$$

In this case of *transpiration* cooling, the heat transfer into the wall is reduced by the decrease of h_c due to blowing. If the coolant gas is supplied from a reservoir at temperature T_o, as shown in Fig. 2.19, then an energy balance between the u- and o-surfaces requires

$$-k\frac{dT}{dy}\bigg|_o + n_{1,o}h_{1,o} = -k\frac{dT}{dy}\bigg|_u + n_{1,u}h_{1,u}$$

But $n_{1,o} = n_{1,u} = \dot{m}''$, and if the o-surface is located sufficiently far from the u-surface, the temperature gradient $(dT/dy)|_o$ will be negligible. Then

$$k\frac{dT}{dy}\bigg|_u = \dot{m}''(h_{1,u} - h_{1,o})$$

and substituting in Eq. (2.124) gives

$$h_c(T_e - T_s) = \dot{m}''(h_{1,u} - h_{1,o}) \tag{2.125}$$

If the specific heat of the coolant is assumed constant, this result simplifies to

$$h_c(T_e - T_s) = \dot{m}''c_{p1}(T_s - T_o) \tag{2.126}$$

which states that the heat transfer to the wall is equal to the coolant flow rate times its enthalpy change as it flows from the reservoir to the wall surface.

Notice that we did not assume all properties constant in the Couette–flow heat transfer analysis: the specific heats of species 1 and 2 were allowed to be different. Figure 2.20 shows h_c/h_c^* plotted from Eq. (2.119) for various values of c_{p1}, where it is seen that the effect of blowing in reducing heat transfer increases with increasing c_{p1}: one can think of a high-specific-heat gas being more effective in "blocking" the heat from reaching the wall. However, high-specific-heat gases such as helium also have a low density and high thermal conductivity. Thus, for real air boundary layer flows with helium transpiration, there are additional effects due to density and

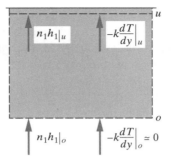

Figure 2.19 Energy balance on a coolant reservoir for transpiration cooling with injection of a single chemical species.

conductivity (and viscosity) variations across the flow. Also, since $PV = c\mathcal{R}T$, the number of moles of gas stored in a tank of given volume and at a given pressure is independent of species. Thus, a larger tank is required to store a given mass of low-molecular-weight gas, and, in a design trade-off, use of air as the injectant may be advantageous even though the blocking effect per unit mass injection rate is less than that for a low-molecular-weight gas.

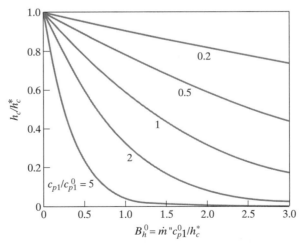

Figure 2.20 The effect of injectant specific heat on heat transfer for injection of single chemical species: a plot of the Couette–flow model result, Eq. (2.119), written as $h_c/h_c^* = (c_{p1}/c_{p1}^0)B_h^0 / \exp[(c_{p1}/c_{p1}^0)B_h^0] - 1$; $B_h^0 = \dot{m}'' c_{p1}^0/h_c^*$.

Problem Solving

In summary, the Couette–flow model has established the effect of finite mass transfer rates on the mass transfer conductance, the wall shear stress, and the heat transfer coefficient, namely,

$$\frac{g_{m1}}{g_{m1}^*} = \frac{B_{m1}}{\exp B_{m1} - 1}; \quad B_{m1} = \frac{\dot{m}''}{g_{m1}^*} \tag{2.127a}$$

or

$$\frac{g_{m1}}{g_{m1}^*} = \frac{\ln(1 + \mathcal{B}_{m1})}{\mathcal{B}_{m1}}; \quad \mathcal{B}_{m1} = \frac{m_{1,e} - m_{1,s}}{m_{1,s} - 1} \tag{2.127b}$$

$$\frac{\tau_s}{\tau_s^*} = \frac{B_f}{\exp B_f - 1}; \quad B_f = \frac{\dot{m}'' u_e}{\tau_s^*} \tag{2.128}$$

$$\frac{h_c}{h_c^*} = \frac{B_h}{\exp B_h - 1}; \quad B_h = \frac{\dot{m}'' c_{p1}}{h_c^*} \tag{2.129}$$

The behavior of Eq. (2.127) is shown in Fig. 2.21. Notice that for strong suction $B_{m1} \to -\infty$, while $\mathcal{B}_{m1} \to -1$. Suction occurs in such situations as condensation from a steam–air mixture, and combustion of a solid fuel when the products of combustion are nonvolatile. Of course, if the flow is a binary gas mixture, species 1 cannot be selectively removed through a porous wall; that is, there is no suction problem corresponding to transpiration cooling.

In solving engineering problems, for which the assumption of low mass transfer rate cannot be justified, the procedure is as follows. Values of g_m^*, τ_s^*, and h_c^* are obtained from correlations for an impermeable wall with the same geometry and flow condition, as was done for low mass transfer rate theory in Section 1.4. The correlations in Table 4.10 of *Heat Transfer* could have been written as starred quantities (for example, Nu*, C_f^*) but such a complication was unnecessary in Chapter 4.

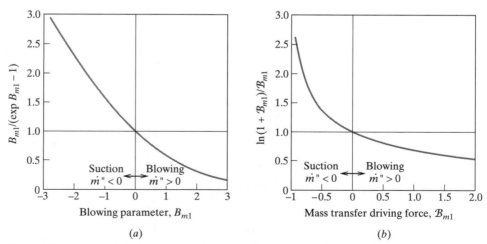

Figure 2.21 Plots of (*a*) Eq. (2.127*a*) and (*b*) Eq. (2.127*b*), showing the effects of suction and blowing on the Couette–flow blowing factor.

Equations (2.127)–(2.129) are then used to correct for high mass transfer rates. The blowing factors are based on the Couette–flow model and are only approximately valid for real flows; nevertheless, they are sufficient for many purposes. More accurate blowing factors for specific flows based on more exact analysis can be found in Section 2.4.2. The foregoing procedure works well for forced-flow laminar and turbulent boundary layers and for turbulent duct flows. It is less satisfactory for natural-convection boundary layers and laminar duct flows. Note that the procedure does not apply to the natural-convection enclosure flows of items 14 through 17 in Table 4.10, since the heat transfer coefficient is defined for transfer across the cavity, and not for transfer from a surface into fluid.

Variable-Property Effects

Accounting for variable-property effects is not straightforward. A major effect, that of the transferred species specific heat on heat transfer, is accounted for in the blowing factor defined by Eq. (2.129). If we ignore the effect of composition on other properties, namely, density, viscosity, and thermal conductivity, then a simple procedure is to simply evaluate g_m^*, τ_s^*, and h_c^* as if we were using low mass transfer rate theory. For example, in the case of an external flow, the properties will be those of a fluid of free-stream composition evaluated at the mean film temperature. The error in results obtained in this manner depends both on the extent to which variable-property effects are accurately accounted for and on the accuracy of the Couette–flow blowing factors for the particular flow under consideration. These errors may tend to cancel in some situations, and add in others. More accurate procedures for dealing with variable-property effects will be dealt with in Section 2.4.2.

Example 2.7   Transpiration Cooling

Air at 1500 K and 1 atm pressure flows at 10 m/s over a porous flat plate that is to be transpiration cooled. Calculate the rate at which coolant must be injected through the plate at a location 0.2 m from the leading edge if the plate surface is to be maintained at 500 K, and the coolant is supplied from a reservoir beneath the plate at 300 K, if (i) the coolant is air, (ii) the coolant is helium.

Solution

Given: Air at 1500 K flowing over a transpiration cooled flat plate.
Required: Coolant injection rate at $x = 0.2$ m to maintain the plate surface at 500 K.
Assumptions:

1. Couette–flow blowing factors are adequate.
2. The injectant specific heat is constant.
3. Radiation heat transfer is negligible.

The required injection rate is given by Eq. (2.126),

$$h_c(T_e - T_s) = \dot{m}'' c_{p1}(T_s - T_o)$$

where the heat transfer coefficient blowing factor is obtained from Eq. (2.119) as

$$\frac{h_c}{h_c^*} = \frac{\dot{m}'' c_{p1}/h_c^*}{\exp(\dot{m}'' c_{p1}/h_c^*) - 1}$$

Combining these two equations and solving for $\dot{m}''$ gives

$$\dot{m}'' = \frac{h_c^*}{c_{p1}} \ln\left(1 + \frac{T_e - T_s}{T_s - T_o}\right) \tag{1}$$

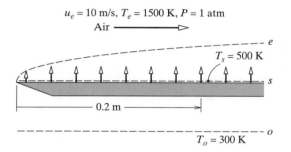

In determining h_c^* we evaluate air properties at the mean film temperature, $T_r = 500 + (1/2)(1500 - 500) = 1000$ K. From Table A.7, $\rho = 0.354$ kg/m³, $\nu = 117.3 \times 10^{-6}$ m²/s, $c_p = 1130$ J/kg K, Pr $= 0.70$. The Reynolds number is

$$\text{Re}_x = \frac{u_e x}{\nu} = \frac{(10)(0.2)}{(117.3 \times 10^{-6})} = 17,050 \quad \text{(laminar flow)}$$

From Eq. (4.56) of *Heat Transfer* the local Stanton number and heat transfer coefficient are

$$\text{St}_x^* = 0.332\text{Re}_x^{-1/2}\text{Pr}^{-2/3}$$

$$= (0.332)(17,050)^{-1/2}(0.70)^{-2/3} = 3.23 \times 10^{-3}$$

$$h_c^* = \rho c_p u_e \text{St}_x^* = (0.354)(1130)(10)(3.23 \times 10^{-3}) = 12.9 \text{ W/m}^2 \text{ K}$$

i. *Air injection.* Substituting in Eq. (1) with the assumed constant c_{p1} evaluated at $(1/2)(T_e + T_o) = 900$ K as 1111 J/kg K gives the injection rate as

$$\dot{m}'' = \frac{h_c^*}{c_{p1}} \ln\left(1 + \frac{T_e - T_s}{T_s - T_o}\right)$$

$$= \frac{12.9}{1111} \ln\left(1 + \frac{1500 - 500}{500 - 300}\right) = 0.0208 \text{ kg/m}^2 \text{ s}$$

ii. *Helium injection.* From Table A.7, $c_{p1} = 5200$ J/kg K, and the required injection rate is

$$\dot{m}'' = \frac{12.9}{5200} \ln\left[1 + \frac{1500 - 500}{500 - 300}\right] = 4.44 \times 10^{-3} \text{ kg/m}^2 \text{ s}$$

Comments

1. Owing to the high specific heat of helium, the helium injection rate is only 21% of that required for air.
2. Check h_c^* using MCONV.
3. To maintain an isothermal plate, Eq. (4.56) in *Heat Transfer* shows that $\dot{m}''$ must vary as $x^{-1/2}$. ▲

Example 2.8   Sweat Cooling

A wall exposed to intense radiative and convective heating is to be protected by sweat cooling. Water is injected through a porous stainless steel surface at a rate just sufficient to keep the surface wetted. Dry air at 840 K and 1 atm pressure flows past the surface at 100 m/s. Heat transfer experiments with air for the same geometry and flow conditions indicate a local Stanton number of 0.00429. (i) Determine the water supply rate to maintain the surface at 360 K. (ii) If the irradiation is 461 kW/m^2, determine the water supply temperature required to maintain a steady state.

Solution

Given: A wall heated by radiation and convection to be sweat cooled with water.
Required: Water supply rate and temperature.
Assumptions:

1. The Couette–flow blowing factors are adequate.
2. The flow is a turbulent boundary layer.

Since the wall temperature is specified, the heat and mass transfer problems are uncoupled. We first solve the mass transfer problem to determine the water supply rate, and then solve the heat transfer problem to determine the water supply temperature required to maintain steady conditions.

i. The water supply rate must balance the evaporation rate; from Eq. (2.100) the evaporation rate is

$$\dot{m}'' = g_{m1}\mathcal{B}_{m1}; \quad \mathcal{B}_{m1} = \frac{m_{1,e} - m_{1,s}}{m_{1,s} - 1}$$

where, from Eq. (2.101), the blowing factor is

$$\frac{g_{m1}}{g_{m1}^*} = \frac{\ln(1 + \mathcal{B}_{m1})}{\mathcal{B}_{m1}}$$

First we calculate the mass transfer driving force $\mathcal{B}_{m1}$. At 360 K, Table A.12a gives the saturation vapor pressure of water vapor as 0.6213×10^5 Pa. For 1 atm $= 1.0133 \times 10^5$ Pa total pressure,

$$
\begin{aligned}
m_{1,s} &= \frac{P_{1,s}}{P_{1,s} + (M_2/M_1)(P - P_{1,s})} \\
&= \frac{0.6213}{0.6213 + (29/18)(1.0133 - 0.6213)} = 0.496 \\
\mathcal{B}_{m1} &= \frac{m_{1,e} - m_{1,s}}{m_{1,s} - 1} = \frac{0 - 0.496}{0.496 - 1} = 0.984
\end{aligned}
$$

Next, we must evaluate the mass transfer conductance g_{m1}^*. The data suggest a turbulent boundary layer, for which Eq. (4.64) gives $St_m^*/St^* = (Sc/Pr)^{-0.57}$. Properties are evaluated at the mean film temperature $T_r = (1/2)(360 + 840) = 600$ K. Using AIRSTE to obtain $Pr = 0.69$ for air and $Sc = 0.54$ for a dilute H_2O-air mixture gives an estimate of St_m^* as

$$St_m^* = 0.00429(0.54/0.69)^{-0.57} = 0.00493$$

For air, Table A.7 gives $\rho = 0.589$ kg/m^3, $c_p = 1038$ J/kg K. Thus, the zero-mass-transfer-limit mass transfer conductance is

$$g_{m1}^* = \rho u_e St_m^* = (0.589)(100)(0.00493) = 0.291 \text{ kg/m}^2 \text{ s}$$

and the blowing factor is

$$\frac{g_{m1}}{g_{m1}^*} = \frac{\ln(1 + \mathcal{B}_{m1})}{\mathcal{B}_{m1}} = \frac{\ln(1 + 0.984)}{0.984} = 0.696$$

Thus, the evaporation rate is

$$\dot{m}'' = g_{m1}\mathcal{B}_{m1} = g_{m1}^* \frac{\ln(1 + \mathcal{B}_{m1})}{\mathcal{B}_{m1}} \mathcal{B}_{m1} = (0.291)(0.696)(0.984)$$

$$= 0.199 \text{ kg/m}^2 \text{ s}$$

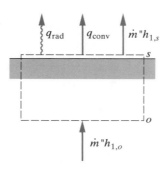

ii. The heat transfer problem is conveniently solved by making an energy balance on a control volume located between the s- and o-surfaces: the o-surface is located in the water supply reservoir where conduction is negligible. Since $n_{1,s} = \dot{m}''$,

$$q_{\text{rad}} + q_{\text{conv}} + \dot{m}'' h_{1,s} = \dot{m}'' h_{1,o}$$

Assuming a constant specific heat for the water, c_{pl}, and writing $q_{\text{conv}} = h_c(T_s - T_e)$,

$$\dot{m}''[h_{\text{fg}} + c_{pl}(T_s - T_o)] = -q_{\text{rad}} + h_c(T_e - T_s) \tag{1}$$

To evaluate the heat transfer coefficient h_c, we first evaluate h_c^*:

$$h_c^* = \rho u_e c_p \text{St}^* = (0.589)(100)(1038)(0.00429) = 262 \text{ W/m}^2 \text{ K}$$

The heat transfer blowing parameter B_h is

$$B_h = \dot{m}'' c_{p1}/h_c^* = (0.199)(2003)/(262) = 1.52$$

where c_{p1} is the specific heat of water vapor, and is also evaluated at $T_r = 600 \text{ K}$. The corrected heat transfer coefficient is then

$$h_c = h_c^* \frac{B_h}{\exp B_h - 1} = (262)\frac{1.52}{\exp(1.52) - 1} = (262)(0.426)$$

$$= 111 \text{ W/m}^2 \text{ K}$$

From Table A.12a, $h_{\text{fg}} = 2.291 \times 10^6$ J/kg at 360 K, and from Table A.8, $c_{pl} = 4178$ J/kg K at a guessed average temperature of $\sim$330 K. Substituting in Eq. (1) gives

$$(0.199)[2.291 \times 10^6 + (4178)(360 - T_o)] = 461 \times 10^3 + 111(840 - 360)$$

Solving, $T_o = 290$ K.

Comments

1. If we assume that the radiation is absorbed and emitted between the s- and u-surfaces (see Section 1.5.1), an energy balance between the u- and o-surfaces gives the conduction heat flux across the u-surface as

$$q_{\text{cond},u} = -k \frac{dT}{dy}\Big|_u = \dot{m}'' c_{pl}(T_o - T_s)$$

$$= (0.199)(4178)(290 - 360)$$

$$= -58.2 \text{ kW/m}^2$$

2. Due to blowing, g_{m1} is 70% of g_{m1}^*, whereas h_c is only 43% of h_c^*: this behavior is due to the relatively high specific heat of water vapor.

3. An alternative way to pose this problem is to specify the water supply temperature and the irradiation. Then the wall temperature is an unknown, and the heat and mass transfer problems must be solved simultaneously, requiring an iterative process. See, for example, Exercise 2–31.   ▲

Example 2.9   Effect of an Air Leak on Condenser Performance

Steam at 336 K and 0.115 atm containing 0.62% air by mass flows down at 1 m/s over a horizontal brass condenser tube of 19.1 mm O.D. and 16.1 mm I.D. Coolant water flows through the tube at a bulk velocity of 1.47 m/s. At a location where the bulk coolant temperature is 283 K, estimate the steam condensation rate and compare it with the value for pure steam.

Solution

Given: Steam containing a small amount of air flowing down over a horizontal condenser tube.
Required: Effect of air on steam condensation rate.
Assumptions:
1. Couette–flow blowing factors are adequate.
2. The presence of the condensate film can be ignored in calculating the steam flow mass transfer conductance.
3. Average values of the transfer coefficients can be used.
4. The isothermal-tube film condensation correlation is adequate.

First we draw an equivalent circuit for this transfer process. It is a hybrid one, involving both heat and mass transfer processes, as shown. The transfer coefficients are all average values. The circuit between T_c and T_s is conventional: $h_{c,i}$ is the inside heat transfer coefficient, L/kA_m is the brass-tube wall resistance, and h_o is the heat transfer coefficient for film condensation. Heat is transferred to the condensate surface both as sensible heat with thermal resistance $1/h_cA_o$ for flow over a cylinder, and as latent heat with the corresponding mass transfer resistance $1/g_{m1}A_o$. In most situations the sensible heat transfer can be neglected compared with the latent heat transfer. Calculation of the thermal circuit between T_c and T_s is straightforward. For evaluation of fluid properties, reference temperatures of 290 K and 300 K are guessed for $h_{c,i}$ and h_o, respectively. Then, using the correlation Eq. (4.44) of *Heat Transfer* for turbulent flow in a tube, $h_{c,i} = 5610$ W/m^2 K, and for laminar film condensation on a horizontal tube,

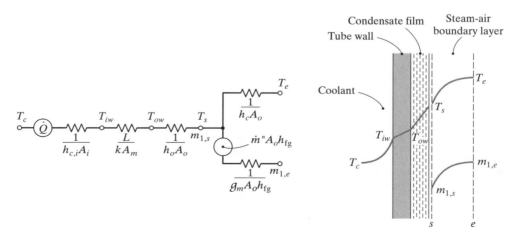

Eq. (7.41) of *Heat Transfer* gives $h_o = 17{,}300(T_s - T_{ow})^{-1/4}$ W/m^2 K. Summing the resistances in the circuit gives a total resistance per unit outside area between T_c and T_s of

$$\frac{1}{U} = \frac{1}{h_{c,i}(A_i/A_o)} + \frac{1}{(k/L)[(A_i + A_o)/2A_o]} + \frac{1}{h_o}, \quad \text{(for } U \text{ based on } A_o\text{)}$$

$$= \frac{1}{(5610)(0.843)} + \frac{1}{(111/1.5 \times 10^{-3})(0.921)} + \frac{1}{(17{,}300)(T_s - T_{ow})^{-1/4}}$$

$$= \frac{1}{4730} + \frac{1}{68{,}200} + \frac{1}{17{,}300(T_s - T_{ow})^{-1/4}}$$

$$U = \frac{10^5}{22.6 + 5.78(T_s - T_{ow})^{1/4}} \tag{1}$$

Next we calculate g_{m1}^*, evaluating properties using the free-stream composition, $m_1 = 0.9938 \simeq 1.0$ (i.e., pure steam) and a guessed reference temperature of 320 K. At 0.115 atm $= 0.1165 \times 10^5$ Pa, the steam is supersaturated, and ρ can be calculated from the ideal gas formula,

$$\rho = PM_1/\mathcal{R}T = (0.1165 \times 10^5)(18)/[(8314)(320)] = 0.0788 \text{ kg/m}^3$$

To calculate ν we first find $\mu = 9.89 \times 10^{-6}$ kg/m s for saturated steam at 320 K from Table A.7; then, since μ is independent of pressure, $\nu = \mu/\rho = 9.89 \times 10^{-6}/0.0788 = 125.5 \times 10^{-6}$ m^2/s for the supersaturated steam. To calculate $\mathcal{D}_{12}$ we use the formula given in Table A.17a, $\mathcal{D}_{12} = 1.97 \times 10^{-5}(1/0.115)(320/256)^{1.685} = 249 \times 10^{-6}$ m^2/s. Hence, Sc $= \nu/\mathcal{D}_{12} = 125.5/249 = 0.504$. The Reynolds number for the steam flow is

$$\text{Re}_D = VD/\nu = (1)(19.1 \times 10^{-3})/(125.5 \times 10^{-6}) = 152.2$$

We will ignore the presence of the condensate film in determining g_{m1}^*: the film is thin compared with the tube diameter, and its surface velocity is small compared with the steam velocity. For forced convection over a cylinder, the mass transfer analog to Eq. (4.71a) of *Heat Transfer* is

$$\overline{\text{Sh}}_D^* = 0.3 + \frac{0.62\text{Re}^{1/2}\text{Sc}^{1/3}}{[1 + (0.4/\text{Sc})^{2/3}]^{1/4}}$$

$$= 0.3 + \frac{(0.62)(152.2)^{1/2}(0.504)^{1/3}}{[1 + (0.4/0.504)^{2/3}]^{1/4}} = 5.51$$

The zero-mass-transfer-limit mass transfer conductance is then

$$g_{m1}^* = \rho \mathcal{D}_{12}\text{Sh}^*/D$$

$$= (0.0788)(249 \times 10^{-6})(5.51)/(19.1 \times 10^{-3}) = 5.66 \times 10^{-3} \text{ kg/m}^2 \text{ s}$$

The rate of latent heat transfer is $\dot{m}'' h_{fg}$; from Eqs. (2.100) and (2.101),

$$\dot{m}'' h_{fg} = g_{m1}^* \ln(1 + \mathcal{B}_{m1})h_{fg} = 5.66 \times 10^{-3} \ln\left(1 + \frac{0.9938 - m_{1,s}}{m_{1,s} - 1}\right) h_{fg} \tag{2}$$

If we neglect sensible heat transfer from the steam to the condensate surface, our thermal circuit shows that the current flowing is

$$\dot{Q} = UA_o(T_s - T_c) = -\dot{m}''h_{fg}A_o \tag{3}$$

Substituting Eqs. (1) and (2) in Eq. (3) gives

$$\frac{\dot{Q}}{A_o} = \frac{(T_s - 283)(10^5)}{22.6 + 5.78(T_s - T_{ow})^{1/4}}$$

$$= -5.66 \times 10^{-3} \ln\left(1 + \frac{0.9938 - m_{1,s}}{m_{1,s} - 1}\right) h_{fg} \tag{4}$$

Equation (4) must be solved by iteration using steam tables for $P_{1,s} = P_{1,\text{sat}}(T_s)$ and $h_{fg}(T_s)$, together with the auxiliary relations

$$m_{1,s} = \frac{P_{1,s}}{P_{1,s} + (29/18)(P - P_{1,s})}; \quad T_s - T_{ow} = \left(\frac{\dot{Q}/A_o}{17{,}300}\right)^{4/3} \tag{5, 6}$$

The result (after some tedious calculations) is $\dot{Q}/A_o = 65{,}500$ W/m^2, with $T_s = 303.7$ K and $m_{1,s} = 0.274$. The condensation rate per unit length of tube is

$$(\dot{Q}/A_o)A_o/h_{fg} = (65{,}500)(\pi)(19.1 \times 10^{-3})/(2.43 \times 10^6)$$

$$= 1.62 \times 10^{-3} \text{ kg/s m}$$

Intermediate temperatures in the circuit can be now calculated and the guessed temperatures for fluid properties revised, if necessary.

We can now check our assumption that the sensible heat transfer is small compared with the latent heat transfer. For pure steam at 320 K, $k = 0.0210$ W/m K, $c_p = 1890$ J/kg K, and Pr = 0.89. Using Eq. (4.71a), $\overline{\text{Nu}}_D^* = 6.86$, and hence

$$h_c^* = (k/D)\text{Nu}_D^* = (0.0210/1.91 \times 10^{-3})(6.86) = 7.54 \text{ W/m}^2 \text{ K}$$

$$\dot{m}'' = -(\dot{Q}/A_o)/h_{fg} = -(65{,}500/2.43 \times 10^6) = -0.0270 \text{ kg/m}^2 \text{ s}$$

To correct h_c^* for suction, we use Eq. (2.129):

$$B_h = \dot{m}''c_{p1}/h_c^* = (-0.0270)(1890)/(7.54) = -6.77$$

$$\frac{h_c}{h_c^*} = \frac{B_h}{\exp B_h - 1} = \frac{-6.77}{\exp(-6.77) - 1} = 6.78$$

Notice that the actual heat transfer coefficient is 6.8 times the value for no mass transfer, due to strong suction.

$$h_c = (6.78)(7.54) = 51.1 \text{ W/m}^2 \text{ K}$$

$$h_c(T_e - T_s) = 51.1(336 - 303.7) = 1650 \text{ W/m}^2$$

The sensible heat transfer proves to be only 2.5% of the latent heat transfer, and was justifiably neglected.

For pure condensing steam at 0.115 atm, T_s is the corresponding saturation value; from Table A.12a, $T_s = 322.0$ K. Again an iterative solution is required. Equating the heat flow across the film and across the wall into the coolant,

$$\frac{\dot{Q}}{A_o} = h_o(T_s - T_{ow})$$

$$= \left(\frac{1}{h_{c,i}(A_i/A_o)} + \frac{1}{(k/L)[(A_i + A_o)/2A_o]} \right)^{-1} (T_{ow} - T_c) \qquad (7)$$

which reduces to

$$17{,}300(322.0 - T_{ow})^{3/4} = 4423(T_{ow} - 283)$$

The solution is $(322.0 - T_{ow}) = 12.7$ K, and $\dot{Q}/A_o = 17{,}300(322.0 - T_{ow})^{3/4} = 116{,}000$ W/m^2. Thus, the heat transfer rate is reduced to $65{,}500/116{,}000 = 56\%$ of the pure steam value by only 0.62% air present in the bulk steam.

Comments

1. Strictly speaking, a new reference temperature for property evaluation should be used in solving Eq. (7).

2. Since power-plant condensers operate at pressures well below atmospheric pressure, air leaks are common. Such condensers are equipped with steam ejectors that continuously pump a steam–air mixture from dead spaces in the condenser to prevent a buildup of air.

3. Notice that the Schmidt number for the bulk steam–air mixture is 0.504, not the value of 0.61 for steam in dilute concentration.

4. The presence of the condensate film was neglected in calculating Re_D, $\overline{\mathrm{Sh}}_D^*$, and $\overline{\mathrm{Nu}}_D^*$. ▲

2.4.2 Improved Blowing Factors

The Couette flow–based blowing factors developed in Section 2.4.1 are useful for routine engineering problem solving. However, often more accurate results are required, for which these blowing factors may be inadequate. Again we will restrict our attention to binary inert mixtures with transfer of a single species, and our concern is with the effects of the type of flow and variable properties on the blowing factors. Figure 2.22 shows the effects of pressure gradient and variable properties on blowing factors for some laminar boundary layers. Figure 2.23 shows the effect of variable properties for a turbulent boundary layer flow on a flat plate. Clearly, deviations from the Couette–flow blowing factors can be substantial.

A simple procedure for correlating the effects of flow type and variable properties is to use weighting factors in the exponential functions suggested by a constant-

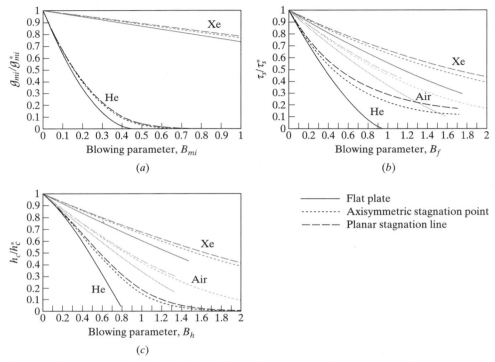

Figure 2.22 Numerical results for the effect of pressure gradient and variable properties on blowing factors for laminar boundary layers: low-speed air flow over a cold wall $(T_s/T_e = 0.1)$ with foreign gas injection. (*a*) Mass transfer conductance. (*b*) Wall shear stress. (*c*) Heat transfer coefficient [9, 10].

property Couette–flow model. Equations (2.127)–(2.129) are generalized as follows. Denoting the injected species as species i, we have

$$\frac{g_{mi}}{g_{mi}^*} = \frac{a_{mi} B_{mi}}{\exp(a_{mi} B_{mi}) - 1}; \quad B_{mi} = \frac{\dot{m}''}{g_{mi}^*} \tag{2.130a}$$

or

$$\frac{g_{mi}}{g_{mi}^*} = \frac{\ln(1 + a_{mi} \mathcal{B}_{mi})}{a_{mi} \mathcal{B}_{mi}}; \quad \mathcal{B}_{mi} = \frac{\dot{m}''}{g_{mi}} = \frac{m_{i,e} - m_{i,s}}{m_{i,s} - 1}$$

$$\frac{\tau_s}{\tau_s^*} = \frac{a_{fi} B_f}{\exp(a_{fi} B_f) - 1}; \quad B_f = \frac{\dot{m}'' u_e}{\tau_s^*} \tag{2.130b}$$

$$\frac{h_c}{h_c^*} = \frac{a_{hi} B_h}{\exp(a_{hi} B_h) - 1}; \quad B_h = \frac{\dot{m}'' c_{pe}}{h_c^*} \tag{2.130c}$$

Notice that g_{mi}^*, τ_s^*, h_c^*, and c_{pe} are evaluated using properties of the free-stream gas at the mean film temperature, and that Eq. (2.130c) is the Couette–flow model result for all properties constant; in particular, the injected and free-stream specific heats are equal. The weighting factor a may be found from exact numerical solutions

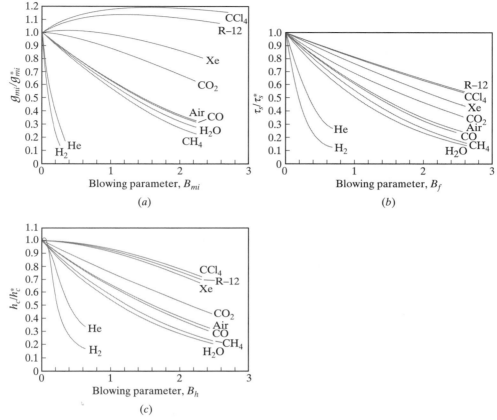

Figure 2.23 Numerical results for the effect of variable properties on blowing factors for a low-speed turbulent air boundary layer on a cold flat plate ($T_s/T_e = 0.2$) with foreign gas injection. (*a*) Mass transfer conductance. (*b*) Wall shear stress. (*c*) Heat transfer coefficient [12].

of boundary layer equations or from experimental data. Some results for laminar and turbulent boundary layers follow.

1. *Laminar boundary layers.* We will restrict our attention to low-speed air flows, for which viscous dissipation and compressibility effects are negligible, and use exact numerical solutions of the self-similar laminar boundary layer equations [9, 10, 11]. Least-squares curve fits of the numerical data were obtained using Eqs. (2.130*a*)–(2.130*c*). Then the weighting factors for axisymmetric stagnation-point flow with a cold wall ($T_s/T_e = 0.1$) were correlated as

$$a_{mi} = 1.65(M_{air}/M_i)^{10/12} \tag{2.131a}$$

$$a_{fi} = 1.38(M_{air}/M_i)^{5/12} \tag{2.131b}$$

$$a_{hi} = 1.30(M_{air}/M_i)^{3/12}[c_{pi}/(2.5\mathcal{R}/M_i)] \tag{2.131c}$$

Notice that $c_{pi}/(2.5\mathcal{R}/M_i)$ is unity for monatomic species. For the planar stagnation line and the flat plate, and other values of the temperature ratio T_s/T_e, the values of the species weighting factors are divided by the values given by Eqs. (2.131) to give correction factors G_{mi}, G_{fi}, and G_{hi}, respectively. These correction factors are listed in Table 2.1.

The exponential relation blowing factors cannot accurately represent some of the more anomalous effects of blowing. For example, when a light gas such as H_2 is injected, Eq. (2.130c) indicates that the effect of blowing is always to reduce heat transfer, due to both the low density and high specific heat of hydrogen. However, at very low injection rates, the heat transfer is actually increased, due to the high thermal conductivity of H_2. For a mixture, $k \simeq \sum x_i k_i$, whereas $c_p = \sum m_i c_{pi}$. At low rates of injection, the mole fraction of H_2 near the wall is much larger than its mass fraction; thus, there is a substantial increase in the mixture conductivity near the wall, but only a small change in the mixture specific heat. An increase in heat transfer results. At higher injection rates, the mass fraction of H_2 is also large, and the effect of high mixture specific heat dominates to cause a decrease in heat transfer.

Table 2.1 G correction factors for foreign gas injection into laminar air boundary layers.

Geometry	Species	G_{mi} T_s/T_e			G_{fi} T_s/T_e			G_{hi} T_s/T_e		
		0.1	0.5	0.9	0.1	0.5	0.9	0.1	0.5	0.9
Flat plate	H	1.14	1.36	1.47	1.30	1.64	1.79	1.15	1.32	—
	H_2	1.03	1.25	1.36	1.19	1.44	1.49	1.56	1.17	1.32
	He	1.05	1.18	1.25	1.34	1.49	1.56	1.18	1.32	—
	Air	—	—	—	1.21	1.27	1.27	1.17	1.21	—
	Xe	1.21	1.13	1.15	1.38	1.35	1.34	1.19	1.18	—
	CCl_4	1.03	0.95	1.00	1.00	1.03	1.03	1.04	1.04	—
Axisymmetric	H	1.00	1.04	1.09	1.00	0.62	0.45	1.00	0.94	0.54
stagnation point	H_2	1.00	1.06	1.06	1.00	0.70	0.62	1.00	1.00	1.01
	He	1.00	1.04	1.03	1.00	0.66	0.56	1.00	1.00	0.95
	C	1.00	1.01	1.00	1.00	0.79	0.69	1.00	0.99	0.87
	CH_4	1.00	1.01	1.00	1.00	0.88	0.84	1.00	1.00	1.00
	O	1.00	0.98	0.97	1.00	0.79	0.70	1.00	0.98	0.95
	H_2O	1.00	1.01	1.00	1.00	0.82	0.73	1.00	1.00	0.99
	Ne	1.00	1.00	0.98	1.00	0.83	0.75	1.00	0.97	0.95
	Air	—	—	—	1.00	0.87	0.82	1.00	0.99	0.97
	A	1.00	0.97	0.94	1.00	0.93	0.91	1.00	0.96	0.95
	CO_2	1.00	0.97	0.95	1.00	0.96	0.94	1.00	0.99	0.97
	Xe	1.00	0.98	0.96	1.00	0.96	1.05	1.00	1.06	0.99
	CCl_4	1.00	0.90	0.83	1.00	1.03	1.07	1.00	0.96	0.93
	I_2	1.00	0.91	0.85	1.00	1.02	1.05	1.00	0.97	0.94
Planar stagnation	He	0.96	0.98	0.98	0.85	0.53	0.47	0.93	0.91	0.92
line	Air	—	—	—	0.94	0.84	0.81	0.94	0.94	—
	Xe	0.92	0.87	0.83	0.90	0.93	0.95	0.93	0.93	—

Based on numerical data of Wortman [11]. Correlations developed by Dr. D. W. Hatfield.

2. *Turbulent boundary layers.* Here we restrict our attention to air flow along a flat plate for Mach numbers up to 6, and use numerical solutions of boundary layer equations with a mixing-length turbulence model [12, 13]. Appropriate species weighting factors for $0.2 < T_s/T_e < 2$ are:

$$a_{mi} = 0.79(M_{\text{air}}/M_i)^{1.33} \tag{2.132a}$$

$$a_{fi} = 0.91(M_{\text{air}}/M_i)^{0.76} \tag{2.132b}$$

$$a_{hi} = 0.86(M_{\text{air}}/M_i)^{0.73} \tag{2.132c}$$

In using Eq. (2.130), the limit values for $\dot{m}'' = 0$ are evaluated at the same location along the plate. Whether the injection rate is constant along the plate or varies as $x^{-0.2}$ to give a self-similar boundary layer has little effect on the blowing factors. Thus, Eq. (2.132) has quite general applicability. Notice that the effects of injectant molecular weight are greater for turbulent boundary layers than for laminar ones, and is due to the effect of fluid density on turbulent transport. Also, the injectant specific heat does not appear in a_{hi} as it did for laminar flows. In general, c_{pi} decreases with increasing M_i and is adequately accounted for in the molecular weight ratio.

Reference State Schemes

The reference state approach, in which constant-property solutions are used with properties evaluated at some reference state, is an alternative method for handling variable-property effects. In principle, the reference state is independent of the precise property data used and of the combination of injectant and free-stream species. A reference state approach for the flat plate laminar boundary layer will be given in Section 2.5.

Example 2.10   Transpiration Cooling

(i) Rework Example 2.7 for helium injection using the improved blowing factors.
(ii) Determine the helium concentration adjacent to the wall.

Solution

Given: Air flow at 1500 K over a transpiration-cooled flat plate.
Required: (i) He injection rate at $x = 0.2$ m to maintain $T_s = 500$ K with $T_o = 300$ K. (ii) The mass fraction $m_{\text{He},s}$.
Assumptions:

1. The boundary layer is laminar.

2. Equation (2.131) and Table 2.1 can be used for the blowing factors.

i. Combining Eqs. (2.126) and (2.130c) and solving for the injection rate gives

$$\dot{m}'' = \frac{h_c^*}{c_{p\text{air}}a_{h\text{He}}} \ln\left[1 + \left(\frac{a_{h\text{He}}c_{p\text{air}}}{c_{p\text{He}}}\right)\left(\frac{T_e - T_s}{T_s - T_o}\right)\right] \tag{1}$$

From Example 2.7, $h_c^* = 12.9$ W/m^2 K, $c_{pHe} = 5200$ J/kg K, and from Table A.7, $c_{pair} = 1130$ J/kg K at the mean film temperature of 1000 K. From Eq. (2.131c) and Table 2.1 for a flat plate and $T_s/T_e = 0.33$,

$$a_{hHe} = (1.26)(1.30)(M_{air}/M_{He})^{3/12}[c_{pHe}/(2.5\mathcal{R}/M_{He})]$$
$$= (1.26)(1.30)(29/4)^{3/12}[5200/(2.5)(8314/4)]$$
$$= (1.26)(1.30)(1.64)(1.00) = 2.69$$

Substituting in Eq. (1),

$$\dot{m}'' = \frac{12.9}{(1130)(2.69)} \ln\left[1 + \left(\frac{(2.69)(1130)}{5200}\right)\left(\frac{1500 - 500}{500 - 300}\right)\right]$$
$$= 5.80 \times 10^{-3} \text{ kg/m}^2 \text{ s}$$

ii. Equation (2.130a) can be rearranged as

$$\dot{m}'' = \left(\frac{g_{mHe}^*}{a_{mHe}}\right) \ln\left(1 + a_{mHe}\frac{m_{He,e} - m_{He,s}}{m_{He,s} - 1}\right) \qquad (2)$$

which must be solved for $m_{He,s}$. From Eq. (4.56) of *Heat Transfer*,

$$\text{St}_{mx}^* = \text{St}_x^*(\text{Sc}/\text{Pr})^{-2/3} \qquad (3)$$

For air at the mean film temperature of 1000 K, Table A.7 and Table A.17a give $\rho = 0.354$ kg/m^3, $\nu = 117.3 \times 10^{-6}$ m^2/s, Pr $= 0.7$, and $\mathcal{D}_{He,air} = 526 \times 10^{-6}$ m^2/s. Hence,

$$\text{Sc} = \nu/D = 117.3 \times 10^{-6}/526 \times 10^{-6} = 0.223$$

Substituting in Eq. (3) with $\text{St}_x^* = 3.23 \times 10^{-3}$ from Example 2.7,

$$\text{St}_{mx}^* = (3.23 \times 10^{-3})(0.223/0.70)^{-2/3} = 6.92 \times 10^{-3}$$
$$g_{mHe}^* = \rho u_e \text{St}_{mx}^* = (0.354)(10)(6.92 \times 10^{-3}) = 2.45 \times 10^{-2} \text{ kg/m}^2 \text{ s}$$

From Eq. (2.131a) and Table 2.1 for a flat plate and $T_s/T_e = 0.33$,

$$a_{mHe} = (1.13)(1.65)(M_{air}/M_{He})^{10/12}$$
$$= (1.13)(1.65)(29/4)^{10/12} = 9.72$$

Substituting in Eq. (2) with $m_{He,e} = 0$,

$$5.80 \times 10^{-3} = \left(\frac{2.45 \times 10^{-2}}{9.72}\right) \ln\left(1 + 9.72\frac{0 - m_{He,s}}{m_{He,s} - 1}\right)$$

Solving, $m_{He,s} = 0.480$.

Comments

1. The injection rate using the improved blowing factor is $(5.80 - 4.44)/4.44 =$ 31% greater than given by the Couette–flow model blowing factor.

2. Use of the Couette–flow model blowing factor for mass transfer to calculate $m_{He,s}$ will give a very poor result for helium injection (since it does not account for density variations). ▲

2.5 LAMINAR BOUNDARY LAYERS

In Section 5.4 of *Heat Transfer* momentum and heat transfer were analyzed for the forced-flow laminar boundary layer on a flat plate. In Section 2.5.1 we extend that analysis to mass transfer. We first examine solutions valid in the limit of zero mass transfer rate, for which the analysis is identical to that of Section 1.4 for heat transfer and yields analogous results. Next we turn to exact self-similar solutions and extend the analysis of Section 5.4.4 to account for blowing or suction at the wall, which allows us to obtain mass transfer solutions valid for high mass transfer rates. In Section 2.5.2 we derive an appropriate form of the energy conservation equation for mixtures and, upon assuming constant properties, obtain two forms of the equation, one in terms of temperature and the other in terms of enthalpy. The advantages of the enthalpy form for simultaneous heat and mass transfer are discussed, and self-similar solutions are obtained for heat transfer that are exactly analogous to those obtained in Section 2.5.1 for mass transfer.

2.5.1 Mass Transfer in Forced Flow along a Flat Plate

Governing Equations
Figure 2.24 shows a schematic of the laminar boundary layer for forced flow along a flat plate, with a constant free-stream velocity u_e and constant mass fraction of species $1, m_{1,e}$. For simplicity we assume steady flow, a binary inert mixture, and constant fluid properties. Appropriate forms of the mass and momentum conservation equations were derived in Section 5.4.1 of *Heat Transfer* as

$$\frac{\partial u}{\partial x} + \frac{\partial v}{\partial y} = 0 \tag{5.37}$$

$$u\frac{\partial u}{\partial x} + v\frac{\partial u}{\partial y} = \nu\frac{\partial^2 u}{\partial y^2} \tag{5.38}$$

The validity of Eq. (5.38) for flow of mixtures with species diffusion is discussed in Section 2.8. The species conservation equation for this flow can be obtained directly from the general species conservation equation derived in Section 2.2.3. However, we will parallel the analysis in Section 5.4.1 and derive the equation from first principles. Figure 2.25 shows an elemental control volume located in the boundary layer. Since there are no chemical reactions, the species conservation principle requires that the

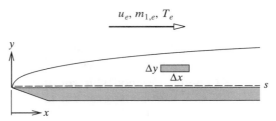

Figure 2.24 Schematic of a laminar boundary layer on a flat plate showing an elemental control volume $\Delta x \Delta y$.

net outflow of species 1 by convection and diffusion equal zero:

$$\rho u m_1 \Delta y|_{x+\Delta x} - \rho u m_1 \Delta y|_x + \rho v m_1 \Delta x|_{y+\Delta y} - \rho v m_1 \Delta x|_y$$

$$- \rho \mathcal{D}_{12} \frac{\partial m_1}{\partial y} \Delta x|_{y+\Delta y} + \rho \mathcal{D}_{12} \frac{\partial m_1}{\partial y} \Delta x|_y = 0$$

Dividing by $\rho \Delta x \Delta y$ and rearranging,

$$\frac{u m_1|_{x+\Delta x} - u m_1|_x}{\Delta x} + \frac{v m_1|_{y+\Delta y} - v m_1|_y}{\Delta y} = \mathcal{D}_{12} \left[\frac{(\partial m_1/\partial y)_{y+\Delta y} - (\partial m_1/\partial y)_y}{\Delta y} \right]$$

Letting $\Delta x, \Delta y \to 0$,

$$\frac{\partial}{\partial x}(u m_1) + \frac{\partial}{\partial y}(v m_1) = \mathcal{D}_{12} \frac{\partial^2 m_1}{\partial y^2} \tag{2.133}$$

or

$$m_1 \frac{\partial u}{\partial x} + u \frac{\partial m_1}{\partial x} + m_1 \frac{\partial v}{\partial y} + v \frac{\partial m_1}{\partial y} = \mathcal{D}_{12} \frac{\partial^2 m_1}{\partial y^2} \tag{2.134}$$

Multiplying the continuity equation, Eq. (5.37), by m_1 and subtracting from Eq. (2.134) gives the desired form of the species conservation equation:

$$u \frac{\partial m_1}{\partial x} + v \frac{\partial m_1}{\partial y} = \mathcal{D}_{12} \frac{\partial^2 m_1}{\partial y^2} \tag{2.135}$$

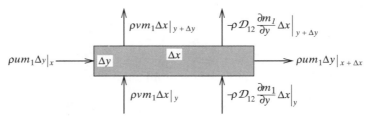

Figure 2.25 Application of the species conservation principle to an elemental control volume in a laminar boundary layer on a flat plate.

In deriving Eq. (2.135), diffusion in the x direction, $-\rho \mathcal{D}_{12}\partial m_1/\partial x$, has been ignored, since, like conduction $-k\partial T/\partial x$ and the shear stress $-\mu \partial u/\partial x$, it is negligible in a thin boundary layer. Scaling arguments to justify this assumption are given in Section 5.8 of *Heat Transfer*. Notice that Eq. (2.135) is of analogous form to the energy equation for low-speed flow of a constant-property fluid, Eq. (5.42), with m_1 corresponding to T and $\mathcal{D}_{12}$ to α. Appropriate boundary conditions for Eq. (2.135) are

$$y = 0: \; m_1 = m_{1,s} \tag{2.136a}$$

$$y \to \infty, x = 0: \; m_1 = m_{1,e} \tag{2.136b}$$

The integral form of the species convection equation can be derived as was done in Section 5.4.3 for the energy equation. But with the analogy between heat and mass transfer clearly established for the differential form of the equation, we will simply write down the result. From Eqs. (5.61) through (5.63),

$$\frac{d\Delta_{m2}}{dx} = \text{St}_{mx} \tag{2.137}$$

where Δ_{m2} is the convection thickness, defined as

$$\Delta_{m2} = \int_0^\infty \left(\frac{u}{u_e}\right)\left(\frac{m_1 - m_{1,e}}{m_{1,s} - m_{1,e}}\right)dy \tag{2.138}$$

and St_{mx} is the local mass transfer Stanton number,

$$\text{St}_{mx} = \frac{\mathcal{g}_{mx}}{\rho u_e} = \frac{-\rho \mathcal{D}_{12}(\partial m_1/\partial y)|_0}{\rho u_e(m_{1,s} - m_{1,e})} \tag{2.139}$$

Low Mass Transfer Rate Solutions

We can often obtain the solution of a mass transfer problem from the solution of an analogous heat transfer problem. In the limit of zero mass transfer rate, we have a simple form of the analogy. The heat transfer solutions obtained in Section 5.4 were for an impermeable wall—that is, $v_s = 0$. In general, the flux of species 1 across the s-surface has both convective and diffusive components,

$$n_{1,s} = m_{1,s}\rho v_s + j_{1,s} \tag{2.140}$$

If we set $v_s = 0$, we obtain $n_{1,s} = j_{1,s} = -\rho \mathcal{D}_{12}(\partial m_1/\partial y)|_0$, which is the basis of low mass transfer rate theory. Thus, to obtain low mass transfer rate solutions, we simply replace the Prandtl number by the Schmidt number and the Nusselt number by the Sherwood number in the corresponding heat transfer solution (as we did for correlations in Section 1.4). The plug flow model and unheated starting length solutions obtained in Section 5.4 then become, from Eq. (5.50),

$$\text{Sh}_x = 0.564\text{Re}_x^{1/2}\text{Sc}^{1/2}(= 0.564\text{Pe}_{mx}^{1/2}) \tag{2.141}$$

and from Eq. (5.68),

$$\text{Sh}_x = \frac{0.331\text{Re}_x^{1/2}\text{Sc}^{1/3}}{[1 - (\xi/x)^{3/4}]^{1/3}} \tag{2.142}$$

The plug flow model is not of much practical use in mass transfer because it is accurate only for very small values of the Schmidt number, and values less than about 0.2 are seldom encountered (see Table 1.2). In the case of heat transfer, liquid metals are characterized by very small Prandtl numbers, and the plug flow model has proven useful. Equation (2.142) applies to discontinuous mass transfer, such as a plate that is bare for $0 < x < \xi$, and coated with naphthalene for $x > \xi$. The plate would also have to be isothermal to have $m_{1,s}$ constant for $x > \xi$. However, discontinuous mass transfer is more often encountered in applications such as transpiration cooling, for which mass transfer (injection) rates tend to be very high and low mass transfer rate theory is invalid.

Self-Similar Solutions

Self-similar solutions of the flat plate laminar boundary layer energy equation were obtained in Section 5.4.4 of *Heat Transfer*. These solutions were obtained for an impermeable wall—that is, $v_s = 0$; thus, the analogous mass transfer solutions are valid for low mass transfer rates. We will now obtain self-similar solutions for $v_s \neq 0$ that will be appropriate for high mass transfer rates. The forms of the governing equations to be solved are the Blasius equation, Eq. (5.81), and the mass transfer analog to Eq. (5.91), namely,

$$f''' + ff'' = 0 \tag{5.81}$$

$$\phi'' + \mathrm{Sc}\,f\phi' = 0; \quad \phi = \frac{m_1 - m_{1,e}}{m_{1,s} - m_{1,e}} \tag{2.143}$$

Primes denote differentiation with respect to the similarity variable $\eta = y(u_e/2\nu x)^{1/2}$, f is the dimensionless stream function, and $f' = u/u_e$ is the dimensionless streamwise velocity. Appropriate boundary conditions are

$$\eta = 0: \quad f' = 0, \quad \phi = 1 \tag{2.144a}$$

$$\eta \to \infty: \quad f' = 1, \quad \phi = 0 \tag{2.144b}$$

Notice that $m_{1,s}$ = constant is required to obtain Eq. (2.143), in the same way T_s = constant was required to obtain Eq. (5.91). If $m_{1,s}$ is not constant, self-similar solutions cannot be obtained.

One further boundary condition is required. Substituting $f'(0) = 0$ in Eq. (5.77) gives

$$v_s = -(u_e\nu/2x)^{1/2}f(0) \tag{2.145}$$

In Section 5.4.4 we set $v_s = 0$ for an impermeable wall, requiring $f(0) = 0$. For finite mass transfer rates we require solutions for $f(0) \neq 0$. The last boundary condition is then

$$\eta = 0: \quad f = \text{Constant} \tag{2.146}$$

where a constant value is required to give f as a function of η only, which is the condition for self-similar solutions. From Eq. (2.145), $v_s \propto x^{-1/2}$ is required for self-similar solutions. However, we shall see later that this condition is not as restrictive as it might first appear. Solution of the Blasius equation for $f(0) \neq 0$ is straightforward and can be effected by the iteration procedure given in Section 5.4.4. For a given value of $f(0)$, the profile $f(\eta)$ is obtained and used in the integration of Eq. (2.143) to give the profile $\phi(\eta)$ and $\phi'(0)$ for a given value of Sc. The desired result is $\phi'(0)$ as a function of $f(0)$ with Sc as a parameter. The mass transfer Stanton number is obtained from

$$\rho u_e \mathrm{St}_{mx} = g_{mx} = \frac{-\rho \mathcal{D}_{12}(\partial m_1/\partial y)|_0}{m_{1,s} - m_{1,e}} = -\rho \mathcal{D}_{12}\phi'(0)(u_e/2\nu x)^{1/2}$$

Rearranging gives the local Stanton and Sherwood numbers:

$$\mathrm{St}_{mx} = \frac{-\phi'(0)}{2^{1/2}\mathrm{Re}_x^{1/2}\mathrm{Sc}} \tag{2.147a}$$

$$\mathrm{Sh}_x = \mathrm{Re}_x\mathrm{Sc}\,\mathrm{St}_{mx} = \frac{-\phi'(0)\mathrm{Re}_x^{1/2}}{2^{1/2}} \tag{2.147b}$$

Of course, Sh_x could have been obtained by analogy from Eq. (5.96).

Figure 2.26 shows dimensionless velocity profiles $f'(\eta)$ for various values of $f(0)$. Negative values of $f(0)$ correspond to blowing (positive mass transfer rates). At $-f(0) = 0.875$, $f''(0) = 0$, indicating a wall shear stress of zero: the boundary layer is then said to have been *blown off* the wall. The boundary layer equations become invalid at boundary layer blow-off. Positive values of $f(0)$ correspond to suction (negative mass transfer rates). As $f(0)$ increases, the boundary layer thins, and $f''(0)$ and the corresponding wall shear stress increase. The boundary layer equations are valid no matter how large $f(0)$ becomes. Figure 2.27 shows concentration profiles $\phi(\eta)$ for various values of $f(0)$, with $\mathrm{Sc} = 0.6$ and $\mathrm{Sc} = 10$. The effects of blowing and suction on those profiles are similar to those for the velocity profile. The effects of blowing to decrease $\phi'(0)$ and suction to increase $\phi'(0)$ are clearly seen. However, for $\mathrm{Sc} = 10$ the effect of blowing in reducing $\phi'(0)$ is more pronounced than for $\mathrm{Sc} = 0.6$.

As was done in the Couette–flow analysis of Section 2.4.1, it proves convenient to normalize the mass transfer conductance by its value in the limit of zero mass transfer in order to obtain the mass transfer from a relation of the form $\dot{m}'' = g_m^*(g_m/g_m^*)\mathcal{B}_m$ [cf. Eq. (2.99)].

$$\frac{g_{mx}}{g_{mx}^*} = \frac{\mathrm{St}_{mx}}{\mathrm{St}_{mx}^*} = \frac{\phi'(0)}{\phi'(0)|_{f(0)=0}} \tag{2.148}$$

where St_{mx}^* is obtained from the analog of Eq. (5.97) as

$$\mathrm{St}_{mx}^* = 0.332\mathrm{Re}_x^{-1/2}\mathrm{Sc}^{-2/3}; \quad \mathrm{Sc} > 0.5 \tag{2.149}$$

The ratio g_{mx}/g_{mx}^* is then a function of Sc and $f(0)$, and from Eq. (2.145) with $\rho v_s = \dot{m}''$,

$$f(0) = -\frac{\dot{m}''}{\rho u_e} 2^{1/2}\,\mathrm{Re}_x^{1/2} \tag{2.150}$$

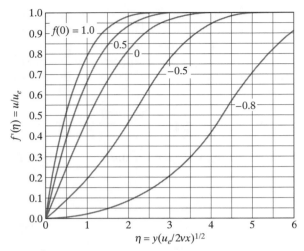

Figure 2.26 Effect of mass transfer on self-similar velocity profiles for a laminar boundary layer on a flat plate: $f(0) = -(\dot{m}''/\rho u_e)(2\mathrm{Re}_x)^{1/2}$.

Figure 2.28 shows g_{mx}/g_{mx}^* plotted versus $f(0)$ with Sc as a parameter. The effect of Sc in this plot is large, and for blowing the curve shapes differ for Sc < 1 versus Sc > 1. A plot of the form of Fig. 2.28 is useful when the mass transfer rate is known at the relevant stage of a calculation—for example, in the case of transpiration cooling

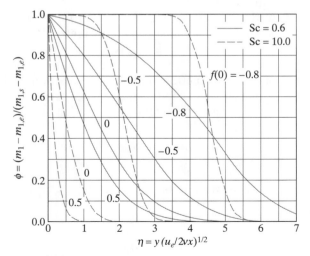

Figure 2.27 Effect of mass transfer on self-similar concentration profiles for a laminar boundary layer on a flat plate for Schmidt numbers equal to 0.6 and 10.0: $f(0) = -(\dot{m}''/\rho u_e)(2\mathrm{Re}_x)^{1/2}$.

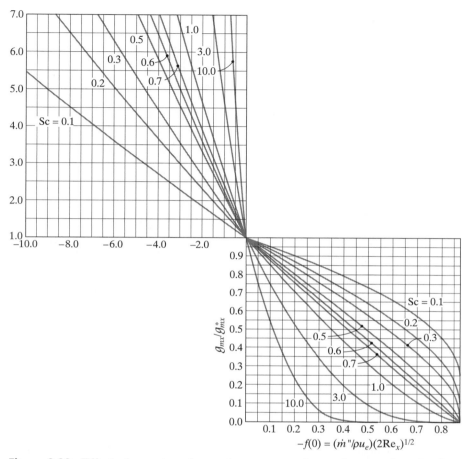

Figure 2.28 Effect of mass transfer on the mass transfer conductance for a laminar
boundary layer on a flat plate: g_{mx}/g_{mx}^* versus dimensionless blowing rate
$-f(0) = (\dot{m}''/\rho u_e)(2\mathrm{Re}_x)^{1/2}$.

with a given injection rate. With $\dot{m}''$ known, $f(0)$ can be calculated from Eq. (2.150)
and Fig. 2.28 used to obtain g_{mx}/g_{mx}^*. Plots similar to Fig. 2.28 are presented in
the pioneering paper of Hartnett and Eckert [14]. An equivalent but different plot
is shown in Fig. 2.29, where the abscissa is the blowing parameter $B_m = \dot{m}''/g_m^*$;
substituting from Eqs. (2.147) into Eq. (2.150) and rearranging gives

$$B_m = \frac{\mathrm{Sc}\, f(0)}{\phi'(0)|_{f(0)=0}} \tag{2.151}$$

The abscissa in Fig. 2.29 is a function of Schmidt number, and the effect of Sc is much
less in Fig. 2.29 than in Fig. 2.28, especially for suction.

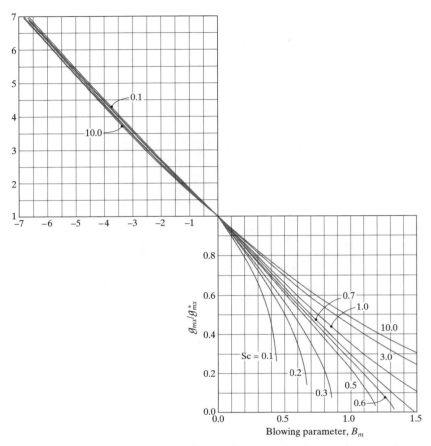

Figure 2.29 Effect of mass transfer on the mass transfer conductance for a laminar boundary layer on a flat plate: g_{mx}/g^*_{mx} versus blowing parameter $B_m = \dot{m}''/g^*_{mx}$.

For problems involving a liquid, species 1, evaporating into a gas, species 2—for example, sweat cooling—the surface temperature T_s might be specified; hence $m_{1,s} = m_{1,s}(T_s, P)$ would be known and the evaporation rate $\dot{m}''$ unknown. Hartnett and Eckert [14] suggest an iterative procedure for such problems, making use of a plot like Fig. 2.28. However, such an iterative procedure can be performed once and for all and the results presented in a convenient form. We proceed by making the usual assumption that the gas is insoluble in the liquid; then

$$n_{2,s} = \rho v_s m_{2,s} - \rho \mathcal{D}_{12} \frac{\partial m_2}{\partial y}\bigg|_0 = 0$$

Substituting $m_2 = 1 - m_1$, $\partial m_2/\partial y = -\partial m_1/\partial y$ gives

$$v_s = \frac{-\mathcal{D}_{12}(\partial m_1/\partial y)|_0}{1 - m_{1,s}} \tag{2.152}$$

From Eq. (2.145), $v_s = -(u_e\nu/2x)^{1/2}f(0)$; from Eq. (5.73b), $\partial/\partial y = (u_e/2\nu x)^{1/2}$ $\partial/\partial\eta$, and from the definition of ϕ, $\partial m_1/\partial y = (m_{1,s} - m_{1,e})\,\partial\phi/\partial y$. Substituting in Eq. (2.152) and rearranging,

$$f(0) = \mathcal{B}_m\frac{\phi'(0)}{\mathrm{Sc}}; \quad \mathcal{B}_m = \frac{m_{1,e} - m_{1,s}}{m_{1,s} - 1} \tag{2.153}$$

By combining Eqs. (2.153), (2.150), and (2.147a), we recover $\dot{m}'' = g_m\mathcal{B}_m$, which was given as Eq. (2.93). The mass transfer driving force $\mathcal{B}_m$ arises naturally in formulating the relation that couples the two governing differential equations, Eqs. (5.81) and (2.143). To every solution $\phi'(0)$ for given $f(0)$ and Sc we can calculate a corresponding value of $\mathcal{B}_m$ from Eq. (2.153) and prepare a plot of g_{mx}/g_{mx}^* versus $\mathcal{B}_m$ with Sc as a parameter, as shown in Fig. 2.30. In problem solving with $m_{1,s}$, and hence $\mathcal{B}_m$, known, Fig. 2.30 gives g_{mx}/g_{mx}^* directly. Then the evaporation rate is obtained from $\dot{m}'' = g_{mx}^*(g_{mx}/g_{mx}^*)\mathcal{B}_m$.

Notice that the effect of mass transfer on the wall shear stress can also be obtained from Fig. 2.28. For a Schmidt number of unity, the momentum and species equations,

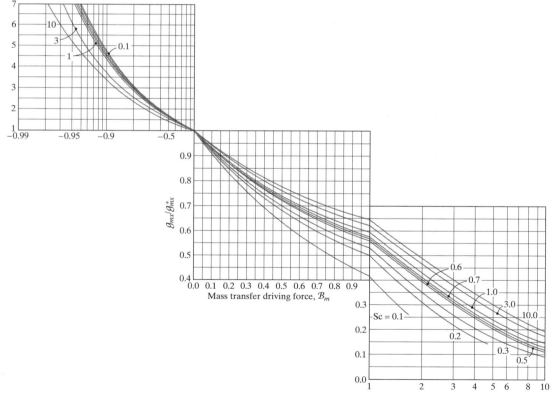

Figure 2.30 Effect of mass transfer on the mass transfer conductance for a laminar boundary layer on a flat plate: g_{mx}/g_{mx}^* versus mass transfer driving force $\mathcal{B}_m = (m_{1,e} - m_{1,s})/(m_{1,s} - 1) = \dot{m}''/g_{mx}$.

Eqs. (5.81) and (2.143), are of identical form with f' corresponding to ϕ. Thus, the curve for $Sc = 1$ in Fig. 2.28 also gives $f''(0)/f''(0)|_{f(0)=0} = \tau_s/\tau_s^*$.

Evaluation of Properties

Examples 2.11 and 2.12 illustrate the practical use of the two types of plot for g_{mx}/g_{mx}^*. However, the implications of the constant-property assumption requires careful consideration. Example 2.11 considers a situation where the properties are essentially constant, so that the solution procedure is exact. But such situations are the exception rather than the rule in engineering practice. Indeed, we recognized this fact when introducing improved blowing factors for use with the Couette–flow model in Section 2.4.2. If we wish to use constant-property solutions for engineering problem solving, properties must be evaluated at an appropriate reference state. Simple reference state schemes involve use of the mean film composition and temperature (a "1/2 rule"), or the 1/3 rule of Hubbard et al., given as Eqs. (2.65). Knuth [15, 16] developed a reference state scheme based on a Couette–flow model of laminar and turbulent flows, in which thermodynamic properties were allowed to vary with composition but transport properties were held constant. For species 1 injected into species 2, the reference mass fraction and temperature are

$$
m_{1,r} = 1 - \frac{M_2}{M_2 - M_1} \frac{\ln(M_e/M_s)}{\ln(m_{2,e}M_e/m_{2,s}M_s)} \tag{2.154}
$$

$$
T_r = 0.5(T_e + T_s) + 0.2r^*(u_e^2/2c_{pr})
$$

$$
+ 0.1\left[B_{hr} + (B_{hr} + B_{mr})\frac{c_{p1} - c_{pr}}{c_{pr}}\right](T_s - T_e) \tag{2.155}
$$

where r^* is the recovery factor for an impermeable wall [see Section 5.2.2 of *Heat Transfer* and cf. Eq. (5.13)]. Use of this scheme requires the calculation of transport properties of gas mixtures. A method based on the kinetic theory of gases is given in Section 1.7. Based on the limited numerical data available at that time, this reference state was shown by Knuth to work well for the laminar boundary layer on a flat plate. Unfortunately, Knuth's scheme is difficult to implement in a hand calculation. Knuth's scheme usually gives values that are between those given by the 1/2 and 1/3 rules. Example 2.12 illustrates use of the 1/2 rule for a situation where property variations are large; Exercises 2–47 and 2–48 require Knuth's scheme to be used. It is also worthwhile to note that some workers have combined the reference state approach for like gas injection, with molecular-weight-dependent empirical factors to account for foreign gas injection. Reasonably good correlations of numerical data for laminar boundary layers have been obtained in this manner [17, 18, 19].

An issue related to variable-property effects concerns the Schmidt-number effect seen in Figs. 2.28 through 2.30. This effect does not have much physical significance. Both Schmidt-number and variable-property effects depend on the combination of species under consideration. For gases, variable properties generally have a larger influence on g_{mx}/g_{mx}^* than does the Schmidt number, as can be seen by comparing Figs. 2.22a and 2.29.

Example 2.11   Diffusion of Nitric Oxide in an Air Boundary Layer

Air at 300 K and 1 atm pressure flows at 10 m/s along a porous flat plate, through which air containing 1% NO by mass is injected at a velocity of 3.53×10^{-3} $(x)^{-1/2}$ m/s. At 0.2 m from the leading edge, calculate the mass transfer conductance for NO, and the NO concentration at the s-surface. Take $\mathrm{Re_{tr}} = 200{,}000$.

Solution

Given: Air containing a small amount of NO injected into an air boundary layer on a flat plate.

Required: At $x = 0.2$ m, g_{mx} for NO, and $m_{\mathrm{NO},s}$.

Assumptions:

1. Constant fluid properties.
2. $\mathrm{Re_{tr}} = 200{,}000$

Since NO is in small concentration and its molecular weight (30) is close to that of air (29), the assumption of constant properties is valid. Air properties at 300 K and 1 atm from Table A.7 are $\rho = 1.177$ kg/m^3 and $\nu = 15.66 \times 10^{-6}$ m^2/s; from Table A.17a, $\mathcal{D}_{\mathrm{NO\ air}} = 0.180 \times 10^{-4}$ m^2/s. The Reynolds and Schmidt numbers are

$$\mathrm{Re}_x = u_e x/\nu = (10)(0.2)/(15.66 \times 10^{-6}) = 127{,}700 < 200{,}000, \text{ laminar}$$

$$\mathrm{Sc} = \nu/\mathcal{D}_{12} = (15.66 \times 10^{-6})/(0.180 \times 10^{-4}) = 0.870$$

The zero-mass-transfer-limit Stanton number is obtained from Eq. (2.149) as

$$\mathrm{St}^*_{mx} = 0.332 \mathrm{Re}_x^{-1/2} \mathrm{Sc}^{-2/3} = (0.332)(127{,}700)^{-1/2}(0.870)^{-2/3} = 1.02 \times 10^{-3}$$

and the zero-mass-transfer-limit conductance is

$$g^*_{mx} = \rho u_e \mathrm{St}^*_{mx} = (1.177)(10)(1.02 \times 10^{-3}) = 1.20 \times 10^{-2} \text{ kg/m}^2 \text{ s}$$

In order to obtain the actual Stanton number from Fig. 2.28, we require $f(0)$; from Eq. (2.150),

$$f(0) = -\frac{\dot{m}''}{\rho u_e} 2^{1/2} \mathrm{Re}_x^{1/2} = -\left(\frac{v_s}{u_e}\right)(2\mathrm{Re}_x)^{1/2}$$

$$v_s = 3.53 \times 10^{-3}(x)^{-1/2} = (3.53 \times 10^{-3})(0.2)^{-1/2} = 7.89 \times 10^{-3} \text{ m/s}$$

$$f(0) = -(7.89 \times 10^{-3}/10)(2 \times 127{,}700)^{1/2} = -0.399$$

From Fig. 2.28, for $\mathrm{Sc} = 0.87$, $f(0) = -0.40$, we obtain $g_{mx}/g^*_{mx} = 0.47$. Thus,

$$g_{mx} = (g_{mx}/g^*_{mx})g^*_{mx} = (0.47)(1.20 \times 10^{-2}) = 5.64 \times 10^{-3} \text{ kg/m}^2 \text{ s}$$

To obtain the mass fraction of NO at the s-surface, we proceed as follows. Denoting NO as species 1, and using Eq. (2.140),

$$n_{1,s} = m_{1,s}\rho v_s + j_{1,s} = m_{1,s}\dot{m}'' + g_{m1}(m_{1,s} - m_{1,e})$$

$$\dot{m}'' = \rho v_s = (1.177)(7.89 \times 10^{-3}) = 9.29 \times 10^{-3} \text{ kg/m}^2 \text{s}$$

Since the injected air contains 1% NO by mass, the absolute flux of NO across the s-surface is

$$n_{1,s} = n_{1,u} = (0.01)(\dot{m}'') = (0.01)(9.29 \times 10^{-3}) = 9.29 \times 10^{-5} \text{ kg/m}^2 \text{s}$$

Substituting above,

$$9.29 \times 10^{-5} = 9.29 \times 10^{-3} m_{1,s} + (5.64 \times 10^{-3})(m_{1,s} - 0)$$

Solving, $m_{1,s} = 6.22 \times 10^{-3}$ (0.622%).

Comments

1. This example is not an engineering problem: it was constructed to illustrate some important aspects of mass transfer in a laminar boundary layer.
2. Notice that v_s proportional to $x^{-1/2}$ gives a constant value of $f(0)$, as required for a self-similar solution. Also, $m_{1,s}$ is constant along the plate as a result.
3. This is a problem in which *two* species are transferred; nearly all the problems we have considered up to now involved transfer of one species only. We see that, through careful use of Eq. (2.140), handling of more than one transferred species is straightforward. Section 2.6 shows how the mass transfer driving force $\mathcal{B}_m$ defined by Eq. (2.93) can be generalized to situations where more than one species is transferred (see also Exercise 2–28).   ▲

Example 2.12   Sweat Cooling of a Flat Plate: Water Supply Rate

A flat plate exposed to intense radiative and convective heating is to be sweat cooled with water. Dry air at 840 K and 1 atm flows at 10 m/s along the plate as a laminar boundary layer. If the plate surface is to be maintained at 360 K, calculate the water supply rate 0.2 m from the leading edge (i) using the improved blowing factors of Section 2.4.2, and (ii) using the constant-property self-similar solution and the 1/2 rule for evaluating properties.

Solution

Given: Air flowing along a radiatively heated, sweat-cooled flat plate.
Required: The water evaporation rate 0.2 m from the leading edge by two methods.
Assumptions: The thermal resistance of the thin film of water on the plate is negligible.

i. *Using the improved blowing factors.* From Eqs. (2.93) and (2.130a),

$$\dot{m}'' = g_{mx}\mathcal{B}_m; \quad \frac{g_m}{g_m^*} = \frac{\ln(1 + a_{m1}\mathcal{B}_m)}{a_{m1}\mathcal{B}_m}$$

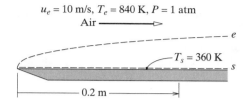

We first evaluate $\mathcal{B}_m$. From PSYCHRO, at $T = 360$ K, $P = 1.013 \times 10^5$ Pa, $m_{1,s} = m_{1,\text{sat}} = 0.496$. Thus,

$$\mathcal{B}_m = \frac{m_{1,e} - m_{1,s}}{m_{1,s} - 1} = \frac{0 - 0.496}{0.496 - 1} = 0.984$$

The reference temperature for evaluating g_{mx}^* is the mean film temperature, $T_r = 360 + (1/2)(840 - 360) = 600$ K. From Table A.7 for dry air, $\rho = 0.589$ kg/m^3, $\mu = 29.74 \times 10^{-6}$ kg/m s. Also, from Table A.17a, $\mathcal{D}_{12} = 92.8 \times 10^{-6}$ m^2/s, and then Sc = 0.54. The Reynolds number is

$$\text{Re}_x = u_e \rho x / \mu = (10)(0.589)(0.2)/29.74 \times 10^{-6} = 3.961 \times 10^4$$

From Eq. (2.149), the local zero-mass-transfer-limit Stanton number and conductance are

$$\text{St}_{mx}^* = 0.332\text{Re}_x^{-1/2}\text{Sc}^{-2/3} = (0.332)(3.961 \times 10^4)^{-1/2}(0.54)^{-2/3}$$
$$= 2.52 \times 10^{-3}$$

$$g_{mx}^* = \rho u_e \text{St}_{mx}^* = (0.589)(10)(2.52 \times 10^{-3})$$
$$= 1.48 \times 10^{-2} \text{ kg/m}^2 \text{ s}$$

Using Eq. (2.131a),

$$a_{m1} = 1.65(M_{\text{air}}/M_1)^{10/12} G_{m1}$$

Table 2.1 does not contain G factors for H_2O injection on a flat plate. However, comparing the factors for a flat plate and axisymmetric stagnation point, and interpolating between $T_s/T_e = 0.1$ and 0.5, a rough estimate of the G factor is

$$G_{m1} \simeq (1.00)(1.1) = 1.1$$

$$a_{m1} \simeq (1.65)(29/18)^{10/12}(1.1) = 2.7$$

$$\frac{g_{mx}}{g_{mx}^*} = \frac{\ln(1 + 2.7 \times 0.984)}{2.7 \times 0.984} = 0.488$$

$$\dot{m}'' = g_{mx}^*(g_{mx}/g_{mx}^*)\mathcal{B}_m = (1.48 \times 10^{-2})(0.488)(0.984)$$
$$= 7.11 \times 10^{-3} \text{ kg/m}^2 \text{ s}$$

ii. *Using the self-similar solution and the 1/2 rule.* The reference state for calculating g_{mx}^* is $m_{1,r} = (1/2)(0.496 + 0) = 0.248$, $T_r = 360 + (1/2)(840 - 360) = 600$ K. Using AIRSTE, $\rho = 0.512$ kg/m^3, $\mu = 2.71 \times 10^{-5}$ kg/m s, $Sc = 0.564$. The Reynolds number, local Stanton number, and conductance are

$$\text{Re}_x = u_e \rho x / \mu = (10)(0.512)(0.2)/2.71 \times 10^{-5}$$
$$= 3.78 \times 10^4$$
$$\text{St}_{mx}^* = 0.332 \text{Re}_x^{-1/2} Sc^{-2/3} = (0.332)(3.78 \times 10^4)^{-1/2}(0.564)^{-2/3}$$
$$= 2.50 \times 10^{-3}$$
$$g_{mx}^* = \rho u_e \text{St}_m^* = (0.512)(10)(2.50 \times 10^{-3})$$
$$= 1.28 \times 10^{-2} \text{ kg/m}^2 \text{ s}$$

From Fig. 2.30 for $Sc = 0.564$ and $\mathcal{B}_m = 0.984$, $g_{mx}/g_{mx}^* = 0.57$. Thus,

$$\dot{m}'' = g_{mx}^*(g_{mx}/g_{mx}^*)\mathcal{B}_m = (1.28 \times 10^{-2})(0.57)(0.984)$$
$$= 7.18 \times 10^{-3} \text{ kg/m}^2 \text{ s}$$

Comments

1. The difference between the two results for the evaporation rate is 1.0%, which is remarkably small considering that we made approximations in each case (G_{m1} was an estimate, and the 1/2 rule is not exact).

2. For a given water reservoir temperature T_o, there is a unique radiative heat flux q_{rad} consistent with a surface temperature of 360 K (see Example 2.13). ▲

2.5.2 Simultaneous Heat and Mass Transfer in Forced Flow along a Flat Plate

We commence by deriving the boundary layer form of the energy conservation equation for a mixture. We assume steady, low-speed, laminar flow and follow the approach used in Section 2.3.5. Referring to Fig. 2.31, the steady-flow energy equation, Eq. (1.4) of *Heat Transfer*, requires that the net outflow of enthalpy equal the heat conducted into the volume:

$$\sum n_{ix} h_i \Delta y|_{x+\Delta x} - \sum n_{ix} h_i \Delta y|_x + \sum n_{iy} h_i \Delta x|_{y+\Delta y} - \sum n_{iy} h_i \Delta x|_y$$
$$= -k \frac{\partial T}{\partial y} \Delta x|_y + k \frac{\partial T}{\partial y} \Delta x|_{y+\Delta y}$$

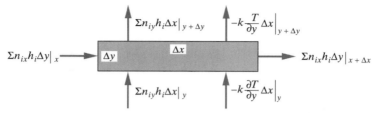

Figure 2.31   Application of the energy conservation principle to an elemental control volume in a flat-plate laminar boundary layer flow of a mixture.

Dividing by $\Delta x \Delta y$ and rearranging,

$$\frac{\sum n_{ix}h_i|_{x+\Delta x} - \sum n_{ix}h_i|_x}{\Delta x} + \frac{\sum n_{iy}h_i|_{y+\Delta y} - \sum n_{iy}h_i|_y}{\Delta y}$$

$$= \frac{(k\partial T/\partial y)_{y+\Delta y} - (k\partial T/\partial y)_y}{\Delta y}$$

Letting $\Delta x, \Delta y \to 0$,

$$\frac{\partial}{\partial x}\left(\sum n_{ix}h_i\right) + \frac{\partial}{\partial y}\left(\sum n_{iy}h_i\right) = \frac{\partial}{\partial y}\left(k\frac{\partial T}{\partial y}\right) \qquad (2.156)$$

In Eq. (2.156), k is the *mixture* conductivity. As in the derivation of Eq. (5.42), conduction in the x-direction has been ignored because it is negligible for a thin boundary layer; also, for this low-speed flow along a flat plate, there are no terms that account for viscous dissipation and compression heating.

Equation (2.156) is further rearranged as follows. By definition, $n_{ix} = m_i\rho u + j_{ix}$; hence,

$$\frac{\partial}{\partial x}\left(\sum n_{ix}h_i\right) = \frac{\partial}{\partial x}\sum(m_i\rho u + j_{ix})h_i$$

$$= \frac{\partial}{\partial x}\left(\rho u \sum m_i h_i + \sum j_{ix}h_i\right)$$

$$\frac{\partial}{\partial x}\left(\sum n_{ix}h_i\right) \simeq \frac{\partial}{\partial x}(\rho u h) \qquad (2.157a)$$

since $\sum m_i h_i = h$, the mixture enthalpy, and the streamwise diffusion fluxes j_{ix} are negligible in a thin boundary layer. Similarly,

$$\frac{\partial}{\partial y}\left(\sum n_{iy}h_i\right) = \frac{\partial}{\partial y}\left(\rho v h + \sum j_{iy}h_i\right) \qquad (2.157b)$$

The second term on the right-hand side of Eq. (2.157b) is not negligible, since it involves diffusion fluxes across the boundary layer. Substituting back in Eq. (2.156) and rearranging,

$$\frac{\partial}{\partial x}(\rho u h) + \frac{\partial}{\partial y}(\rho v h) = \frac{\partial}{\partial y}\left[k\frac{\partial T}{\partial y} - \sum j_{iy}h_i\right] \qquad (2.158)$$

Expanding the left-hand side of Eq. (2.158) and using the continuity equation in the usual manner gives

$$\rho u \frac{\partial h}{\partial x} + \rho v \frac{\partial h}{\partial y} = \frac{\partial}{\partial y}\left[k\frac{\partial T}{\partial y} - \sum j_{iy}h_i \right] \qquad (2.159)$$

The second term on the right-hand side of Eq. (2.159) is called the *interdiffusion* contribution to energy transport in a mixture, because it accounts for enthalpy transport due to interdiffusion of chemical species in the mixture.[5] For a binary mixture this term becomes

$$-\sum j_{iy}h_i = \rho \mathcal{D}_{12}\frac{\partial m_1}{\partial y}h_1 + \rho \mathcal{D}_{12}\frac{\partial m_2}{\partial y}h_2 = \rho \mathcal{D}_{12}(h_1 - h_2)\frac{\partial m_1}{\partial y} \qquad (2.160)$$

since $\partial m_2/\partial y = -\partial m_1/\partial y$. If $h_1 = h_2$, the interdiffusion term is seen to be zero.

Within the limitation of the constant properties assumption, there are two possible approaches to a heat transfer analysis.

Approach 1: Temperature formulation. We assume an inert mixture with all species specific heats equal, $c_{p1} = c_{p2} = \cdots = c_p$. Then, using the same enthalpy datum state for all species gives $\sum j_{iy}h_i = h_i \sum j_{iy} = 0$, $\partial h = c_p \partial T$, and Eq. (2.159) becomes

$$\rho u c_p \frac{\partial T}{\partial x} + \rho v c_p \frac{\partial T}{\partial y} = \frac{\partial}{\partial y}\left(k\frac{\partial T}{\partial y} \right)$$

Finally, the mixture conductivity k is assumed constant to recover Eq. (5.42),

$$u \frac{\partial T}{\partial x} + v \frac{\partial T}{\partial y} = \alpha \frac{\partial^2 T}{\partial y^2} \qquad (5.42)$$

which is the energy equation for a constant-property pure fluid. The interdiffusion term in the energy equation for a mixture has been eliminated by assuming an inert mixture with equal species specific heats. Boundary conditions for Eq. (5.42) are

$$y = 0: \ T = T_s \qquad (5.88a)$$

$$y \to \infty, x = 0: \ T = T_e \qquad (5.88b)$$

and T_s, T_e are constants for a self-similar solution. As noted before, Eqs. (5.42) of *Heat Transfer* and (2.135) of the present volume are of the same form; that is, the heat and mass transfer problems are analogous. In fact, if $\alpha = \mathcal{D}_{12}$, the solutions are identical, and the normalized temperature profile $(T - T_e)/(T_s - T_e)$ is identical to the normalized mass fraction profile $(m_1 - m_{1,e})/(m_{1,s} - m_{1,e})$. The ratio $\mathcal{D}_{12}/\alpha = \rho c_p \mathcal{D}_{12}/k$ is the Lewis number, which was introduced in Section 1.5.2. When the normalized profiles are identical, it follows from Eqs. (5.96) and (2.147b) that the Nusselt number equals the Sherwood number (or, equivalently, the heat transfer and mass transfer Stanton numbers are equal) when the Lewis number is unity. In most

---

[5] Energy transport due to interdiffusion must not be confused with *diffusional conduction*, which is an energy transport similar to heat conduction, but due to concentration gradients rather than a temperature gradient. Diffusional conduction is also called the *diffusion thermo-effect* or *Dufour effect*, and has been ignored in our analysis since it is usually small compared with ordinary conduction.

gas mixtures the Lewis number is indeed close to unity, but for liquid solutions it can be quite different from unity.

Approach 2: Enthalpy formulation. We assume effective binary diffusion and that the Lewis numbers of all species present are equal to unity, $Le_i = \mathcal{D}_{im}/\alpha = 1$. Notice that in the definition of Lewis number, $\alpha = k/\rho c_p$ is a *mixture property*, and $\mathcal{D}_{im}$ is the species-dependent effective binary diffusion coefficient for species i in the mixture. Implicit in the requirement that $Le_i = 1$ is that all the $\mathcal{D}_{im}$ are equal. Then, proceeding as we did in Section 2.3.5, we obtain

$$\frac{k}{c_p}\frac{\partial h}{\partial y} = k\frac{\partial T}{\partial y} - \sum j_{iy}h_i, \quad \text{for } Le_i = 1 \tag{2.161}$$

Substituting in Eq. (2.159),

$$\rho u\frac{\partial h}{\partial x} + \rho v\frac{\partial h}{\partial y} = \frac{\partial}{\partial y}\left(\frac{k}{c_p}\frac{\partial h}{\partial y}\right) \tag{2.162}$$

and if we now assume that k/c_p is constant,

$$u\frac{\partial h}{\partial x} + v\frac{\partial h}{\partial y} = \alpha\frac{\partial^2 h}{\partial y^2} \tag{2.163}$$

Equation (2.163) is of the same mathematical form as Eq. (5.42); however, it is of more general validity, provided the Lewis number is close to unity, as it is for most gas mixtures. Chemical reactions are allowed, and the interdiffusion term has been retained (albeit approximately). Equation (2.163) is particularly useful for combustion of gaseous or volatile fuels. Thus, we will develop the remainder of our heat transfer analysis using the enthalpy formulation. Appropriate boundary conditions for Eq. (2.163) are

$$y = 0: \ h = h_s \tag{2.164a}$$

$$y \rightarrow \infty, x = 0: \ h = h_e \tag{2.164b}$$

A relation coupling the momentum and energy equations can be derived that is similar to the relation coupling the momentum and species equations, Eq. (2.153). We introduce a new concept for this purpose, the *transferred state enthalpy*, h_t, defined by the relation

$$\dot{m}''h_t = \sum n_i h_i|_s - k\frac{\partial T}{\partial y}\Big|_s \tag{2.165}$$

The right-hand side of Eq. (2.165) is the energy flux across the s-surface expressed as a sum of absolute convection and conduction components. The concept of a transferred state enthalpy was introduced by Spalding [20, 21] and allows us to conveniently display the analogy between heat and mass transfer. In general, the transferred state is a fictitious state that simply satisfies Eq. (2.165); for the special cases of transpiration or sweat cooling, h_t proves to be equal to the enthalpy in the reservoir. [However, see Eq. (2.194) if there is thermal radiation.] Introducing $n_{i,s} = m_{i,s}\dot{m}'' + j_{i,s}$ into

Eq. (2.165) gives

$$\dot{m}''h_t = \dot{m}''h_s + \sum j_i \, h_i|_s - k\frac{\partial T}{\partial y}\bigg|_s$$

$$= \dot{m}''h_s - \frac{k}{c_p}\frac{\partial h}{\partial y}\bigg|_s, \quad \text{for Le}_i = 1$$

Introducing a dimensionless enthalpy $\Phi = (h - h_e)/(h_s - h_e)$ and solving for $\dot{m}''$ gives

$$\dot{m}'' = -\frac{h_e - h_s}{h_s - h_t}\frac{k}{c_p}\frac{\partial \Phi}{\partial y}\bigg|_s \tag{2.166}$$

or, in terms of self-similar variables using Eqs. (5.77) and (5.73b),

$$f(0) = \mathcal{B}_h\frac{\Phi'(0)}{\text{Pr}}; \quad \mathcal{B}_h = \frac{h_e - h_s}{h_s - h_t} \tag{2.167}$$

where $\mathcal{B}_h$ is the heat transfer driving force and is analogous to the mass transfer driving force $\mathcal{B}_m$.[6] Equation (2.167) is of the same form as Eq. (2.153).

In order to obtain a complete analogy between heat and mass transfer, the Stanton number must be defined in terms of the sum of the conduction and interdiffusion energy fluxes across the s-surface, and the enthalpy difference across the boundary layer. Using the *heat transfer conductance* g_h introduced in Section 1.4.4,

$$\rho u_e \text{St}_x = g_{hx} = \frac{-k(\partial T/\partial y)|_s + \sum_i j_i h_i|_s}{h_s - h_e} \tag{2.168}$$

$$= \frac{-(k/c_p)(\partial h/\partial y)|_s}{h_s - h_e}, \quad \text{for Le}_i = 1$$

$$= -\frac{k}{c_p}\Phi'(0)(u_e/2\nu x)^{1/2}$$

Recall that assuming $\text{Le}_i = 1$ does not eliminate the interdiffusion term. Rearranging gives the local Stanton number as

$$\text{St}_x = \frac{-\Phi'(0)}{2^{1/2}\text{Re}_x^{1/2}\,\text{Pr}} \tag{2.169}$$

which is analogous to Eq. (2.147a). Also,

$$\frac{g_{hx}}{g_{hx}^*} = \frac{\text{St}_x}{\text{St}_x^*} = \frac{\Phi'(0)}{\Phi'(0)|_{f(0)=0}} \tag{2.170}$$

and from Eq. (5.97),

$$\text{St}_x^* = 0.332\text{Re}_x^{-1/2}\text{Pr}^{-2/3}, \quad \text{Pr} > 0.5 \tag{2.171}$$

---

[6] Actually $\mathcal{B}_m$ contains 1 in the denominator, rather than a quantity corresponding directly to h_t. Section 2.6 will define $\mathcal{B}_m$ in terms of a *transferred state mass fraction* $m_{i,ts} = n_{i,s}/\dot{m}''$. In the special case of transfer of a single inert species, $m_{i,ts} = 1$, giving the result used here.

As was the case for mass transfer, g_{hx}/g_{hx}^* can be given as a function of $f(0)$ (or, equivalently, $B_h = \dot{m}''/g_{hx}^*$) or of $\mathcal{B}_h$, with Pr as a parameter. The plots are then identical to Figs. (2.28) through (2.30). Notice that Eqs. (2.145), (2.167), and (2.169) combine to give

$$\dot{m}'' = g_{hx}\mathcal{B}_h; \quad \mathcal{B}_h = \frac{h_e - h_s}{h_s - h_t}; \quad g_{hx} = \rho u_e St_x \qquad (2.172)$$

A troublesome issue concerning this analysis is the use of Prandtl number in Eqs. (2.169) through (2.171). Since we assumed unity Lewis numbers ($Pr = Sc_i$) to obtain Eq. (2.169), why not use a Schmidt number in these equations? If all the Le_i were exactly equal to unity, there would be no problem; but in reality $Le_i \neq 1$, and thus a choice must be made. We choose to use Pr because Eq. (2.169) is the solution of the *energy* equation, which is used to determine *heat transfer*. In particular, we would like our solution to be exact in the limit $\dot{m}'' \to 0$, for which we have a heat transfer problem only: then Eq. (2.171) gives the heat transfer Stanton number exactly.

When using the enthalpy formulation, we see that the analogy between mass and heat transfer results in an analogy between the diffusion flux across the s-surface for mass transfer and the sum of the conduction and interdiffusion fluxes across the s-surface for heat transfer. The driving potential for this sum of conduction and interdiffusion fluxes is the enthalpy difference across the boundary layer $(h_s - h_e)$. For *inert* systems we can derive another form of the analogy between mass and heat transfer, as follows. For an inert mixture, we are free to choose the enthalpy datum state for each species such that the enthalpy of each is zero at the surface temperature T_s. Then Eq. (2.168) becomes

$$\rho u_e St_x = g_{hx} = \frac{-k(\partial T/\partial y)|_s}{-h_e^{T_s}} \qquad (2.173)$$

and

$$h_e^{T_s} = \sum m_{i,e} h_{i,e}^{T_s}$$
$$= \sum m_{i,e}(h_{i,e} - h_{i,s}) \quad \text{independent of datum state}$$
$$= \sum m_{i,e} h_{i,e} - \sum m_{i,e} h_{i,s}$$
$$h_e^{T_s} = h_e - h_{es} \qquad (2.174)$$

where $h_{es} = \sum m_{i,e} h_{i,s}$ is the enthalpy of a mixture of e-state composition at the s-state temperature. Notice that the right-hand side of Eq. (2.174) is independent of datum states, since it involves only enthalpy differences of pure components. Substituting in Eq. (2.173),

$$\rho u_e St_x = g_{hx} = \frac{-k(\partial T/\partial y)|_s}{h_{es} - h_e} \qquad (2.175)$$

Thus, we see that the heat transfer Stanton number defined by Eq. (2.168) also gives the conduction heat flux across the s-surface in an inert system, provided that the driving potential is taken to be $(h_{es} - h_e)$, and not $(h_s - h_e)$. This result using the enthalpy formulation is different from the corresponding result obtained using the temperature formulation. In the temperature formulation we assumed all species specific heats equal, $c_{p1} = c_{p2} = \cdots = c_p$. From Eq. (5.96),

$$\mathrm{St}_x = \frac{-k(\partial T/\partial y)|_s}{\rho u_e c_p (T_s - T_e)} \tag{2.176}$$

whereas from Eq. (2.175),

$$\mathrm{St}_x = \frac{-k(\partial T/\partial y)|_s}{\rho u_e c_{pe} (T_s - T_e)} \tag{2.177}$$

if we take c_p independent of temperature to facilitate the comparison. The temperature formulation requires that we use a specific heat that is some average over the species involved; in contrast, the enthalpy formulation requires that we use the specific heat of the free-stream mixture. The reason for this difference is that the temperature formulation eliminated the interdiffusion term at the outset, whereas the enthalpy formulation retained the interdiffusion term.

The role played by the enthalpy difference $(h_{es} - h_e)$ can be further illuminated if we consider the evaluation of $\mathcal{B}_h$ for transpiration cooling. If an inert gas, species 1, is injected into a flow of inert gas, species 2, then, referring to Eq. (2.165) and Fig. 2.32, an energy balance between the s- and o-surfaces gives

$$\dot{m}'' h_t = \sum n_i h_i |_s - k \frac{\partial T}{\partial y}|_s = n_1 h_{1,o} = \dot{m}'' h_{1,o}$$

Substituting in Eq. (2.172),

$$\mathcal{B}_h = \frac{h_e - h_s}{h_s - h_{1,o}} \tag{2.178}$$

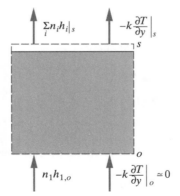

Figure 2.32 Transpiration cooling: the energy conservation principle applied to a control volume located between the s-surface and an o-surface in the coolant reservoir.

Choosing enthalpy datum states such that $h_1 = h_2 = 0$ at $T = T_s$, and proceeding as before,

$$\mathcal{B}_h = \frac{h_e^{T_s}}{-h_{1,o}^{T_s}} = \frac{h_e - h_{es}}{h_{1,s} - h_{1,o}} \tag{2.179}$$

Substituting back in Eq. (2.172),

$$\dot{m}'' = g_h \mathcal{B}_h = g_h \frac{(h_e - h_{es})}{h_{1,s} - h_{1,o}} \tag{2.180}$$

Equation (2.179) gives a driving force for transpiration cooling that is easy to evaluate, since it does not involve the unknown s-surface composition. Further insight can be obtained by rearranging Eq. (2.180) as

$$g_h(h_e - h_{es}) = \dot{m}''(h_{1,s} - h_{1,o})$$

For specific heats independent of temperature and using Eq. (2.177),

$$g_h c_{pe}(T_e - T_s) = -k\frac{\partial T}{\partial y}\bigg|_s = \dot{m}'' c_{p1}(T_s - T_o) \tag{2.181}$$

which states that the conduction heat flux across the s-surface equals $\dot{m}''$ times the increase in enthalpy of the coolant. Thus, even in a mixture where there is an interdiffusion contribution to energy transfer across the s-surface, the heat transfer to the wall for transpiration cooling is actually only the conduction component. In the notation of the steady-flow energy equation, Eq. (1.4) of *Heat Transfer*, the left-hand side of Eq. (2.181) is $\dot{Q}$ and the right-hand side is $\dot{m}\Delta h$ (per unit area).

Examples 2.13 and 2.14 illustrate the use of the enthalpy formulation–based constant-property solutions. As for mass transfer in Section 2.5.1, consideration of real engineering problems requires the use of a suitable reference state for evaluation of properties, and Knuth's scheme or a 1/2 or 1/3 rule may be used. In problems involving the combustion of volatile hydrocarbon fuels, the scheme of Law and Williams given in Section 2.3.5 may also be useful.

Example 2.13   Sweat Cooling of a Flat Plate: Heat Transfer

If, in Example 2.12, the water is supplied from a reservoir at 300 K, determine the radiative heat flux absorbed by the plate. Use the constant-property self-similar solution with the 1/2 rule for evaluating properties.

Solution

Given: Air flowing along a radiatively heated, sweat-cooled flat plate.
Required: The radiative heat flux 0.2 m from the leading edge.
Assumptions:

1. The thermal resistance of the thin film of water on the plate is negligible.
2. A laminar boundary layer.

3. The 1/2 rule is adequate for evaluating properties.

4. Lewis number unity.

5. Species specific heats are independent of temperature.

From Example 2.12, $m_{1,s} = 0.496$, $m_{1,r} = 0.248$, $T_r = 600$ K. Using AIRSTE, $\rho = 0.512$ kg/m^3, $\mu = 2.71 \times 10^{-5}$ kg/m s, Sc $= 0.564$, $c_p = 1275$ J/kg K, $k = 0.0454$ W/m K, Pr $= 0.763$. Also, $\text{Re}_x = 3.78 \times 10^4$. The heat transfer Stanton number and conductance are

$$\text{St}_x^* = 0.332\text{Re}_x^{-1/2}\text{Pr}^{-2/3} = (0.332)(3.78 \times 10^4)^{-1/2}(0.763)^{-2/3} = 2.05 \times 10^{-3}$$

$$g_{hx}^* = \rho u_e \text{St}_x^* = (0.512)(10)(2.05 \times 10^{-3}) = 1.05 \times 10^{-2} \text{ kg/m}^2 \text{ s}$$

Since we know $\dot{m}'' = 7.18 \times 10^{-3}$ kg/m^2 s from Example 2.12, Fig. 2.28 will be used to obtain g_h/g_h^*:

$$-f(0) = \frac{\dot{m}''}{\rho u_e}2^{1/2}\text{Re}_x^{1/2} = \frac{7.18 \times 10^{-3}}{(0.512)(10)}(2 \times 3.78 \times 10^4)^{1/2} = 0.386$$

Then, by analogy from Fig. 2.28 for Pr(Sc) $= 0.763$ and $-f(0) = 0.386$, $g_{hx}/g_{hx}^* = 0.52$. Since the system is inert, we can conveniently use Eq. (2.175), which gives

$$k\frac{\partial T}{\partial y}\bigg|_s = g_{hx}(h_e - h_{es}) \simeq g_{hx}^*(g_{hx}/g_{hx}^*)\bar{c}_{pair}(T_e - T_s)$$

From Table A.7, $\bar{c}_{pair} = (1/2)(1098 + 1007) = 1052$ J/kg K.

$$k\frac{\partial T}{\partial y}\bigg|_s = (1.05 \times 10^{-2})(0.52)(1052)(840 - 360) = 2757 \text{ W/m}^2$$

Referring to the figure, an energy balance between the s- and o-surfaces with $n_1 = \dot{m}''$ gives

$$\dot{m}''h_1\big|_s - k\frac{\partial T}{\partial y}\bigg|_s + q_{\text{rad}} = \dot{m}''h_1|_o$$

since only species 1 is transferred. Rearranging,

$$k\frac{\partial T}{\partial y}\bigg|_s - q_{\text{rad}} = \dot{m}''(h_{1,s} - h_{1,o})$$

$$= \dot{m}''[h_{\text{fg}} + c_{pl}(T_s - T_o)]$$

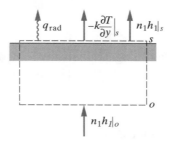

From Tables A.8 and A.12a, $c_{pl} = 4178$ J/kg K, $h_{fg} = 2.291 \times 10^6$ J/kg; thus,

$$k \frac{\partial T}{\partial y}\Big|_s - q_{rad} = (7.18 \times 10^{-3})[2.291 \times 10^6 + 4178(360 - 300)]$$

$$2757 - q_{rad} = 1.825 \times 10^4$$

Solving, $q_{rad} = 2757 - 1.825 \times 10^4 = -1.55 \times 10^4$ W/m^2 (15.5 kW/m^2)

Comments

1. Solve this problem using the improved blowing factors of Section 2.4.2 and compare results.

2. In this problem T_s was specified and q_{rad} was unknown. If we specify q_{rad} and make T_s the unknown, a complicated iteration process is required. Alternatively, q_{rad} could be varied parametrically to obtain a graph of T_s versus q_{rad}. Either way, sweat-cooling problem solving is not for the fainthearted!

3. Use an air enthalpy table to evaluate $(h_e - h_{es})$ and compare the result with the approximation obtained using $(h_e - h_{es}) \simeq \bar{c}_{pair}(T_e - T_s)$.

4. Although the analytical result used was obtained assuming Le $= 1$(Pr $=$ Sc), we were careful to use Pr and the heat transfer conductance (rather than Sc and the mass transfer conductance) for these *heat transfer* calculations. We thus ensured that our procedure is exact in the limit of zero mass transfer.

5. Notice that q_{rad} must also vary as $x^{-1/2}$ in order to have a self-similar boundary layer.   ▲

Example 2.14   Heat Transfer into a Graphite Heat Shield

Graphite is often used as a heat shield material due to its high sublimation temperature ($> 3500°$C). In a particular application, a graphite heat shield is exposed to air at 2000 K and 2 atm flowing at 30 m/s. The flow can be modeled as a laminar boundary layer on a flat plate. Calculate the heat flux into the heat shield at a location 3 cm from the leading edge when the surface temperature has reached 1500 K. The surroundings are black at 1400 K, and graphite has an emittance of 0.88.

Solution

Given: A graphite surface exposed to a hot air flow.
Required: Heat flux at $x = 0.03$ m when the surface temperature is 1500 K.
Assumptions:

1. Diffusion-controlled oxidation to carbon monoxide.

2. Carbon vapor pressure negligible (heterogeneous reaction).

3. Species Lewis numbers unity.

4. Radiation can be modeled as interchange between a small gray body and large black surroundings.

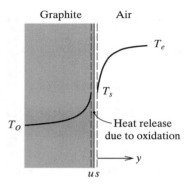

The combustion process is the same as was analyzed in Section 2.3.2: at the specified temperatures, the product of combustion is carbon monoxide, and the reaction is diffusion-controlled with $m_{O_2,s} \simeq 0$.

$$2C + O_2 \rightarrow 2CO$$

$$1\text{kg} + r\text{kg} \rightarrow (1 + r)\text{kg}$$

where $r = 32/24 = 1.33$ is the stoichiometric ratio of the reaction. In Section 1.5.3 we made an ad hoc estimate of the heat transfer during combustion of carbon: here a more rigorous approach is used. The surface energy balance is

$$-k\frac{\partial T}{\partial y}\bigg|_u + \sum n_i h_i|_u = -k\frac{\partial T}{\partial y}\bigg|_s + \sum n_i h_{i,s} + q_{\text{rad}}$$

as shown in the figure.

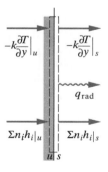

To obtain a better grasp on the physical significance of the terms in this energy balance, it is rearranged as

$$k\frac{\partial T}{\partial y}\bigg|_u = k\frac{\partial T}{\partial y}\bigg|_s + \left(\sum n_i h_i|_u - \sum n_i h_i|_s\right) - q_{\text{rad}} \qquad (1)$$

$$\underbrace{\phantom{k\frac{\partial T}{\partial y}}}_{\substack{\text{conduction into} \\ \text{heat shield}}} \qquad \underbrace{\phantom{k\frac{\partial T}{\partial y}}}_{\substack{\text{convection from} \\ \text{air flow}}} \qquad \underbrace{}_{\substack{\text{heat release due} \\ \text{to combustion}}} \qquad \underbrace{}_{\substack{\text{radiation loss} \\ \text{to surroundings}}}$$

The term accounting for heat release due to combustion is evaluated as follows:

$$\sum n_i h_i |_u = n_{C,u} h_{C,u} = \dot{m}'' h_{C,u}$$

where C denotes solid carbon.

$$\sum n_i h_i |_s = n_{CO,s} h_{CO,s} + n_{O_2,s} h_{O_2,s} = \dot{m}''[(1+r)h_{CO,s} - r h_{O_2,s}]$$

Thus, the rate of heat release is

$$\left(\sum n_i h_i |_u - \sum n_i h_i |_s\right) = \dot{m}''[h_{C,u} + r h_{O_2,s} - (1+r)h_{CO,s}] = \dot{m}'' \Delta h_c$$

where Δh_c is the heat of combustion per kilogram of solid carbon at temperature T_s. Also, for a small gray surface in large black surroundings, we can write $q_{\rm rad} \simeq \sigma \varepsilon (T_s^4 - T_w^4)$. Substituting in Eq. (1),

$$k \frac{\partial T}{\partial y}\bigg|_u = k \frac{\partial T}{\partial y}\bigg|_s + \dot{m}'' \Delta h_c - \sigma \varepsilon (T_s^4 - T_w^4) \tag{2}$$

We next calculate the mass transfer rate $\dot{m}''$,

$$\dot{m}'' = g_{mO_2} \mathcal{B}_{mO_2}$$

where, from Eq. (2.41), $\mathcal{B}_{mO_2} = 0.173$. The 1/2 rule will be used to evaluate properties: $T_r = (1/2)(T_e + T_s) = (1/2)(2000 + 1500) = 1750$ K. Again from Section 2.3.2, $m_{CO,s} = 0.345$. Denoting O_2, CO, and N_2 as species 1, 2, and 3, respectively, $m_{1,r} = (1/2)(0.231 + 0) = 0.116$; $m_{2,r} = (1/2)(0 + 0.345) = 0.173$; $m_{3,r} = (1/2)(0.769 + 0.655) = 0.711$. GASMIX gives the following properties at 2 atm: $\rho = 0.3959$ kg/m^3, $\mu = 5.83 \times 10^{-5}$ kg/m s, $c_p = 1257$ J/kg K, $Sc_{O_2,m} \simeq 0.73$, Pr $= 0.702$.

The Reynolds number is

$$Re_x = u_e \rho x / \mu = (30)(0.3959)(0.03)/(5.83 \times 10^{-5}) = 6112$$

and from Eq. (2.149) the mass transfer Stanton number is

$$St^*_{mx} = 0.332 Re_x^{-1/2} Sc^{-2/3} = (0.332)(6112)^{-1/2}(0.73)^{-2/3} = 5.24 \times 10^{-3}$$

$$g^*_{mx} = \rho u_e St^*_m = (0.3959)(30)(5.24 \times 10^{-3}) = 6.22 \times 10^{-2} \text{ kg/m}^2 \text{ s}$$

From Fig. 2.30 with $\mathcal{B}_m = 0.173$, Sc $= 0.73$, $g_{mx}/g^*_{mx} = 0.88$. Thus, $g_{mx} = (0.88)(6.22 \times 10^{-2}) = 5.47 \times 10^{-2}$ kg/m^2 s. Hence,

$$\dot{m}'' = g_m \mathcal{B}_m = (5.47 \times 10^{-2})(0.173) = 9.47 \times 10^{-3} \text{ kg/m}^2 \text{ s}$$

Since the boundary layer is inert, we can use Eq. (2.175) to obtain the s-surface conduction,

$$k \frac{\partial T}{\partial y}\bigg|_s = g_{hx}(h_e - h_{es}) \simeq g_{hx} \bar{c}_{p\,\text{air}}(T_e - T_s) \tag{3}$$

where $\bar{c}_{p\,\text{air}}$ is evaluated at $(T_e + T_s)/2$. Using Eqs. (2.149) and (2.171),

$$g_{hx}^* = g_{mx}^*(\text{Pr}/\text{Sc})^{-2/3} = (6.22 \times 10^{-2})(0.702/0.730)^{-2/3}$$
$$= 6.38 \times 10^{-2}\ \text{kg/m}^2\,\text{s}$$

Since we know $\dot{m}''$, we can use Fig. 2.28 to obtain g_{hx}/g_{hx}^*:

$$-f(0) = (\dot{m}''/\rho u_e)2^{1/2}\text{Re}_x^{1/2}$$
$$= [9.47 \times 10^{-3}/(0.3959)(30)](2)^{1/2}(6112)^{1/2} = 8.82 \times 10^{-2}$$

Then, by analogy from Fig. 2.28 for $\text{Pr(Sc)} = 0.7$ and $-f(0) = 8.82 \times 10^{-2}$, $g_{hx}/g_{hx}^* = 0.88$. Hence, $g_{hx} = (0.88)(6.38 \times 10^{-2}) = 5.61 \times 10^{-2}\,\text{kg/m}^2\,\text{s}$.

For $T_e = 2000$ K, $T_s = 1500$ K, Table A.7 gives $\bar{c}_{p\,\text{air}} = (1/2)(1244 + 1202) = 1223$ J/kg K. Substituting in Eq. (3),

$$k\left.\frac{\partial T}{\partial y}\right|_s = (5.61 \times 10^{-2})(1223)(2000 - 1500) = 3.43 \times 10^4\ \text{W/m}^2$$

An estimate of the heat of combustion Δh_c can be obtained from the data in Table A.25 by subtracting (M_{CO}/M_C) of the value for CO from the value for C (values in Table A.25 are for a CO_2 product per kilogram of fuel):

$$\Delta h_c \simeq 32.78 \times 10^6 - (28/12)(2.11 \times 10^6) \simeq 9.2 \times 10^6\ \text{J/kg}$$

Finally, substituting into Eq. (2) gives the conduction flux into the heat shield:

$$k\left.\frac{\partial T}{\partial y}\right|_u = 3.43 \times 10^4 + (9.47 \times 10^{-3})(9.2 \times 10^6) - (5.67)(0.88)(15^4 - 14^4)$$

$$= 3.43 \times 10^4 + 8.71 \times 10^4 - 6.09 \times 10^4$$

$$= 6.05 \times 10^4\ \text{W/m}^2(61\ \text{kW/m}^2)$$

Comments

1. The conduction heat flux calculated above would be used as a boundary condition in a finite-difference numerical solution of the heat conduction equation to obtain the transient response of the heat shield.

2. In this situation the surface energy balance is dominated by the radiation heat flux. As T_s increases above 1500 K, q_{rad} increases rapidly and $k\,\partial T/\partial y|_u = 0$ at a value of T_s below 1600 K.

3. Notice that since $\text{Pr} \simeq \text{Sc}$, the blowing corrections for mass and heat transfer are identical.

4. The surface recession can cause the heat shield shape to change. Such shape changes are usually tolerated to take advantage of the high sublimation temperature of graphite.   ▲

2.6 A GENERAL PROBLEM-SOLVING PROCEDURE FOR CONVECTIVE HEAT AND MASS TRANSFER

In Sections 2.3, 2.4, and 2.5, a number of specific problems involving high mass transfer rates were analyzed. These problems were selected because of their fundamental importance and engineering significance. Section 2.3 introduced diffusion-induced convection (Stefan flow) in the context of some practical evaporation and combustion problems. Section 2.4 used the Couette flow as a model of real convective flows to obtain engineering methods that account for high mass transfer rates. In Section 2.5 convection in a laminar boundary layer on a flat plate was rigorously analyzed, extending the classical momentum and heat transfer analyses to include mass transfer. Each analysis was self-contained, and little was done to generalize the applicability of the results. However, the student should have discerned that the analyses had much in common, and that appropriate generalization should yield results applicable to a wide range of physical situations. For example, the Shvab–Zeldovich transformation used in Section 2.3.3 to analyze carbon particle combustion can also be used to analyze general chemical reactions in the laminar boundary layer of Section 2.5.1. The assumption of unity Lewis number that was used to simplify the energy equation for the laminar boundary layer in Section 2.5.2 can also be used to obtain an alternative result to Eq. (2.63) for droplet evaporation.

In this section we organize the results obtained already to give a general problem-solving procedure for steady convective heat and mass transfer. To develop the procedure, we first carefully discuss the concept of a *transferred state* in Section 2.6.1. The transferred state for enthalpy was introduced in Section 2.5.2, and we now introduce the transferred state for chemical species and elements. The actual problem-solving procedure is presented in Section 2.6.2, and in Section 2.6.3 examples of its application to more complicated problems are given. These problems involve such features as transfer of more than one species, chemical reactions, simultaneous heat and mass transfer, and coupled mass transfer in two adjacent phases.

The development that follows is based mainly on the contributions of D. B. Spalding [20, 21, 22].

2.6.1 The Transferred State

Mass Transfer

To introduce the concept of a *transferred state* for mass transfer, it is convenient to return to Section 2.4.1, where the mass transfer driving force was defined for a binary inert mixture, with transfer of a single species. We now extend that definition to a multicomponent mixture with transfer of more than one species. The mass transfer conductance for species i is

$$g_{mi} = \frac{j_{i,s}}{m_{i,s} - m_{i,e}} \tag{2.182}$$

Notice again that g_{mi} is defined in terms of the *diffusive* flux $j_{i,s}$. Since g_{mi} is not well behaved if species i reacts in the flow, we will restrict our attention to species that are inert in the flow (but that may be involved in heterogeneous reactions at the wall). The *absolute* flux of species i across the s-surface is, by definition,

$$n_{i,s} = m_{i,s} n_s + j_{i,s} \tag{2.183}$$

where $n_s = \dot{m}''$, the mass transfer rate. We define the transferred-state mass fraction $m_{i,ts}$ as

$$m_{i,ts} = \frac{n_{i,s}}{\dot{m}''} \tag{2.184}$$

for transfer across the s-surface. If we imagine the mass transfer process as a stream of matter flowing through the s-surface, $m_{i,ts}$ is the fraction of the stream that is species i. Substituting Eqs. (2.182) and (2.184) into Eq. (2.183),

$$m_{i,ts}\dot{m}'' = m_{i,s}\dot{m}'' + g_{mi}(m_{i,s} - m_{i,e})$$

Solving for $\dot{m}''$,

$$\dot{m}'' = g_{mi} \frac{m_{i,e} - m_{i,s}}{m_{i,s} - m_{i,ts}}$$

or

$$\dot{m}'' = g_{mi} \mathcal{B}_{mi}; \quad \mathcal{B}_{mi} = \frac{m_{i,e} - m_{i,s}}{m_{i,s} - m_{i,ts}} \tag{2.185}$$

where $\mathcal{B}_{mi}$ is the mass transfer driving force for species i. We can also define a transferred-state mass fraction for transfer across the u-surface as

$$m_{i,tu} = \frac{n_{i,u}}{\dot{m}''} \tag{2.186}$$

In the special case that species i does not react at the wall—that is, between the u- and s-surfaces—$n_{i,u} = n_{i,s}$ and

$$m_{i,tu} = m_{i,ts} \equiv m_{i,t} \tag{2.187}$$

Thus, for problems involving inert species—for example, evaporation and condensation—we need not distinguish between the ts- and tu-states, since they are the same, and we simply write $m_{i,t}$ for a transferred-state mass fraction.

We have, of course, implicitly used the transferred-state mass fraction concept in many of the analyses in Sections 2.3 through 2.5. For example, when only a single component is transferred in an inert binary system, as occurs for evaporation of water into air, we can write $n_{1,u} = \dot{m}''$, since only water crosses the u-surface. Thus, $m_{1,tu} = 1$ and, from Eq. (2.187), $m_{1,t} = 1$. Substituting in Eq. (2.185) recovers Eq. (2.93):

$$\dot{m}'' = g_{m1} \mathcal{B}_{m1}; \quad \mathcal{B}_{m1} = \frac{m_{1,e} - m_{1,s}}{m_{1,s} - 1} \tag{2.93}$$

Figure 2.33 shows heterogeneous oxidation of carbon, as analyzed in Section 2.3.2. At the high temperatures under consideration, the reaction was

$$2C + O_2 \rightarrow 2\,CO$$

$$2 \times 12\,\text{kg} + 32\,\text{kg} \rightarrow 2 \times 28\,\text{kg}$$

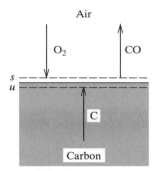

Figure 2.33 Surface
species balance for
high-temperature
combustion of carbon.

Since only carbon crosses the u-surface, $n_{C,u} = \dot{m}''$, $n_{O_2,u} = 0$, $n_{CO,u} = 0$, and

$$m_{C,tu} = 1; \quad m_{O_2,tu} = 0; \quad m_{CO,tu} = 0$$

To obtain $m_{O_2,ts}$ we note that for each kilogram of carbon crossing the u-surface in the positive direction, 4/3 kg of oxygen crosses the s-surface in the negative direction; thus

$$\frac{n_{O_2,s}}{n_{C,u}} = -\frac{4}{3}$$

But $n_{C,u} = m_{C,tu}\dot{m}'' = (1)\dot{m}'' = \dot{m}''$; hence

$$m_{O_2,ts} = \frac{n_{O_2,s}}{\dot{m}''} = -\frac{4}{3}$$

Substituting in Eq. (2.185) gives

$$\dot{m}'' = g_{mO_2} \frac{m_{O_2,e} - m_{O_2,s}}{m_{O_2,s} + 4/3}$$

which can be compared to Eqs. (2.39) and (2.41). (As before, if the reaction is diffusion-controlled, we can set $m_{O_2,s} \simeq 0$.) We see that use of the transferred-state mass fraction allows us to handle the interface boundary condition in a systematic manner.

We also define a transferred-state mass fraction for chemical element (α). Recall that in Section 2.3.3, we defined the absolute flux of element (α) as $\mathbf{n}_{(\alpha)} = \sum_i m_{(\alpha)i}\mathbf{n}_i$, where $m_{(\alpha)i}$ is the mass fraction of element (α) in species i. Then

$$m_{(\alpha),t} = \frac{n_{(\alpha),s}}{\dot{m}''} = \frac{n_{(\alpha),u}}{\dot{m}''} \qquad (2.188)$$

We need not distinguish between ts- and tu-states since $n_{(\alpha),s} = n_{(\alpha),u}$; chemical elements cannot be created or destroyed between the s- and u-surfaces. As an example, consider combustion of carbon in air. Irrespective of the reaction stoichiometry, $n_{(C),u} = \dot{m}''$, $n_{(O),u} = 0$; hence

$$m_{(C),t} = \frac{n_{(C),u}}{\dot{m}''} = 1; \quad m_{(O),t} = \frac{n_{(O),s}}{\dot{m}''} = \frac{n_{(O),u}}{\dot{m}''} = 0$$

—a result that was used in Section 2.3.3. Notice that we can allow chemical reactions to occur both in the flow and at the wall when the analysis is based on chemical elements.

Heat Transfer

The transferred-state enthalpy h_t was introduced in Section 2.5.2 and was defined as

$$\dot{m}'' h_t = \sum n_i h_i|_s - k \left.\frac{\partial T}{\partial y}\right|_s \tag{2.165}$$

That is, if energy transfer across the s-surface by convection and conduction is imagined to be due to convection only by a stream of matter crossing the interface at a rate $\dot{m}''$, then h_t is the enthalpy of the stream. In Section 2.5.2, h_t was introduced to obtain a heat transfer result that was of the same form as the mass transfer result of Section 2.5.1. We also saw in Section 2.5.2 that problem solving required evaluating h_t for a specific physical situation by performing an energy balance between the s-surface and the u-surface, or a reservoir o-surface. In transpiration or sweat cooling with negligible surface radiation, h_t proved to be equal to the reservoir enthalpy h_o. For the noncondensable gas problem in a condenser, h_t is evaluated by performing a further energy balance between the u-surface and the bulk coolant.

2.6.2 The Problem-Solving Procedure

Most (but not all) of the results obtained in Sections 2.3 through 2.5 can be cast in the form

$$\dot{m}'' = g\mathcal{B} \tag{2.189}$$

where $\mathcal{B}$ is a *driving force* based on the mass fraction of a mass species or chemical element, or on enthalpy, and g is the corresponding conductance. In general, we can write

$$\mathcal{B} = \frac{\mathcal{P}_e - \mathcal{P}_s}{\mathcal{P}_s - \mathcal{P}_{ts}} \tag{2.190}$$

where the **conserved property** $\mathcal{P} = m_i, m_{(\alpha)}$, or h. When i is inert in the phase under consideration, or when $\mathcal{P} = m_{(\alpha)}$ or h, $\mathcal{P}_{ts}$ can be replaced by $\mathcal{P}_t$ because the ts- and t-states are identical. Equation (2.189) forms the basis of our problem-solving procedure. In problems involving transfer of more than one species, simultaneous consideration of two adjacent phases, or simultaneous heat and mass transfer, two or more equations of the form of Eq. (2.189) must be solved simultaneously. Also, auxiliary relations such as vapor pressure or solubility data are often required. Two essentially distinct issues are involved in using Eq. (2.189): one is the evaluation of the driving force $\mathcal{B}$, and the other is the evaluation of the conductance g.

Choice of Driving Force

The choice of driving force depends on the type of mass transfer problem under consideration and the information available. The situation may involve evaporation, combustion of a solid fuel, combustion of a volatile fuel, catalysis, and so on. The problem considered most often in Sections 2.3 through 2.5 was mass transfer of a single inert species, either in connection with evaporation, condensation, or transpiration cooling. The driving force was based on mass fraction of the transferred species, $\mathcal{P} = m_1$, with $m_{1,t} = 1$.

If the results of the analysis of carbon combustion in Section 2.3.3 are recast in the form of Eq. (2.189), the driving force is seen to be based on a linear combination of element mass fractions, $\mathcal{P} = \zeta m_{(O)} + \xi m_{(C)}$. In deriving this driving force, the diffusion coefficients of all species containing the elements were assumed equal. Similarly, Section 2.5.2 showed that for an enthalpy-based driving force to be valid, the Lewis numbers, $\mathrm{Le}_i = (k/c_p)/\rho \mathcal{D}_{im}$, of all species in the mixture must equal unity. Since k, c_p, and ρ are all mixture properties, this condition requires all the diffusion coefficients to be equal. In addition, it is easy to show that an enthalpy-based driving force is also valid for a nonreacting mixture of species with all species specific heats equal. The transfer number obtained for combustion of a liquid fuel droplet in Section 2.3.5 was, in fact, an enthalpy-based driving force with the Lewis numbers assumed to be unity, and as a further simplification, the species specific heats were assumed equal and constant.

Evaluation of the Conductance

Evaluation of the conductance g, corresponding to the chosen driving force $\mathcal{B}$, requires identification of the flow configuration, evaluation of a zero mass transfer limit conductance g^*, and its correction using an appropriate blowing factor. If $\mathcal{B}$ is based on the mass fraction of species i, then the corresponding conductance is a mass transfer conductance. For example, for a laminar boundary layer on a flat plate, the local value of g^*_{mi} is obtained from Eq. (2.149) as

$$g^*_{mxi} = \rho u_e \mathrm{St}^*_{mxi} = 0.332 \, \mathrm{Re}_x^{-1/2} \, \mathrm{Sc}_i^{-2/3}; \quad \mathrm{Sc}_i > 0.5 \qquad (2.191)$$

where $\mathrm{Sc}_i = \nu/\mathcal{D}_{im}$. If $\mathcal{B}$ is based on chemical element (α), then the local value of $g^*_{m(\alpha)}$ is obtained from Eq. (2.191) by replacing Sc_i by $\mathrm{Sc}_{(\alpha)} = \nu/\mathcal{D}$, where $\mathcal{D}$ is some average of the $\mathcal{D}_{im}$'s for the species containing element (α). If an enthalpy-based driving force is obtained for an inert mixture by assuming equal species specific heats, then the corresponding zero-mass-transfer-limit conductance is g^*_h. Again, for a laminar boundary layer on a flat plate, the local value of g^*_h is obtained from Eq. (2.171) as

$$g^*_{hx} = \rho u_e \mathrm{St}^*_x = 0.332 \, \mathrm{Re}_x^{-1/2} \mathrm{Pr}^{-2/3}; \quad \mathrm{Pr} > 0.5 \qquad (2.192)$$

On the other hand, if an enthalpy-based driving force is obtained by assuming unity Lewis numbers, there is the implication $\mathrm{Sc}_i = \mathrm{Pr}$ for all species i and hence $g^*_{mi} = g^*_h$. Since usually $\mathrm{Sc}_i \neq \mathrm{Pr}$, evaluation of g^*_h poses a problem: we usually base it on Pr in order to have an exact result for heat transfer in the limit $\dot{m}'' \to 0$.

Blowing factors to correct g^* were obtained from various sources in Sections 2.4 and 2.5. In Example 2.8 the blowing factor was obtained from a constant-property

Couette–flow model. In Example 2.12, part (i), an improved blowing factor that accounted for variable-property effects and flow configuration was used, and in part (ii) the blowing factor was obtained from an exact self-similar solution for the constant-property laminar boundary layer. It is also possible to use experimental data for blowing factors when available; for example, some such data are available for the combustion of liquid hydrocarbon fuel droplets.

The results for the spherically symmetric flows analyzed in Section 2.3 can also be recast in a form consistent with the foregoing procedure for evaluating the conductance. For example, the analysis of combustion of liquid fuel droplets in Section 2.3.5 gave Eq. (2.74), which can be rearranged as

$$\dot{m}'' = g_h \mathcal{B}_h; \quad \mathcal{B}_h = \frac{h_e - h_s}{h_s - h_t} \tag{2.193a}$$

$$g_h^* = \frac{k/c_p}{R}; \quad \frac{g_h}{g_h^*} = \frac{\ln(1 + \mathcal{B}_h)}{\mathcal{B}_h} \tag{2.193b}$$

since $h_u = h_t$ as a result of taking $dT/dr|_u = 0$.

Surface Radiation

It is possible to allow for surface radiation when using the general problem-solving procedure with an enthalpy-based driving force. To do so, we must recognize that the transferred-state enthalpy h_t is always evaluated by performing an energy balance between the s-surface and either the u-surface or a reservoir o-surface. For example, consider simple transpiration cooling with an inert species 1 injected into a flow of species 2, as shown in Fig. 2.34a. An energy balance between the s- and o-surfaces with $n_{1,s} = \dot{m}''$ gives

$$\dot{m}'' h_{1,s} - k \left.\frac{\partial T}{\partial y}\right|_s + q_{\text{rad}} = \dot{m}'' h_{1,o}$$

Using the definition of h_t from Eq. (2.165) and rearranging,

$$h_t = h_{1,o} - (q_{\text{rad}}/\dot{m}'') \tag{2.194}$$

and the enthalpy driving force becomes

$$\mathcal{B}_h = \frac{h_e - h_s}{h_s - h_t} = \frac{h_e - h_s}{(h_s - h_{1,o}) + (q_{\text{rad}}/\dot{m}'')}$$

As usual, we choose enthalpy datum states such that $h_1 = h_2 = 0$ at $T = T_s$, to obtain

$$\mathcal{B}_h = \frac{h_e^{T_s}}{-h_{1,o}^{T_s} + (q_{\text{rad}}/\dot{m}'')} = \frac{h_e - h_{es}}{(h_{1,s} - h_{1,o}) + (q_{\text{rad}}/\dot{m}'')}$$

Finally, assuming constant species specific heats,

$$\dot{m}'' = g_h \frac{c_{pe}(T_e - T_s)}{c_{p1}(T_s - T_o) + (q_{\text{rad}}/\dot{m}'')} \tag{2.195}$$

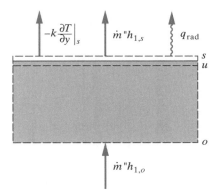

Figure 2.34 *(a)* The surface energy balance with radiation. Transpiration cooling.

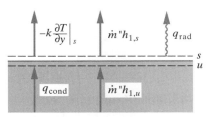

Figure 2.34 *(b)* The surface energy balance with radiation. Textile drying.

We see that radiation heat transfer away from the surface (positive q_{rad}) reduces $\mathcal{B}_h$ and hence reduces the injection rate required to maintain a specified surface temperature.

As a second example, consider drying of a wet textile by combined convective and radiative heating, as shown in Fig. 2.34*b*. An energy balance between the *s*- and *u*-surfaces with $n_{1,s} = \dot{m}''$ gives[7]

$$\dot{m}'' h_{1,s} - k \left. \frac{\partial T}{\partial y} \right|_s + q_{rad} = \dot{m}'' h_{1,u} + q_{cond}$$

$$\dot{m}'' h_t = \dot{m}'' h_{1,u} + q_{cond} - q_{rad}$$

and the driving force becomes

$$\mathcal{B}_h = \frac{h_e - h_s}{h_s - h_{1,u} - (q_{cond}/\dot{m}'') + (q_{rad}/\dot{m}'')}$$

Again taking $h_1 = h_2 = 0$ at $T = T_s$,

$$\mathcal{B}_h = \frac{h_e - h_{es}}{h_{fg} - (q_{cond}/\dot{m}'') + (q_{rad}/\dot{m}'')} \tag{2.196}$$

In the *constant drying period* the textile is nearly isothermal and $q_{cond} \simeq 0$. Then, assuming constant species specific heats,

$$\dot{m}'' = g_h \frac{c_{pe}(T_e - T_s)}{h_{fg} + (q_{rad}/\dot{m}'')} \tag{2.197}$$

---

[7] Strictly speaking, we should subscript q_{cond} as $q_{cond,u}$ and q_{rad} as $q_{rad,s}$, but to be consistent with Section 1.5.1, we take these subscripts as understood.

In this drying process, q_{rad} is negative, and for a given value of T_s the effect of the radiation flux is to increase the evaporation rate, as expected and desired. Equation (2.197) can be rewritten as

$$g_h c_{pe}(T_e - T_s) = \dot{m}'' h_{fg} + q_{rad}$$

or

$$k \left. \frac{\partial T}{\partial y} \right|_s + (-q_{rad}) = \dot{m}'' h_{fg} \tag{2.198}$$

which clearly shows the energy balance implicit in Eq. (2.197).

Variable-Property Effects

A number of schemes are available to account for the effects of variable properties. The schemes recommended in Sections 2.3 through 2.5 are summarized below.

A. Improved blowing factors. For forced-convection laminar or turbulent air boundary layers with injection of a single inert species, the improved blowing factors of Section 2.4.2 should be used. Recall that in this scheme, air properties at the mean film temperature are used to calculate g^*. Also notice that the heat transfer blowing factor gives the conduction component of the s-surface energy flux.

B. General reference states. When constant-property analytical formulas, numerical data, or experimental data are available for the flow configuration of concern, a reference state scheme can be used in which all properties are evaluated at the reference state. Examples include the 1/2 rule (mean film temperature and composition), Hubbard's 1/3 rule [(Eqs. 2.65)], or Knuth's Couette–flow model–based scheme [(Eqs. 2.154 and 2.155)].

C. Specific prescriptions. Sometimes specific prescriptions for the problem of concern are available based on experimental data or exact numerical solutions. Examples include the scheme of Law and Williams for liquid hydrocarbon fuel droplet combustion given in Section 2.3.5, and the laminar boundary layer scheme of Simon et al. [19].

D. Couette–flow model. If nothing better is available, the simple Couette–flow model results of Section 2.4.1 should be used. Two possible strategies are (1) calculate all properties at one of the general reference states listed under scheme *B* above, and (2) evaluate g^* using the free-stream composition and mean film temperature, and use Eq. (2.119) to account for the effect of unequal specific heats. Since these schemes are rather crude, every effort should be made to obtain appropriate experimental data to validate the results.

Validity of the Procedure

On a first reading, the student may well obtain the impression that this general problem-solving procedure is an ad hoc proposal and does not have a rigorous analytical basis. It is straightforward to provide such a basis, but in so doing one simply repeats elements of the analyses already presented in Sections 2.3 through 2.5. Spalding [20, 21] and Kays and Crawford [22] give a more complete analytical basis for the procedure. In general, the procedure applies to steady, low-speed laminar or turbulent flows of any configuration. The effects of thermal, forced, and pressure diffusion, and diffusional conduction, must be negligible, and multicomponent diffusion must be modeled as effective binary diffusion. Absorption or emission of radiation in the fluid must be negligible. High-speed flows can be treated as a special case if a unity Prandtl number is assumed (see Exercise 2–51). The fluid should be an ideal gas mixture or ideal liquid solution.

Notice that the heat transfer results for the evaporating droplet, Eq. (2.63), and the simple Couette–flow model, Eq. (2.119), cannot be recast in the general form of Eq. (2.189), namely, $\dot{m}'' = g\mathcal{B}$. For these simple flows we were able to account for unequal species specific heats and took advantage of this fact to obtain useful analytical results. Thus, in using these results, care must be taken to adapt the general problem-solving procedure accordingly.

2.6.3 Applications of the Procedure

There were many examples of problem solving in Sections 2.3 through 2.5; usually the problems were relatively simple. Here, more complicated problems are considered, involving such features as transfer of more than one species, simultaneous heat and mass transfer, complex chemical reactions, and coupled mass transfer in two adjacent phases. The emphasis is on formulating the set of equations to be solved, and evaluating the driving forces.

1. *Reaction of carbon with a steam–air mixture.* In a number of technological processes, carbon is oxidized by a mixture of air and steam (see Example 2.4). The surface reactions are

$$CO_2 + C \rightarrow 2\,CO$$
$$H_2O + C \rightarrow CO + H_2$$

At typical process temperatures, these reactions are very rapid and are diffusion-controlled with $m_{CO_2,s} = m_{H_2O,s} = 0$. If we are prepared to assume equal effective binary diffusion coefficients of all species containing the elements (C) and (O), then we can follow the procedure of Section 2.3.3 and choose as conserved property $\mathcal{P}$ the combination of elemental mass fractions $[(12/16)m_{(O)} - m_{(C)}]$, as was done in Example 2.4. Recall that use of this combination avoids the need to know $m_{CO,s}$. Then

$$\mathcal{P}_e = \left(\frac{12}{16}m_{(O)} - m_{(C)}\right)_e = \frac{12}{16}\left(m_{O_2,e} + \frac{16}{18}m_{H_2O,e}\right)$$

$$\mathcal{P}_s = \left(\frac{12}{16}m_{(O)} - m_{(C)}\right)_s = 0$$

$$\mathcal{P}_t = \left(\frac{12}{16}m_{(O)} - m_{(C)}\right)_t = -1$$

$$\dot{m}'' = g_m \mathcal{B}; \quad \mathcal{B} = \frac{\mathcal{P}_e - \mathcal{P}_s}{\mathcal{P}_s - \mathcal{P}_t} = \frac{12}{16}m_{O_2,e} + \frac{12}{18}m_{H_2O,e} \qquad (2.199)$$

The assumption of equal diffusion coefficients is not too bad because we do not have to include $\mathcal{D}_{H_2 m}$ in this assumption ($\mathcal{D}_{H_2 m}$ is much larger than the other effective binary diffusion coefficients in the mixture).

If there are gas-phase reactions, such as $2\,CO + O_2 \rightarrow 2\,CO_2$, Eq. (2.199) remains valid. However, for high bulk gas temperatures there are no gas-phase reactions, and since both O_2 and H_2O are inert in the gas phase, we can then base driving forces on m_{O_2} and m_{H_2O} to write

$$\dot{m}'' = \mathcal{g}_{mO_2} \frac{m_{O_2,e} - m_{O_2,s}}{m_{O_2,s} - m_{O_2,ts}} = \mathcal{g}_{mH_2O} \frac{m_{H_2O,e} - m_{H_2O,s}}{m_{H_2O,s} - m_{H_2O,ts}} \quad (2.200a,\,b)$$

The stoichiometry of the surface reactions is

$$2C + O_2 \rightarrow 2\,CO \quad C + H_2O \rightarrow CO + H_2$$

$$1\,kg + \frac{16}{12}\,kg \rightarrow \quad 1\,kg + \frac{18}{12}\,kg \rightarrow$$

and thus the mass transfer rate is

$$\dot{m}'' = n_{C,u} = -\left(\frac{12}{16}n_{O_2,s} + \frac{12}{18}n_{H_2O,s} \right)$$

Dividing by $\dot{m}''$,

$$1 = -\frac{3}{4}m_{O_2,ts} - \frac{2}{3}m_{H_2O,ts} \quad (2.200c)$$

For fast surface reactions, $m_{O_2,s} = m_{H_2O,s} = 0$, and Eqs. (2.200a,b,c) are three equations in the three unknowns $\dot{m}''$, $m_{O_2,ts}$, and $m_{H_2O,ts}$. Appropriate blowing factors must be used to obtain $\mathcal{g}_m$ from $\mathcal{g}_m^*$.

At 1600 K, binary diffusion coefficients of O_2, CO, and H_2O in air are 1.52×10^{-4} m²/s, 1.54×10^{-4} m²/s, and 2.17×10^{-4} m²/s, respectively. Thus, assuming equal effective binary diffusion coefficients for these species will give a good result using Eq. (2.199), but in principle, Eqs. (2.200) should give a more accurate result because equal diffusion coefficients for O_2 and H_2O are not assumed. However, accurate blowing factors are not easily obtained, because four species are transferred across the s-surface, not just one, as considered in Sections 2.4 and 2.5.

2. *Gas absorption.* Many industrial processes employ absorption into a falling liquid film in order to remove, or *scrub*, an unwanted species from a gas stream. Figure 1.1 shows a packed column scrubber, and Exercise 1–30 considered absorption from a pure gas into a laminar falling film. In Chapter 3, gas scrubbers will be analyzed as two-stream mass exchangers using low mass transfer rate theory. Here we consider the application of high mass transfer rate theory to gas absorption to illustrate how transfer between two phases is handled by our general problem-solving procedure. Figure 2.35 shows conditions at a specific location in a packed column where species 1—for example, SO_2—is being absorbed into water. The local bulk concentrations are $m_{1,e}$ and $m_{1,b}$ in the gas and liquid phases, respectively. If the gas stream is saturated with water vapor

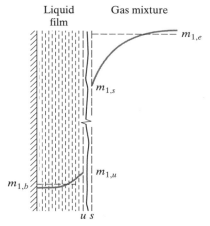

Figure 2.35 Concentration profiles for gas absorption into a falling liquid film in a packed tower.

so that no evaporation takes place, and the carrier gas is highly insoluble in water, then $m_{1,t} \simeq 1$ and the gas absorption rate is

$$\dot{m}'' = \mathcal{g}_{mG} \frac{m_{1,e} - m_{1,s}}{m_{1,s} - 1} = -\mathcal{g}_{mL} \frac{m_{1,b} - m_{1,u}}{m_{1,u} - 1} \qquad (2.201a, \, b)$$

The subscripts G and L on the conductances refer to the gas and liquid phases, respectively. A required auxiliary relation is obtained from solubility data for species 1,

$$m_{1,u} = m_{1,u}(m_{1,s}, P, T) \qquad (2.201c)$$

Equations (2.201) are three equations in the three unknowns $\dot{m}''$, $m_{1,s}$, and $m_{1,u}$. Appropriate correlations for $\mathcal{g}_{mG}^*$ and $\mathcal{g}_{mL}^*$ will be given in Chapter 3.

Gas absorption problems are usually dealt with by chemical engineers, who prefer to work on a molar rather than a mass basis. Thus, in Chapter 3, gas scrubbers are analyzed on a molar basis, rather than the mass basis used in this analysis.

3. *Gas desorption in a direct-contact condenser.* In the open cycle for ocean thermal energy conversion (the *Claude* cycle), steam containing air is condensed onto cold seawater containing dissolved air (see Fig. 7.34 in *Heat Transfer*). Conditions in the condenser allow air to desorb from the seawater as the steam condenses. Concentration and temperature profiles are shown in Fig. 2.36. We denote the bulk liquid state by subscript b, and gas and liquid side conductances by subscripts G and L, respectively. Species 1 and 2 are H_2O and air. We could formulate this problem as one involving simultaneous transfer of two species and heat. But we can simplify the problem by recognizing that the desorption of air affects the process only indirectly, because the air concentration in the

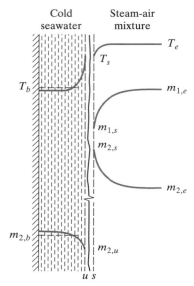

Figure 2.36 Temperature and concentration profiles for direct contact condensation in a condenser for Claude–cycle ocean thermal energy conversion.

liquid phase and the air desorption rate are very small. That is, we can first solve the usual noncondensable gas problem assuming $m_{1,u} = m_{1,t} = 1$, to obtain $\dot{m}''$ and T_s, and then solve the desorption problem. Thus, we write

$$\dot{m}'' = g_{mG}\frac{m_{1,e} - m_{1,s}}{m_{1,s} - 1} = g_{hG}\frac{h_e - h_s}{h_s - h_t} \qquad (2.202a, b)$$

$$m_{1,s} = m_{1,s}(T_s, P) \qquad (2.202c)$$

Evaluation of the heat transfer driving force is simplified by performing an energy balance between the s- and u-surfaces, writing $q_{cond} = h_{cL}(T_b - T_s)$, and then setting enthalpies equal to zero at T_s, to obtain

$$\mathcal{B}_h = \frac{h_e - h_{es}}{h_{fg} - (h_{cL}/\dot{m}'')(T_b - T_s)} = \frac{c_{pe}(T_e - T_s)}{h_{fg} - (h_{cL}/\dot{m}'')(T_b - T_s)}$$

($\dot{m}''$ is negative). The heat transfer coefficient h_{cL} is for heat transfer *into* the bulk liquid: it should not be confused with the heat transfer coefficient of film condensation, which is for heat transfer *across* the liquid film. In a falling-film direct-contact condenser, the film is usually in turbulent flow, and the required h_{cL} can be estimated from the heat transfer analog to the mass transfer correlation given as Eq. (3.45).

Having solved for $\dot{m}''$, m_1, and T_s, we then further simplify the problem by recognizing that air mass-fraction gradients in the liquid phase are very small because the air mass fractions themselves are very small. Hence, if we view the gas desorption process as liquid- and gas-phase resistances in series, the liquid-side resistance dominates, and the gas-side resistance can be assumed to be negligible (see the discussion of the mass transfer Biot number in Section 1.4.4). Thus, we need only consider transfer in the liquid phase, and write

$$\dot{m}'' = -g_{mL}\frac{m_{2,b} - m_{2,u}}{m_{2,u} - m_{2,t}} \qquad (2.203a)$$

$$m_{2,u} = m_{2,u}(m_{2,s}, T_s, P), \quad \text{with } m_{2,s} = 1 - m_{1,s} \qquad (2.203b)$$

Equation $(2.203a)$ is then rearranged as

$$\dot{m}''m_{2,t} = n_{2,u} = \dot{m}''m_{2,u} - g_{mL}(m_{2,u} - m_{2,b}) \qquad (2.204)$$

which can be solved for the air desorption rate $n_{2,u}$. An appropriate correlation for g_{mL} is given by Eq. (3.45).

4. *Transpiration cooling with injectant dissociation.* A porous wall is to be protected from a high-temperature air flow by transpiration of a dimer A_2. We wish to investigate the effect of dissociation of the dimer, $A_2 \rightleftharpoons 2A$, on heat transfer into the wall, and we will evaluate the driving force for three cases:

    i. Species A_2 does not dissociate at all.
    ii. Species A_2 dissociates wholly within the flow—that is, at temperatures higher than the surface temperature T_s.
    iii. Species A_2 dissociates wholly within the porous wall—that is, at temperatures below T_s.

To make the comparison, it will be sufficient to evaluate the heat transfer driving force $\mathcal{B}_h$ obtained by assuming unity Lewis numbers.

Case (i). Species A_2 is inert: thus, Eq. (2.179) for $\mathcal{B}_h$ can be used. Denoting A_2 as species 1,

$$\mathcal{B}_h = \frac{h_e - h_{es}}{h_{1,s} - h_{1,o}} = \frac{c_{pe}(T_e - T_s)}{c_{p1}(T_s - T_o)} \qquad (2.205)$$

where constant species specific heats have been assumed for convenience, subscript o refers to the reservoir state, and radiation has been ignored.

Case (ii). Let A be species 2 and the air (assumed to be inert) species 3. From Eq. (2.190),

$$\mathcal{B}_h = \frac{h_e - h_s}{h_s - h_t}$$

An energy balance between the s- or u-surface and an o-surface in the reservoir shows that $h_t = h_o$; hence,

$$\mathcal{B}_h = \frac{h_e - h_s}{h_s - h_o} \qquad (2.206)$$

By definition, $h_i = h_i^0 + \int_{T^0}^{T} c_{pi}dT$

$$h = \sum m_i h_i, \quad i = 1, 2, 3$$

If we choose the datum temperature $T^0 = T_s$ and $h_i(T_s) = 0$ for species 1 and 3 (i.e., $h_1^0 = h_3^0 = 0$), then

$$h_1 = \int_{T_s}^{T} c_{p1} dT; \quad h_2 = h_2^{T_s} + \int_{T_s}^{T} c_{p2} dT; \quad h_3 = \int_{T_s}^{T} c_{p3} dT$$

and the e-, s-, and o-state enthalpies are

$$h_e = \sum m_{i,e} h_{i,e} = (1)(h_{3,e})$$

$$= \int_{T_s}^{T_e} c_{p3} \, dT \simeq c_{pe}(T_e - T_s) \quad (\text{since } c_{p3} = c_{pe})$$

$$h_s = 0 \quad (\text{since } m_{2,s} = 0: \text{ A}_2 \text{ dissociation takes place wholly}$$
$$\text{within the flow, i.e., at temperatures above } T_s)$$

$$h_o = \int_{T_s}^{T_o} c_{p1} dT = h_{1,o} - h_{1,s} \simeq c_{p1}(T_o - T_s)$$

Substituting in Eq. (2.206),

$$\mathcal{B}_h = \frac{c_{pe}(T_e - T_s)}{c_{p1}(T_s - T_o)}$$

which is the same result as for case (i); hence, the required $\dot{m}''$ and heat transfer to the surface are the same.

Case (iii). Again we choose $T^0 = T_s$, but now we set $h_i(T_s) = 0$ for species 2 and 3:

$$h_1 = h_1^{T_s} + \int_{T_s}^{T} c_{p1} \, dT; \quad h_2 = \int_{T_s}^{T} c_{p2} \, dT; \quad h_3 = \int_{T_s}^{T} c_{p3} \, dT$$

$$h_e \simeq c_{pe}(T_e - T_s), \quad \text{as before}$$

$$h_s = 0 \quad (\text{since } m_{1,s} = 0: \text{ A}_2 \text{ dissociation takes place wholly}$$
$$\text{within the wall, i.e., at temperatures below } T_s)$$

$$h_o = h_1^{T_s} + \int_{T_s}^{T_o} c_{p1} \, dT \simeq h_1^{T_s} + c_{p1}(T_o - T_s)$$

where $h_1^{T_s}$ is the negative of Δh_d, the heat of dissociation of A$_2$ at temperature T_s. Substituting in Eq. (2.206),

$$\mathcal{B}_h = \frac{c_{pe}(T_e - T_s)}{c_{p1}(T_s - T_o) + \Delta h_d} \quad (\Delta h_d \text{ is positive}) \tag{2.207}$$

Thus, if the dissociation takes place within the wall, the heat taken up reduces the driving force and injection rate required to balance the surface heating. In contrast, if the dissociation takes place within the boundary layer, there is no effect on heat transfer to the surface. Of course, dissociation in the boundary layer does lower temperatures, and there are second-order effects via temperature-dependent properties. Also, the Lewis number is seldom exactly equal to unity.

5. *Evaporation of a multicomponent droplet.* Liquid hydrocarbon fuels are usually mixtures of a number of components, with different vapor pressures and other thermophysical properties. When such droplets evaporate, the more volatile components distill off more rapidly than the less volatile components, and the composition of the vapor mixture produced changes with time. Similar phenomena occur in distillation columns of oil-refining operations (and in the production of fine cognac). To illustrate how such problems can be analyzed, we consider the evaporation of an n-component droplet into an inert gas, I. To simplify the problem, we will assume that there is good mixing within the droplet so that the droplet composition and temperature are uniform. When this assumption is made, chemical engineers call the process *batch distillation*; the validity of the assumption will be discussed later. The rate of evaporation is given by

$$\dot{m}'' = g_{mi}\,\mathcal{B}_{mi} = g_h\,\mathcal{B}_h; \quad i = 1, 2, \ldots, n$$

$$\mathcal{B}_{mi} = \frac{m_{i,e} - m_{i,s}}{m_{i,s} - m_{i,t}}; \quad \mathcal{B}_h = \frac{h_e - h_s}{h_s - h_t} \tag{2.208}$$

$$m_{i,s} = m_{i,s}(m_{i,u}, T_s, P)$$

usually in the form of *Raoult's law.* (Henry's law is a limit form of Raoult's law.) The heat transfer driving force assumes a unity Lewis number, and is simplified by first taking the enthalpy datum state such that species enthalpies are zero at temperature T_s; then

$$\mathcal{B}_h = \frac{h_e - h_{es}}{-h_t^{T_s}} \tag{2.209}$$

An energy balance on a control volume between the u- and s-surfaces allows the transferred-state enthalpy to be evaluated:

$$\dot{m}'' h_t^{T_s} = \sum n_i h_{i,u}^{T_s} + q_{\text{cond}}$$

$$h_t^{T_s} = \sum m_{i,t}(-h_{\text{fgi}}) + (q_{\text{cond}}/\dot{m}'')$$

Assuming constant species specific heats and substituting in Eq. (2.209),

$$\mathcal{B}_h \simeq \frac{c_{pe}(T_e - T_s)}{\sum m_{i,t} h_{\text{fgi}} - (q_{\text{cond}}/\dot{m}'')} \tag{2.210}$$

For batch distillation we can use a lumped-capacity model for the droplet of radius R, to write

$$\frac{d}{dt}\left(\frac{4}{3}\pi R^3 \rho_l h\right) = -(4\pi R^2)q_{\text{cond}} \tag{2.211}$$

$$\frac{d}{dt}\left(\frac{4}{3}\pi R^3 \rho_l m_i\right) = -(4\pi R^2)m_{i,t}\dot{m}'', \quad i = 1, 2, \ldots, n \tag{2.212}$$

with initial conditions

$$t = 0: \; R = R_0, T_s = T_0, m_i = m_{i,0} \tag{2.213}$$

Notice that $h = h_u$ and $m_i = m_{i,u}$, since for batch distillation the liquid temperature and composition are assumed to be uniform. Equations (2.211) and (2.212) must be solved numerically, with Eqs. (2.208) solved simultaneously at each time step. However, Eqs. (2.208) are highly nonlinear, and care must be taken to develop a stable and accurate numerical procedure. Gas-phase properties can be evaluated using Hubbard's 1/3 rule extended to multicomponent evaporation.

The batch distillation model is a limiting case corresponding to perfect mixing inside the droplet: the opposite limit is a model that assumes a stationary liquid, and heat and mass transport in the droplet occur by conduction and diffusion. Results for these two limiting cases for a binary droplet are compared by Landis and Mills [23]. If there is relative motion between the droplet and the surrounding gas phase, a flow circulation can be induced in the droplet. Flow circulation models—for example, [24, 25]—give results in between those given by the batch distillation and conduction/diffusion models. However, under some circumstances, droplets may oscillate, with the predominant mode being a shape oscillation. Transport rates in oscillating droplets are much larger than in circulating droplets [26, 27, 28], and in such situations the batch distillation model may be more appropriate. Notice that our analysis did not require an expression for the heat flux across the u-surface, q_{cond}. The foregoing discussion suggests that it is difficult to specify q_{cond} because the liquid-gas interface does not behave like a solid wall. When a droplet oscillates, the flow inside the droplet is complex and poorly understood: such matters are the subject of current research.

2.7 TRANSPORT IN MULTICOMPONENT SYSTEMS

The Chapman–Enskog kinetic theory of gases, introduced in Section 1.7, yields results that are satisfactory for gases at low pressures. The theory describes ordinary diffusion in a multicomponent mixture, as well as thermal, forced, and pressure diffusion. In addition, it describes diffusional conduction (also called the diffusion thermo- or Dufour effect). In Section 1.7, formulas are presented for the calculation of viscosity, thermal conductivity, and binary diffusion coefficients. Here we will be concerned with various forms of the general diffusion and energy flux vectors, and a prescription for calculating thermal diffusion coefficients. The most useful reference work from which the results of the Chapman–Enskog theory may be obtained is *The Molecular Theory of Gases and Liquids,* by J. O. Hirschfelder, C. F. Curtiss, and R. B. Bird [29], hereafter denoted as H.C.B. A nice introduction to the kinetic theory of gases is given by L. C. Woods [30].

Models of transport in liquid solutions are not well developed. Ad hoc empirical models combined with the methodology of irreversible thermodynamics are used to obtain equations governing diffusion in multicomponent concentrated liquid solutions. Some of the main results will be presented here. Section 2.7 concludes with a discussion of forced diffusion in ionized gases and in dilute liquid solutions.

2.7.1 The Chapman–Enskog Kinetic Theory

Important features of this kinetic theory of gases are as follows:

1. The density of the mixture must be low enough for three-body collisions to occur with negligible frequency.

2. The model assumes monoatomic molecules, but little error is introduced by applying the results to polyatomic gases. The momentum flux and diffusion flux are not appreciably affected by the internal degrees of freedom. The energy flux vector, on the other hand, contains both the translational energy and the energy of the internal degrees of freedom; the "Eucken correction" is introduced to take this into account.

3. The flux vectors are general in the sense that they do not explicitly contain the force law that is assumed to characterize the molecular interaction. The various transport coefficients, however, do depend on the particular force law.

4. The solution of the Boltzmann equation involves an expansion in terms of Sonine polynomials. Chapman and Cowling [31] in their solution used an infinite series of these polynomials, with the result that the transport coefficients are expressed in terms of ratios of infinite determinants. However, to obtain numerical values, it is necessary to consider only a few elements of the determinants, since convergence is rapid as additional rows and columns are included. For viscosity, thermal conductivity, and ordinary diffusion, one term does not give a dependence of the coefficients on concentration, whereas two terms show a slight dependence. Thermal diffusion and diffusional conduction appear only when the second term is included, indicating that they are usually second-order effects.

5. The expressions for the flux vectors contain only the first spatial derivatives of temperature, pressure, concentration, and so on. Thus, the results are inapplicable when gradients change abruptly—for example, within a shock wave.

2.7.2 The Diffusion and Energy Flux Vectors

H.C.B. Eq. (8.1.1) gives the mass flux vector relative to the mass average velocity for the species i in a mixture of n components as:

$$
\begin{aligned}
\mathbf{j}_i &= \mathcal{N}_i m_i \hat{\mathbf{v}}_i \\
&= \frac{\mathcal{N}^2}{\rho} \sum_{j=1}^{n} m_i m_j D_{ij} \mathbf{d}_j - D_i^T \nabla \ln T
\end{aligned} \tag{2.214}
$$

where $\mathbf{d}_j$ includes the gradients of the number fraction $\mathcal{N}_j / \mathcal{N}$, pressure P and also the external forces $\mathbf{X}_k$ acting on the molecules,

$$
\mathbf{d}_j = \nabla \frac{\mathcal{N}_j}{\mathcal{N}} + \left(\frac{\mathcal{N}_j}{\mathcal{N}} - \frac{\mathcal{N}_j m_j}{\rho} \right) \nabla \ln P - \frac{\mathcal{N}_j m_j}{P\rho} \left(\frac{\rho}{m_j} \mathbf{X}_j - \sum_{k=1}^{n} \mathcal{N}_k \mathbf{X}_k \right) \tag{2.215}
$$

The mass of a molecule of species k is m_k, and $\mathbf{X}_k$ is the external force per molecule: thus, $\mathbf{X}_k = m_k \mathbf{f}_k$, where $\mathbf{f}_k$ is the force per unit mass. For a gravitational force field, the force per unit mass is constant: substitution in the last term of Eq. (2.215) gives

$$\frac{\mathcal{N}_j m_j}{P \rho} \left(\rho \mathbf{f}_j - \sum_{k=1}^{n} \rho_k \mathbf{f}_k \right) = 0 \quad \text{if } \mathbf{f}_k \text{ is constant.}$$

However, a gravitational field does produce a pressure gradient and thus indirectly yields a contribution to diffusion as pressure diffusion.

The D_{ij} are *multicomponent diffusion coefficients.* Except in binary mixtures, for which $D_{12} = \mathcal{D}_{12}$, $D_{ij} \neq D_{ji}$, $D_{ii} = 0$, and the D_{ij} are concentration dependent. The D_{ij} satisfy the following summation rule:

$$\sum_i (m_i m_h D_{ih} - m_i m_k D_{ik}) = 0 \tag{2.216}$$

H.C.B. relates the D_{ij} to the binary coefficients $\mathcal{D}_{ij}$; as mentioned above, for a binary mixture, $D_{12} = \mathcal{D}_{12}$, and for a ternary mixture

$$D_{12} = \mathcal{D}_{12} \left[1 + \frac{\mathcal{N}_3 [(m_3/m_2) \mathcal{D}_{13} - \mathcal{D}_{12}]}{\mathcal{N}_1 \mathcal{D}_{23} + \mathcal{N}_2 \mathcal{D}_{13} + \mathcal{N}_3 \mathcal{D}_{12}} \right] \tag{2.217}$$

The D_i^T are the multicomponent thermal diffusion coefficients and depend in a complex manner in temperature, concentration, molecular weights, and the force law of the molecular interaction. For a binary mixture it is important to note that the D_i^T defined here does not reduce to the coefficient of Chapman and Cowling [31], but the difference arises only due to a difference in the definitions.

Because the multicomponent diffusion coefficients D_{ij} are concentration dependent, it is convenient in some situations to replace the n equations given by Eq. (2.214) by a set of $(n-1)$ independent equations involving the binary diffusion coefficients $\mathcal{D}_{ij}$:

$$\sum_{j=1}^{n} \frac{\mathcal{N}_i \mathcal{N}_i}{\mathcal{N}^2 \mathcal{D}_{ij}} (\mathbf{v}_j - \mathbf{v}_i) = \mathbf{d}_i - \nabla \ln T \sum_{j=1}^{n} \frac{\mathcal{N}_i \mathcal{N}_i}{\mathcal{N}^2 \mathcal{D}_{ij}} \left(\frac{D_j^T}{\mathcal{N}_j m_j} - \frac{D_i^T}{\mathcal{N}_i m_j} \right) \tag{2.218}$$

These are the *Stefan–Maxwell* equations in their most general form; their derivation is presented in H.C.B. Section 7.4.

The relation $\sum \mathbf{j}_i = 0$ is easily proven; similar summation rates are valid for the ordinary, pressure and forced, and thermal diffusion components separately, which can be shown as follows:

Sum Eq. (2.214) over all species,

$$\sum_{i=1}^{n} \mathbf{j}_i = \frac{\mathcal{N}^2}{\rho} \sum_{i=1}^{n} \sum_{j=1}^{n} m_i m_j D_{ij} \mathbf{d}_j - \sum_{i=1}^{n} D_i^T \nabla \ln T$$

$$= \frac{\mathcal{N}^2}{\rho} \sum_{j=1}^{n} \mathbf{d}_j \sum_{i=1}^{n} m_i m_j D_{ij} - \sum_{i=1}^{n} D_i^T \nabla \ln T$$

$$= \frac{\mathcal{N}^2}{\rho} \sum_{j=1}^{n} \mathbf{d}_j \left(\sum_{i=1}^{n} m_i m_h D_{ih} \right) - \sum_{i=1}^{n} D_i^T \nabla \ln T \quad \text{[using Eq. (2.216)]}$$

$$= \frac{\mathcal{N}^2}{\rho} \left(\sum_{i=1}^{n} m_i m_h D_{ih} \right) \sum_{j=1}^{n} \mathbf{d}_j - \sum_{i=1}^{n} D_i^T \nabla \ln T$$

Now consider each term of $\sum_{i=1}^{n} \mathbf{d}_j$ in turn:

$$\sum_{j} \nabla \frac{\mathcal{N}_i}{\mathcal{N}} = \sum_{j} \nabla x_j = 0, \quad \text{since } \sum_{j} x_j = 1$$

$$\sum_{j} \left(\frac{\mathcal{N}_j}{\mathcal{N}} - \frac{\mathcal{N}_j m_j}{\rho} \right) \nabla \ln P = \nabla \ln P \left[\sum_{j} x_j - \sum_{j} m_j \right]$$

$$= \nabla \ln P (1 - 1)$$

$$= 0$$

$$\sum_{j} \frac{\mathcal{N}_j m_j}{\rho P} \left[\frac{\rho}{m_j} \mathbf{X}_j - \sum_{k} \mathcal{N}_k \mathbf{X}_k \right] = \frac{1}{P} \left[\sum_{j} \mathcal{N}_j \mathbf{X}_j - \sum_{k} \mathcal{N}_k \mathbf{X}_k \sum_{j} \frac{\rho_j}{\rho} \right]$$

$$= \frac{1}{P} \left[\sum_{j} \mathcal{N}_j \mathbf{X}_j - \sum_{k} \mathcal{N}_k \mathbf{X}_k \right]$$

$$= 0$$

Thus, summation rules have been proven for ordinary, pressure, and forced diffusion in turn. For thermal diffusion, the rule follows immediately from the relation $\sum_i D_i^T = 0$, or by subtraction.

H.C.B. Eq. (8.1.23) gives the energy flux relative to the mass average velocity in a multicomponent mixture as

$$\mathbf{q} = -k \nabla T + \frac{5}{2} kT \sum_{i=1}^{n} \mathcal{N}_i \hat{\mathbf{v}}_i + \frac{kT}{\mathcal{N}} \sum_{i=1}^{n} \sum_{j=1}^{n} \frac{\mathcal{N}_j D_i^T}{m_i \mathcal{D}_{ij}} (\hat{\mathbf{v}}_i - \hat{\mathbf{v}}_j) \qquad (2.219)$$

The first term is ordinary thermal conduction. The second term is the interdiffusion energy flux since, $\mathcal{N}_i \hat{\mathbf{v}}_i$ is the diffusion flux of species i relative to the mass average velocity in molecules/unit time-unit area, and each monatomic molecule carries, on an average, a quantity of thermal energy equal to $\frac{5}{2}kT$ (relative to the mass average velocity $\mathbf{v}$). For polyatomic molecules the interdiffusion term becomes $\sum_{i=1}^{n} \mathcal{N}_i m_i h_i \hat{\mathbf{v}}_i$, where h_i is the enthalpy per unit mass of species i. H.C.B. Section 7.6b discusses this modification for polyatomic molecules. The last term in Eq. (2.219) is the diffusional conduction. Notice that the model used in deriving the energy equation for a laminar boundary layer on a flat plate in Section 2.5.2 gives the interdiffusion term directly.

Simplification for a Binary Mixture

We first write down some of the simplifications that obtain for a binary system in terms of microscopic parameters. The diffusion flux as given by Eq. (2.214) reduces to:

$$\mathbf{j}_1 = \mathcal{N}_1 m_1 \hat{\mathbf{v}}_1$$

$$= \frac{\mathcal{N}^2}{\rho} m_1 m_2 \mathcal{D}_{12} \mathbf{d}_2 - D_1^T \nabla \ln T \qquad (2.220)$$

whereas Eq. (2.215) reduces to:

$$\mathbf{d}_2 = \nabla \frac{\mathcal{N}_2}{\mathcal{N}} + \left(\frac{\mathcal{N}_2}{\mathcal{N}} - \frac{\mathcal{N}_2 m_2}{\rho} \right) \nabla \ln P - \frac{\mathcal{N}_1 \mathcal{N}_2}{P\rho} (m_1 \mathbf{X}_2 - m_2 \mathbf{X}_1) \qquad (2.221)$$

The coefficient of $\nabla \ln P$ in this equation is simply the difference between the mole fraction and the mass fraction; hence it follows that

$$\mathbf{d}_1 = -\mathbf{d}_2 \qquad (2.222)$$

Also, since $\mathbf{j}_1 = -\mathbf{j}_2$, it follows that

$$\mathcal{D}_{12} = \mathcal{D}_{21} \quad \text{and} \quad D_1^T = -D_2^T \qquad (2.223)$$

By writing Eq. (2.220) for $\mathbf{j}_1$ and $\mathbf{j}_2$, respectively, and subtracting, the Stefan–Maxwell form is obtained as

$$(\mathbf{v}_1 - \mathbf{v}_2) = -\frac{\mathcal{N}^2}{\mathcal{N}_1 \mathcal{N}_2} \mathcal{D}_{12}[\mathbf{d}_1 + k_T \nabla \ln T] \qquad (2.224)$$

where the *thermal diffusion ratio* k_T has been introduced and is defined as

$$k_T = -\frac{\rho}{\mathcal{N}^2 m_1 m_2} \frac{D_1^T}{\mathcal{D}_{12}} \qquad (2.225)$$

Thus, k_T is a measure of the relative importance of thermal and ordinary diffusion.

The foregoing flux vectors contain microscopic parameters such as the mass of a molecule and number density. For purposes of engineering analysis we prefer to have these vectors expressed in terms of appropriate continuum parameters. The microscopic parameters can be eliminated from Eqs. (2.220), (2.221) and (2.225) by introducing Avogadro's number $\mathcal{A}$ in the following identities:

$$m_1 = M_1/\mathcal{A}; \quad \mathcal{N} = \mathcal{A}c$$

The three equations become, respectively,

$$\mathbf{j}_1 = \rho_1 \hat{\mathbf{v}}_1 = \frac{c^2}{\rho} M_1 M_2 \mathcal{D}_{12} \mathbf{d}_2 - D_1^T \nabla \ln T \tag{2.226}$$

$$\mathbf{d}_2 = \nabla x_2 - (x_2 - m_2) \nabla \ln P - \frac{m_1 m_2}{RT}(\mathbf{f}_2 - \mathbf{f}_1) = -\mathbf{d}_1 \tag{2.227}$$

$$k_T = \frac{\rho}{c^2 M_1 M_2} \frac{D_1^T}{\mathcal{D}_{12}} = \frac{M^2}{M_1 M_2 \rho} \frac{D_1^T}{\mathcal{D}_{12}} \tag{2.228}$$

where $R = \mathcal{R}/M$ and $1/M = m_1/M_1 + m_2/M_2$. If Eq. (2.227) is written for $\mathbf{d}_1$ and substituted in Eq. (2.226), then after some manipulation,

$$\mathbf{j}_1 = -\rho \mathcal{D}_{12} \left[\nabla m_1 + \frac{M_2}{M} \left(1 - \frac{M_1}{M} \right) m_1 \nabla \ln P \right.$$
$$\left. - \frac{M_1 M_2}{M^2} \frac{m_1 m_2}{RT}(\mathbf{f}_1 - \mathbf{f}_2) + \frac{M_1 M_2}{M^2} k_T \nabla \ln T \right] \tag{2.229}$$

By using the relation $\mathbf{J}_1^* = (M/M_1 M_2)\mathbf{j}_1$, the equivalent molar form is obtained,

$$\mathbf{J}_1^* = -c \mathcal{D}_{12} \left[\nabla x_1 + \left(1 - \frac{M_1}{M} \right) x_1 \nabla \ln P \right.$$
$$\left. - \frac{x_2 x_1}{M^2 RT}(M_2 \mathbf{F}_1 - M_1 \mathbf{F}_2) + k_T \nabla \ln T \right] \tag{2.230}$$

where $\mathbf{F}$ is the external force per mole.

To obtain the binary mixture energy flux vector in terms of continuum parameters, the first step is to rewrite Eq. (2.219) for a binary system, and then replace the microscopic parameters by their macroscopic counterparts, yielding final forms in terms of either mass or molar units. For a polyatomic gas Eq. (2.219) becomes

$$\mathbf{q} = -k \nabla T + \mathbf{j}_1(h_1 - h_2) + \frac{kT}{\mathcal{N}} \frac{D_1^T}{\mathcal{D}_{12}} \left[\frac{\mathcal{N}_2}{m_1}(\hat{\mathbf{v}}_1 - \hat{\mathbf{v}}_2) - \frac{\mathcal{N}_1}{m_2}(\hat{\mathbf{v}}_2 - \hat{\mathbf{v}}_1) \right] \tag{2.231}$$

where the relations $\mathbf{j}_1 = \mathcal{N}_1 m_1 \hat{\mathbf{v}}_1 = -\mathbf{j}_2$ and $D_1^T = -D_2^T$ have been used. If the relation $k = \mathcal{R}/\mathcal{A} = RM/\mathcal{A}$ is then introduced, after some manipulation Eq. (2.231) in mass terms becomes

$$\mathbf{q} = -k \nabla T + \mathbf{j}_1(h_1 - h_2) + \mathbf{j}_1 k_T RT \frac{\rho^2}{\rho_1 \rho_2} \tag{2.232}$$

or

$$\mathbf{q} = -k \nabla T + \mathbf{j}_1(h_1 - h_2) + \mathbf{j}_1 \frac{k_T RT}{m_1 m_2} \tag{2.233}$$

An alternative form is obtained by introducing the *thermal diffusion factor* $\alpha_{12} = k_T/x_1 x_2$. With 1 denoting the heavier species, α_{12} is positive—that is, the heavier species diffuses down the temperature gradient. Substitution in Eq. (2.232) gives

$$\mathbf{q} = -k\nabla T + \mathbf{j}_1(h_1 - h_2) + \mathbf{j}_1\alpha_{12}RT\frac{M^2}{M_1 M_2} \tag{2.234}$$

The merit of this last equation is that, whereas k_T is strongly dependent on composition, α_{12} is essentially independent of composition. Thus, data for thermal diffusion coefficients are more conveniently expressed in terms α_{12}.

In molar terms Eq. (2.234) becomes

$$\mathbf{q} = -k\nabla T + \left(\frac{M_2}{M}H_1 - \frac{M_1}{M}H_2\right)\mathbf{J}_1^* + \alpha_{12}RT\mathbf{J}_1^* \tag{2.235}$$

where H is the enthalpy per mole.

Evaluation of the Thermal Diffusion Factor

Data for thermal diffusion coefficients are sparse and unreliable. Owing to its relative insensitivity to concentration, the thermal diffusion factor α_{12} is the preferred form for presenting experimental or theoretical results for thermal diffusion in binary mixtures. Rosner [32] has proposed a simple correlation for α_{12} that has the form

$$\alpha_{12} = \alpha_{12}^\infty\left[1 - \frac{T^*}{T}\right] \tag{2.236}$$

where α_{12}^∞ is the value of α_{12} as T approaches infinity and T^* is a characteristic temperature. Table 2.2 lists values of α_{12}^∞ and T^* for various species in air. Figure 2.37 shows the molecular weight variation of α_{12}^∞ for various species in air. Additional data may be found in References [31, 33, 34].

Table 2.2 Constants for use in Eq. (2.236) for the thermal diffusion factor, from Rosner [32].

Species	M	α_{12}^∞	T^*, K
H	1.01	−0.2901	17.94
H_2	2.016	−0.2881	30.37
He	4.000	−0.2707	−29.95
H_2O	18.02	−0.1424	290.6
CO	28.01	−0.00888	32.02
H_2S	34.08	0.05918	79.38
HCl	36.74	0.0687	90.02
NaOH	40.00	0.1626	394.6
CO_2	44.01	0.1480	49.56
NaCl	58.45	0.3462	420.0
SO_2	64.07	0.2948	104.6
SO_3	80.07	0.3825	138.9
Na_2SO_4	142.04	0.7765	458.6
UF_6	352.02	0.9001	67.12
WCl_6	396.66	1.048	228.6
UI_4	746.65	1.267	329.6

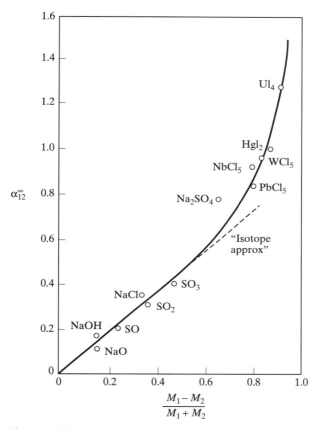

Figure 2.37   Molecular weight dependence of α_{12}^{∞}; species 2 is air.

2.7.3 Diffusion in Liquid Solutions

As mentioned in Section 1.2.3, experiment has shown that Fick's law is valid for diffusion in dilute liquid solutions. Diffusion in liquids in commonly analyzed in a molar basis: since for dilute solutions, with moderate temperature and pressure gradients, c and ρ can be taken as constants, the precise manner in which Fick's law is written is unimportant: commonly it is written

$$\mathbf{J}_i = -\mathcal{D}\nabla c_i \qquad (2.237)$$

where the diffusion coefficient $\mathcal{D}_i$ has units $kmol/m^2s$, and no distinction is made between mass or molar average velocity related quantities.

In concentrated multicomponent solutions, Fick's law is no longer valid. Even in ternary systems, there are two independent concentration gradients, and the diffusion flux of each species can be affected by both concentration gradients. For a concentrated electrolyte at constant temperature and pressure, the appropriate driving potential for combined ordinary and forced diffusion is the electrochemical potential, customarily denoted μ_i, and we write

$$c_i \nabla \mu_i = \sum_j K_{ij}(\mathbf{v}_j - \mathbf{v}_i) \tag{2.238}$$

where K_{ij} are termed friction coefficients. Equation (2.238) is based on a *flow model* of diffusion and expresses the simple idea that the movement of species i is resisted by the other species j, the resistance being proportional to the relative velocity of the two species. The friction coefficients are replaced by binary diffusion coefficients $\mathcal{D}_{ij}$,

$$K_{ij} = -\frac{\mathcal{R}T c_i c_j}{c \mathcal{D}_{ij}} \tag{2.239}$$

to obtain

$$c_i \nabla \mu_i = \mathcal{R}T \sum_j \frac{c_i c_j}{c \mathcal{D}_{ij}}(\mathbf{v}_j - \mathbf{v}_i) \tag{2.240}$$

which are similar to the Stefan–Maxwell equations introduced for multicomponent gas mixtures as Eq. (2.218). By Newton's third law of motion $K_{ji} = K_{ij}$, thus

$$\mathcal{D}_{ij} = \mathcal{D}_{ji} \tag{2.241}$$

The number of independent equations given by Eq. (2.240) is the number of species less one, since summation of Eq. (2.238) over i gives

$$\sum_i c_i \nabla \mu_i = \sum_i \sum_j K_{ij}(\mathbf{v}_j - \mathbf{v}_i) \tag{2.242}$$

At constant pressure and temperature the left-hand side is zero by the Gibbs–Duhem relation, and the right-hand side is zero since $K_{ji} = K_{ij}$.

Equation (2.240) applies at constant temperature and pressure. Newman [35] gives the general form as

$$c_i \left(\nabla \mu_i + \bar{S}_i \nabla T - \frac{M_i}{\rho} \nabla P \right)$$

$$= \mathcal{R}T \sum_j \frac{c_i c_j}{c \mathcal{D}_{ij}} \left[\mathbf{v}_j - \mathbf{v}_i + \left(\frac{D_j^T}{\rho_j} - \frac{D_i^T}{\rho_i} \right) \nabla \ln T \right] \tag{2.243}$$

where $\bar{S}_i$ is the partial molar entropy of species i. The driving forces on the left-hand side will sum to zero even when pressure and temperature vary, and thermal diffusion has been included on the right-hand side of the equation.

2.7.4 Forced Diffusion

Ionized Gases

If an electric field is applied to an ionized gas, the negatively charged electrons diffuse to the anode and the positively charged ions migrate to the cathode. The molar flux due to forced diffusion of a singly ionized species is

$$\mathbf{J}_i^{*F} = \mu_i \mathbf{E} c x_i \tag{2.244}$$

where E [V/m] is the electric field strength or voltage gradient, and μ_i [m^2/V s] is the *mobility* of species i. The Nernst–Einstein relation is used to relate mobility to the binary diffusion coefficient,

$$\mu_i = \frac{e\mathcal{D}_{im}}{kT} \qquad (2.245)$$

where e is the elementary electric charge, 1.60×10^{-19} A s, and k is the Boltzmann constant, 1.38×10^{-23} J/K. Notice that $\mu_i \mathbf{E}$ has the dimensions of velocity; the phenomenon of forced diffusion in an ionized gas is often called cataphoresis, and $\mu_i \mathbf{E}$ is the *cataphoretic velocity*. An analysis of forced diffusion in a helium–cadmium laser is required as Exercise 2–18.

Dilute Liquid Electrolytes

Following the practice of electrochemists, we write the forced diffusion flux of an ionic species in a dilute electrolyte as

$$\mathbf{J}_i^{*F} = -u_i z_i F c_i \nabla \Phi \qquad (2.246)$$

where Φ is the electrostatic potential, and $-\nabla \Phi = \mathbf{E}$ [V/m] is the electric field strength. Also, F is Faraday's constant and equals 9.6487×10^7 coulombs/kg equivalent, z_i is the number of proton charges carried by the ion, and u_i [kmol/m^{2}J s] is the mobility of the ion. Notice that we use μ_i to denote the mobility of a gaseous ion and u_i for a liquid ion: they have different units, and it is advisable to be consistent with common practice in plasma physics and electrochemistry respectively. The product $z_i F$ is the charge per mole on species i, $z_i F(-\nabla \Phi)$ is the force per mole, and $u_i z_i F(-\nabla \Phi)$ is the *migration velocity*. The mobility u_i is related to the diffusion coefficient D_i by the Nernst–Einstein equation

$$\mathcal{D}_i = \mathcal{R}T u_i \qquad (2.247)$$

which is strictly valid only in the limit of an infinitely dilute solution. Table 2.3 gives diffusion coefficients and the number of proton charges of some ions for infinite dilution in water.

Table 2.3 Diffusion coefficients of ions for infinite dilution in water at 25° C^a [35].

Ion	z_i	$\mathcal{D}_i \times 10^9$, m^2/s	Ion	z_i	$\mathcal{D}_i \times 10^9$ m^2/s
H$^+$	1	9.321	OH$^-$	-1	5.26
Na$^+$	1	1.334	Cl$^-$	-1	2.03
K$^+$	1	1.957	NO$_3^-$	-1	1.90
NH$_4^+$	1	1.954	HCO$_3^-$	-1	1.105
Ag^{++}	1	1.648	SO$_4^-$	-2	1.065
Mg^{++}	2	0.706	Fe(CN)$_6^{3-}$	-3	0.896
Ca^{++}	2	0.792	Fe(CN)$_6^{4-}$	-4	0.739
Cu^{++}	2	0.72	ClO$_4^-$	-1	1.79
Zn^{++}	2	0.71	HSO$_4^-$	-1	1.33

^{a}The Wilke-Chang formula, Eq. (1.122), can be used to obtain values at other temperatures ($\mathcal{D}_i \mu / T = $ Constant).

2.8 CONSERVATION EQUATIONS FOR A MULTICOMPONENT GAS MIXTURE

Analysis of mass, momentum, and heat transport in gas mixtures requires a set of *conservation equations* or *equations of change*. The equations in common use are derived from kinetic theory or coexisting continua models of a multicomponent gas mixture. The coexisting continua model was used to derive the species conservation equation in Section 2.2.3 and to derive the momentum and energy conservation equations for a laminar boundary layer of a binary mixture in Section 2.5.2. Our objective here is to derive more general momentum and energy conservation equations.

Kinetic-theory derivations of the conservation equations are given by Chapman and Cowling [31], Hirschfelder, Curtiss, and Bird [29], Woods [30, 36], Vincenti and Kruger [37], and Williams [38]. It is not easy to master these derivations: both the physics and mathematics are challenging, and a number of issues are controversial and have not been satisfactorily resolved [39]. Fortunately, it is quite easy to derive a set of conservation equations that on one hand are not completely general, but on the other hand are general enough for most engineering situations. In Section 2.8.1, such a derivation is given; also special forms that have been used in practice are obtained in Section 2.8.2.

2.8.1 Conservation Equations for a Model Chemically Reacting Multicomponent Gas Mixture

The model considered is not completely general. The only external force allowed is gravity, and stresses and kinetic energy that may be associated with pressure and thermal diffusion are assumed negligible. This model allows the momentum and energy conservation equations to be derived as simple extensions to the equations for a pure fluid. To obtain more general conservation equations we have to derive separate equations for each species that include interactions between species, and then sum over all species to obtain mixture equations. Such a derivation is a challenging task and if not necessary should be avoided [39]. Since we do not include external force fields other than gravity, the equations obtained do not apply to an ionized gas plasma. On the other hand, it is probably safe to say that the literature does not contain a single example of a situation where stresses or kinetic energy associated with pressure or thermal diffusion are significant; thus, neglect of such terms is of no consequence.

Conservation of Species
The species conservation equation derived in Section 2.2.3 is quite general; on a mass basis Eq. (2.13) is

$$\frac{\partial \rho_i}{\partial t} + \nabla \cdot \mathbf{n}_i = \dot{r}_i''' \tag{2.13}$$

Substituting $\mathbf{n}_i = \rho_i \mathbf{v} + \mathbf{j}_i$ and rearranging,

$$\frac{\partial \rho_i}{\partial t} + \nabla \cdot \rho_i \mathbf{v} = -\nabla \cdot \mathbf{j}_i + \dot{r}_i''' \tag{2.248}$$

The diffusion flux vector can be obtained from the Chapman–Enskog kinetic theory of gases; from Eqs. (2.214) and (2.215) written in terms of continuum parameters

$$\mathbf{j}_i = \rho \sum_{j=1}^{n} \frac{M_i M_j}{M^2} D_{ij} \left[\nabla x_j + (x_j - m_j) \nabla \ln P \right] - D_i^T \nabla \ln T \qquad (2.249)$$

where there is no forced diffusion term since, for gravity, the force per unit mass is constant for each species.

Conservation of Momentum

The momentum conservation equation for a pure fluid is derived in Section 5.7.2 of *Heat Transfer*. This derivation can be generalized to apply to our model gas mixture by simply replacing the pure fluid density and viscosity by mixture properties. The density ρ becomes $\sum \rho_i$, and the viscosity is given by a mixture rule—for example, the Wilke formula, Eq. (1.116). As in Section 5.7.2, we will ignore the contributions of terms involving $\nabla \cdot \mathbf{v}$ to the normal stresses. Then Eqs. (5.195)–(5.198) also apply to our model gas mixture, and the momentum conservation equations become

$$\rho \left(\frac{\partial u}{\partial t} + u \frac{\partial u}{\partial x} + v \frac{\partial u}{\partial y} + w \frac{\partial u}{\partial z} \right)$$
$$= -\frac{\partial P}{\partial x} + \frac{\partial}{\partial x} \left(\mu \frac{\partial u}{\partial x} \right) + \frac{\partial}{\partial y} \left(\mu \frac{\partial u}{\partial y} \right) + \frac{\partial}{\partial z} \left(\mu \frac{\partial u}{\partial z} \right) + \rho g_x \qquad (2.250a)$$

$$\rho \left(\frac{\partial v}{\partial t} + u \frac{\partial v}{\partial x} + v \frac{\partial v}{\partial y} + w \frac{\partial v}{\partial z} \right)$$
$$= -\frac{\partial P}{\partial y} + \frac{\partial}{\partial x} \left(\mu \frac{\partial v}{\partial x} \right) + \frac{\partial}{\partial y} \left(\mu \frac{\partial v}{\partial y} \right) + \frac{\partial}{\partial z} \left(\mu \frac{\partial v}{\partial z} \right) + \rho g_y \qquad (2.250b)$$

$$\rho \left(\frac{\partial w}{\partial t} + u \frac{\partial w}{\partial x} + v \frac{\partial w}{\partial y} + w \frac{\partial w}{\partial z} \right)$$
$$= -\frac{\partial P}{\partial z} + \frac{\partial}{\partial x} \left(\mu \frac{\partial w}{\partial x} \right) + \frac{\partial}{\partial y} \left(\mu \frac{\partial w}{\partial y} \right) + \frac{\partial}{\partial z} \left(\mu \frac{\partial w}{\partial z} \right) + \rho g_z \qquad (2.250c)$$

Equations (2.250) can be written in the compact form

$$\rho \frac{D\mathbf{v}}{Dt} = -\nabla P + \nabla \cdot (\mu \nabla \mathbf{v}) + \rho \mathbf{g} \qquad (2.251)$$

Conservation of Energy

The energy equations for a pure fluid are derived in Section 5.7.3. This derivation can be generalized to apply to our model gas mixture by noting the following:

1. The density ρ is the mixture density, $\sum_i \rho_i$
2. The internal energy u is the mixture internal energy, $\sum_i m_i u_i$
3. The enthalpy h is the mixture enthalpy, $\sum_i m_i h_i$
4. The heat transfer into the control volume is the energy flux $\mathbf{q}$ relative to the mass average velocity given by Eq. (2.219). Rewriting this relation in terms of

macroscopic variables for polyatomic molecules,

$$
\mathbf{q} = -k\nabla T + \sum_i \mathbf{j}_i h_i + \frac{RT}{\rho}\sum_i \sum_j \frac{M^2 D_i^T}{M_i M_j \mathcal{D}_{ij}}\left(\frac{m_j}{m_i}\mathbf{j}_i - \mathbf{j}_j\right) \tag{2.252}
$$

5. The heat generation term remains unchanged. Note that heat release due to chemical reactions is included in the internal energy or enthalpy terms.

Then the total energy equation for the mixture can be written down using Eq. (5.202) as

$$
\rho\frac{D}{Dt}\left(u + \frac{1}{2}v^2\right) = -\nabla \cdot \mathbf{q} + \rho\mathbf{g}\cdot\mathbf{v} - \frac{\partial}{\partial x}(\mathbf{v}\cdot\mathbf{F}_x)
$$
$$
- \frac{\partial}{\partial y}(\mathbf{v}\cdot\mathbf{F}_y) - \frac{\partial}{\partial z}(\mathbf{v}\cdot\mathbf{F}_z) + \dot{Q}_v''' \tag{2.253}
$$

where $\mathbf{F}_x = \mathbf{i}(P + \tau_{xx}) + \mathbf{j}(\tau_{xy}) + k(\tau_{xz})$, etc.

The thermal energy equation in terms of internal energy is obtained from Eq. (5.205) as

$$
\rho\frac{Du}{Dt} = -P\cdot\nabla\mathbf{v} - \nabla\cdot\mathbf{q} + \mu\Phi + \dot{Q}_v''' \tag{2.254}
$$

where the dissipation function Φ is given by Eq. (5.206). The thermal energy equation in terms of enthalpy is obtained from (5.207) as

$$
\rho\frac{Dh}{Dt} = \frac{DP}{Dt} - \nabla\cdot\mathbf{q} + \mu\Phi + \dot{Q}_v''' \tag{2.255}
$$

Notice that Eq. (5.208) cannot be generalized by simply replacing $\nabla k\nabla T$ with $-\nabla\cdot\mathbf{q}$ because $dh \neq c_p dT$ for mixture (see Section 2.8.2).

2.8.2 Simplifications for a Steady Boundary Layer Flow of an Inert Binary Mixture

Many engineering analyses have been performed for steady boundary-layer flows of inert binary mixtures. Thus it is useful to develop some forms of the conservation equations that have been used in these analyses.

Laminar Boundary Layers
Using Eq. (2.229) and noting that $dP/dy = 0$ for a boundary layer, the species conservation equation Eq. (2.248) becomes

$$
\rho u\frac{\partial m_1}{\partial x} + \rho v\frac{\partial m_1}{\partial y} = \frac{\partial}{\partial y}\left\{\rho\mathcal{D}_{12}\left[\frac{\partial m_1}{\partial y} + \frac{M_1 M_2}{M^2}\frac{k_T}{T}\frac{\partial T}{\partial y}\right]\right\} \tag{2.256}
$$

The momentum conservation equation Eq. (2.251) becomes

$$
\rho u\frac{\partial u}{\partial x} + \rho v\frac{\partial u}{\partial y} = -\frac{dP}{dx} + \frac{\partial}{\partial y}\left(\mu\frac{\partial u}{\partial y}\right) + \rho g_x \tag{2.257}
$$

Using Eq. (2.234) the thermal energy equation in terms of enthalpy Eq. (2.255) becomes

$$\rho u \frac{\partial h}{\partial x} + \rho v \frac{\partial h}{\partial y} = u \frac{dP}{dx} - \frac{\partial}{\partial y} \left\{ -k \frac{\partial T}{\partial y} - \rho \mathcal{D}_{12} \frac{\partial m_1}{\partial y} \left[(h_1 - h_2) + \alpha_{12} R T \frac{M^2}{M_1 M_2} \right] \right\}$$
$$+ \mu \left(\frac{\partial u}{\partial y} \right)^2 + \dot{Q}_v''' \qquad (2.258)$$

It is useful to have the thermal energy equation in terms of temperature. As in Section 2.3.5, we start by writing $h = \sum_i m_i h_i$, and then

$$dh = \sum m_i dh_i + \sum h_i dm_i$$
$$= \sum m_i c_{pi} dT + \sum h_i dm_i \qquad (2.259)$$
$$dh = c_p dT + (h_1 - h_2) dm_1$$

for a binary mixture, with $c_p = m_1 c_{p1} + m_2 c_{p2}$ as the mixture specific heat. Then the left side of Eq. (2.258) becomes

$$\rho u \frac{\partial h}{\partial x} + \rho v \frac{\partial h}{\partial y} = \rho u c_p \frac{\partial T}{\partial x} + \rho v c_p \frac{\partial T}{\partial y} + \left(\rho u \frac{\partial m_1}{\partial x} + \rho v \frac{\partial m_1}{\partial y} \right)(h_1 - h_2)$$

To obtain a further simplification of the equations we now ignore thermal diffusion. Then, using Eq. (2.256),

$$\rho u \frac{\partial h}{\partial x} + \rho v \frac{\partial h}{\partial y} = \rho u c_p \frac{\partial T}{\partial x} + \rho v c_p \frac{\partial T}{\partial y} + \frac{\partial}{\partial y} \left(\rho \mathcal{D}_{12} \frac{\partial m_1}{\partial y} \right)(h_1 - h_2)$$

Substituting in Eq. (2.258) with $\dot{Q}_v''' = 0$ gives

$$\rho u c_p \frac{\partial T}{\partial x} + \rho v c_p \frac{\partial T}{\partial x} = u \frac{dP}{dx} - \frac{\partial}{\partial y} \left(-k \frac{\partial T}{\partial y} - \rho \mathcal{D}_{12} \frac{\partial m_1}{\partial y}(h_1 - h_2) \right)$$
$$- \frac{\partial}{\partial y} \left(\rho \mathcal{D}_{12} \frac{\partial m_1}{\partial y} \right)(h_1 - h_2) + \mu \left(\frac{\partial u}{\partial y} \right)^2 \qquad (2.260)$$

Introducing Prandtl and Schmidt numbers for the mixture, $Pr = c_p \mu / k$, $Sc = \mu / \rho \mathcal{D}_{12}$, and assuming an inert mixture to eliminate enthalpies of formation gives

$$\rho u \frac{\partial T}{\partial x} + \rho v \frac{\partial T}{\partial y} = \frac{u}{c_p} \frac{dP}{dx} + \frac{\partial}{\partial y} \left(\frac{\mu}{Pr} \frac{\partial T}{\partial y} \right)$$
$$+ \frac{1}{c_p} \left[\frac{\mu}{Pr} \frac{\partial c_p}{\partial y} \frac{\partial T}{\partial y} + \frac{\mu}{Sc}(c_{p1} - c_{p2}) \frac{\partial m_1}{\partial y} \frac{\partial T}{\partial y} \right]$$
$$+ \frac{\mu}{c_p} \left(\frac{\partial u}{\partial y} \right)^2 \qquad (2.261)$$

This form of the energy conservation equation has been widely used—for example, in the analysis of condensation in the presence of a noncondensable gas [40, 41].

Turbulent Boundary Layers

We will restrict our attention to use of an eddy diffusivity model for turbulent transport as described in Section 5.5.1. In order to obtain conservation equations for turbulent flow directly from their laminar counterparts, we define effective transport properties as

$$\mu_{eff} = \mu + \rho \varepsilon_M \tag{2.262a}$$

$$\frac{\mu_{eff}}{Pr_{eff}} = \frac{\mu}{Pr} + \frac{\rho \varepsilon_M}{Pr_t} \tag{2.262b}$$

$$\frac{\mu_{eff}}{Sc_{eff}} = \frac{\mu}{Sc} + \frac{\rho \varepsilon_M}{Sc_t} \tag{2.262c}$$

where ε_M is the eddy viscosity (eddy diffusivity of momentum) and Pr_t and Sc_t are the turbulent Prandtl and Schmidt numbers, respectively. Here we are ignoring forced, thermal, and pressure diffusion and diffusional conduction. Then the boundary layer momentum, mass, and energy conservation equations are obtained directly from Eqs. (2.257), (2.256), and (2.261) as

$$\rho u \frac{\partial u}{\partial x} + \rho v \frac{\partial u}{\partial y} = -\frac{dP}{dx} + \frac{\partial}{\partial y}\left(\mu_{eff}\frac{\partial u}{\partial y}\right) + \rho g_x \tag{2.263}$$

$$\rho u \frac{\partial m_1}{\partial x} + \rho v \frac{\partial m_1}{\partial y} = \frac{\partial}{\partial y}\left[\frac{\mu_{eff}}{Sc_{eff}}\frac{\partial m_1}{\partial y}\right] \tag{2.264}$$

$$\rho u \frac{dT}{\partial x} + \rho v \frac{dT}{\partial y} = \frac{u}{c_p}\frac{\partial P}{\partial x} + \frac{\partial}{\partial y}\left(\frac{\mu_{eff}}{Pr_{eff}}\frac{\partial T}{\partial y}\right)$$

$$+ \frac{1}{c_p}\left[\frac{\mu_{eff}}{Pr_{eff}}\frac{\partial c_p}{\partial y}\frac{\partial T}{\partial y} + \frac{\mu_{eff}}{Sc_{eff}}(c_{p1} - c_{p2})\frac{\partial m_1}{\partial y}\frac{\partial T}{\partial y}\right]$$

$$+ \frac{\mu_{eff}}{c_p}\left(\frac{\partial u}{\partial y}\right)^2 \tag{2.265}$$

2.9 CLOSURE

In this chapter the theory of simultaneous diffusion and convection was developed, with applications including heatpipes and volatile liquid fuel droplet combustion. High mass transfer rate convection was introduced, initially with emphasis on engineering problem solving. Applications included transpiration and sweat cooling, and the effect of air leaks on condenser performance. Rigorous analysis of high mass transfer rate convection requires the numerical solution of coupled nonlinear governing differential equations. As an example, the solution for the constant-property, forced-convection laminar boundary layer on a flat plate was obtained. The advantage of an enthalpy formulation for simultaneous heat and mass transfer was demonstrated, with special emphasis placed on gaining an understanding of the role played by energy transport due to interdiffusion of species in a mixture. The discussion was rather long and involved, but mastery of these concepts is essential before the student proceeds to the research literature.

The chapter continued with a general problem-solving procedure for steady convective heat and mass transfer problems. The procedure is based on a generalization of the various analyses presented earlier in the chapter, and it was not necessary to develop the theory underlying the procedure from first principles. The chapter concluded with results from the Chapman–Enskog kinetic theory of multicomponent gas mixtures, and conservation equations for model multicomponent gas mixtures.

REFERENCES

1. Hubbard, G. L., Denny, V. E., and Mills, A. F., "Droplet evaporation: Effects of transients and variable properties," *Int. J. Heat Mass Transfer*, 18, 1003–1008 (1975).

2. Kent, J. C., "Quasi-steady diffusion controlled droplet evaporation and condensation," *Appl. Sci. Res. Ser. A*, 28, 315–359 (1973).

3. Law, C. K., "Quasi-steady droplet vaporization theory with property variations," *Physics of Fluids*, 18, 1426–1432 (1975).

4. Godsave, G. A. E., "Studies of the combustion of drops in a fuel spray—the burning of single drops of fuel," *Fourth Symposium on Combustion*, Williams & Wilkins, Baltimore, p. 818 (1953).

5. Spalding, D. B., "The combustion of liquid fuels," *Fourth Symposium on Combustion*, Williams & Wilkins, Baltimore, p. 847 (1953).

6. Law, C. K., and Williams, F. A., "Kinetics and convection in the combustion of alkane droplets," *Combustion and Flame*, 19, 393–405 (1972).

7. Nuruzzaman, A. S. M., Martin, G. F., and Hedley, A. B., "Combustion of single droplets and simplified spray systems," *J. Inst. Fuel*, 44, 38–54 (1971).

8. Saitoh, T., Yamazaki, K., and Viskanta, R., "Effect of thermal radiation on transient combustion of a fuel droplet," *J. Thermophysics and Heat Transfer,* 7, 94–100 (1993).

9. Mills, A. F., and Wortman, A., "Two-dimensional stagnation point flows of binary mixtures," *Int. J. Heat Mass Transfer*, 15, 969–987 (1972).

10. Wortman, A., and Mills, A. F., "Accelerating compressible laminar boundary layers of binary gas mixtures," *Archiwum Mechaniki Stosowanej,* 26, 479–497 (1974).

11. Wortman, A., "Mass transfer in self-similar boundary layer flows," Ph.D. dissertation, School of Engineering and Applied Science, University of California, Los Angeles (1969).

12. Landis, R. B., and Mills, A. F., "The calculation of turbulent boundary layers with foreign gas injection," *Int. J. Heat Mass Transfer*, 15, 1905–1932 (1972).

13. Landis, R. B., "Numerical solution of variable property turbulent boundary layers with foreign gas injection," Ph.D. dissertation, School of Engineering and Applied Science, University of California, Los Angeles (1971).

14. Hartnett, J. P., and Eckert, E. R. G., "Mass transfer cooling in a laminar boundary layer with constant fluid properties," *Trans. ASME*, 79, 247–254 (1957).

15. Knuth, E. L., "Use of reference states and constant property solutions in predicting mass-, momentum-, and energy-transfer rates in high speed laminar flows," *Int. J. Heat Mass Transfer,* 6, 1–22 (1963).

16. Knuth, E. L., and Dershin, H., "Use of reference states in predicting transport rates in high-speed turbulent flows with mass transfer," *Int. J. Heat Mass Transfer*, 6, 999–1018 (1963).

17. Baron, J. R., "The binary-mixture boundary layer associated with mass transfer cooling at high speeds," MIT Naval Supersonic Laboratory, Tech. Report 190 (1956).

18. Gross, J. F., Hartnett, J. P., Masson, D. J., and Gazley, C., Jr., "A review of binary laminar boundary layer characteristics," *Int. J. Heat Mass Transfer*, 3, 198–221 (1961).

19. Simon, H. A., Hartnett, J. P., and Liu, C. A., "Transpiration cooling correlations of air and non-air free streams," AIAA Paper no. 68–758, presented at the AIAA Third Thermophysics Conference, Los Angeles, June 1968.

20. Spalding, D. B., *Convective Mass Transfer*, McGraw–Hill, New York (1963).

21. Spalding, D. B., "A standard formulation of the steady convective mass-transfer problem," *Int. J. Heat Mass Transfer*, 1, 192–207 (1960).

22. Kays, W. M., and Crawford, M. E., *Convective Heat and Mass Transfer*, 3d ed., McGraw–Hill, New York (1993).

23. Landis, R. B., and Mills, A. F., "Effect of internal diffusional resistance on the evaporation of binary droplets," Paper B.7.9, Fifth International Heat Transfer Conference, Tokyo (1974).

24. Chiang, C. H., Raja, M. S., and Sirignano, W. A., "Numerical analysis of a convecting, vaporizing fuel droplet with variable properties," *Int. J. Heat Mass Transfer*, 35, 1307–1324 (1992).

25. Renksizbulut, M., and Bussman, M., "Multicomponent droplet evaporation at intermediate Reynolds numbers," *Int. J. Heat Mass Transfer*, 36, 2827–2835 (1993).

26. Hijikata, K., Mori, Y., and Kawaguchi, S., "Direct contact condensation of vapor to falling cooled droplets," *Int. J. Heat Mass Transfer*, 27, 1631–1640 (1984).

27. Brunson, R. J., and Welleck, R. M., "Mass transfer within an oscillating liquid droplet," *Can. J. Chem. Eng.*, 48, 267–274 (1970).

28. Mills, A. F., and Hoseyni, M. S., "Diffusive deposition of aerosols in a rising bubble," *Aerosol Science and Technology*, 8, 103–105 (1988).

29. Hirschfelder, J. O., Curtiss, C. P., and Bird, R.B., *Molecular Theory of Gases and Liquids,* Wiley, New York (1954).

30. Woods, L. C., *An Introduction to the Kinetic Theory of Gases and Magnetoplasmas*, Oxford University Press, Oxford (1993).

31. Chapman, S., and Cowling, T. G., *The Mathematical Theory of Nonuniform Gases*, 2d ed., Cambridge University Press, New York (1964).

32. Rosner, D. E., "Thermal (Soret) diffusion effects in interfacial mass transfer rates," *PCH Physicochemical Hydrodynamics*, 1, 91–123 (1980).

33. Grew, K. E., and Ibbs, T. L., *Thermal Diffusion in Gases*, Cambridge University Press, London (1952).

34. Mason, E. A., Munn, R. S., and Smith, F. J., "Thermal diffusion in gases," in *Advances in Atomic and Molecular Physics*, Vol. 2, ed. D. R. Bates and I. Esterman, Academic Press (1966).

35. Newman, J. S., *Electrochemical Systems*, Prentice Hall, Englewood Cliffs, N.J. (1973).

36. Woods, L. C., *Thermodynamics of Fluid Systems*, Clarendon Press, Oxford (1975).

37. Vincenti, W. G., and Kruger, C. H., *Physical Gas Dynamics*, John Wiley, New York (1965).

38. Williams, F. A., *Combustion Theory*, 2d ed., Addison-Wesley, Reading, Mass. (1985).

39. Mills, A. F., "The use of the diffusion velocity in conservation equations for multicomponent gas mixtures," *Int. J. Heat Mass Transfer*, 41, 1955–1968 (1998).

40. Minkowycz, W. J., and Sparrow, E. M., "Condensation heat transfer in the presence of noncondensables, interfacial resistance, superheating, variable properties and diffusion," *Int. J. Heat Mass Transfer*, 9, 1125–1144 (1966).

41. Denny, V. E., Mills, A. F., and Jusionis, V. J., "Laminar film condensation from a steam–air mixture undergoing forced flow down a vertical surface," *J. Heat Transfer*, 93, 297–304 (1971).

42. Smirnov, V. A., Verevochkin, G. E., and Brdlick, P. M., "Heat transfer between a jet and held plate normal to flow," *Int. J. Heat Mass Transfer*, 2, 1–7 (1961).

43. Rose, J. W., "Condensation of a vapor in the presence of a noncondensing gas," *Int. J. Heat Mass Transfer*, 12, 233–237 (1969).

EXERCISES

2-1 Show that in a binary mixture

i. $\nabla x_1 = \dfrac{M^2}{M_1 M_2} \nabla m_1$

ii. $\mathbf{j}_1 = \dfrac{M_1 M_2}{M} \mathbf{J}_1^*$

2-2 The Chapman–Enskog kinetic theory of gases shows that for ordinary diffusion in a binary mixture, the mass flux relative to the mass average velocity is

$$\mathbf{j}_1 = -\frac{\mathcal{N}^2 m_1 m_2}{\rho} \mathcal{D}_{12} \nabla n_1$$

where m_1 and m_2 are masses of the molecules, and $\mathcal{N}$ and n are number density and number fraction, respectively. Derive from this relation Fick's law in the forms

$$\mathbf{j}_1 = -\rho \mathcal{D}_{12} \nabla m_1; \qquad \mathbf{J}_1^* = -c \mathcal{D}_{12} \nabla x_1$$

2-3 A gas mixture at 1 bar pressure and 300 K contains 20% H_2, 40% O_2, and 40% H_2O by weight. The absolute velocities of each species are 10 m/s, -2 m/s, and 12 m/s, respectively, all in the direction of the z-axis. Calculate $\mathbf{v}$ and $\mathbf{v}^*$ for the mixture and for each species, $\mathbf{n}_i$, $\mathbf{j}_i$, $\mathbf{N}_i$, and $\mathbf{J}_i^*$.

2-4 An end-condenser water heatpipe, 1 m long and with 2 cm inside diameter, is intended to carry a heat load of 100 W when the evaporator temperature is 100°C. If 30 μg of air leak into the heatpipe and the condenser cannot operate below 95°C, determine the new power rating of the heatpipe.

2-5 Water is supplied to a porous plug at the top of a 10 cm–long, 1 cm–diameter tube. The lower end is open, and dry air at 1 atm flows gently over the tube mouth. The water supply rate is controlled to keep the plug surface wet. The complete system is temperature controlled by a thermostat and can operate in the range 290 K to 370 K. Prepare graphs of the quantities $x_{H_2O,s}$, $n_{H_2O,s}$, $j_{H_2O,s}$, $(x_{air})_{lm}$ as a function of temperature and discuss their behavior.

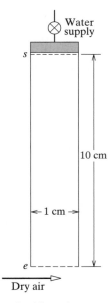

2-6 Water is contained in the bottom of a 15 cm–long vertical tube maintained at 295 K. Pure helium blows gently over the mouth of the tube at 1 atm pressure. At the s-surface, determine the absolute molar flux of the water vapor. Also determine the water vapor diffusion flux relative to the molar average velocity.

2-7 Calculate the Schmidt number of a saturated mixture of water vapor and air at 300 K containing a trace amount of air.

2-8 In a condenser to recover solvent vapors, 9 kmol of steam (species 1) are condensed for each kmol of solvent (species 2). At the condensate surface $x_{2,s} = 0.85$. The total condensation rate of steam and solvent vapors is 0.002 kmol/m^2 s. Using a one-dimensional model, find the mole fractions x_1 and x_2 as functions of distance from the condensate surface. The total pressure is 3 atm, $T_s = 320$ K, and $\mathcal{D}_{12} = 10 \times 10^{-6}$ m^2/s. (*Hint:* Neither component is stationary.)

2-9 Boron is under consideration as fuel for a solid propellant rocket. Boron particles in air ignite at 1900–2300 K, and, provided the temperature is not too high, the oxidation is heterogeneous to form BO and is diffusion controlled with $m_{O_2,s} \simeq 0$. In an experimental study, 50 μm–diameter boron particles are ignited in an air stream at 1000 K and 1 atm. Estimate the lifetime of the particles if the particle temperature is measured to be approximately 3000 K. The density of boron at 3000 K is 2370 kg/m^3.

2-10 Dust particles impinging on a graphite heat shield of a hypersonic vehicle can cause carbon particles to be ejected into the high-temperature boundary layer surrounding the vehicle. Such particles may ignite and burn, and in the temperature range of concern (1600–3000 K) diffusion-controlled oxidation to carbon monoxide may be assumed. For an air temperature of 1500 K, estimate particle temperatures and lifetimes for particles with diameters of 1 and 10 μm (see Example 1.11). Take $\varepsilon = 0.9$ and $\Delta h_c = 9.21 \times 10^6$ J/kg carbon.

2-11 The analysis of Section 2.3.3 gives the burning rate of the carbon particle but does not give the location of the flame. If we assume that the flame is thin enough to be represented by a surface at $r = R_f$, the reaction $2CO + O_2 \rightarrow 2CO_2$ can be treated as a surface reaction occurring between the u' and s' surfaces shown in the figure. To determine the location of the flame, first show that the mass transfer rates across the s and s' surfaces are given by

$$\dot{m}''_s = \frac{R_f}{R_f - R} \frac{\rho \mathcal{D}_{im}}{R} \ln \left[1 + \frac{m_{i,u'} - m_{i,s}}{m_{i,s} - (n_{i,s}/\dot{m}''_s)} \right], \quad i = CO, CO_2$$

$$\dot{m}''_{s'} = \frac{\rho \mathcal{D}_{im}}{R_f} \ln \left[1 + \frac{m_{i,e} - m_{i,s'}}{m_{i,s'} - (n_{i,s'}/\dot{m}''_{s'})} \right], \quad i = O_2, CO_2$$

Then assume equal effective diffusion coefficients for O_2, CO, and CO_2 and apply mass continuity, equilibrium, and stoichiometric relations at the u-, s-, u'-, and s'-surfaces to relate the unknown species fluxes and concentrations. Hence show that $m_{CO,s} = 0.346$, $m_{CO_2,s'} = 0.293$, $R_f/R = 1.922$, and $\dot{m}''_s = 0.160 \rho \mathcal{D}/R$ (as obtained in Section 2.3.3).

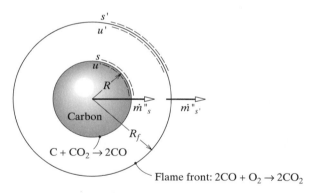

2-12 A spherical droplet of radius R, species 1, evaporates into a stagnant surrounding gas, species 2.

   i. Show that the molar evaporation rate is given by

$$N_{1,s} = \frac{c \mathcal{D}_{12}}{R} \ln \left(1 + \frac{x_{1,e} - x_{1,s}}{x_{1,s} - 1} \right)$$

   by solving the governing differential equation in spherical coordinates. The s- and e-surfaces are at the droplet surface and infinity, respectively.

   ii. Show how the preceding result can be obtained from Eq. (2.29) by using the stagnant film model.

   iii. The corresponding result on a mass basis is Eq. (2.58). Discuss why the molar-based result might be preferred for many calculations.

   iv. For evaporation of water into dry air, plot $N_{1,s} R/c \mathcal{D}_{12}$ versus T_s for $0°C < T_s < 100°C$.

2-13 Following the analysis of droplet combustion in Section 2.3.5, show that the quasi-steady evaporation rate of a spherical droplet into stagnant surrounding gas can be expressed as

$$\dot{m}'' = \frac{k/c_p}{R} \ln \left(1 + \frac{c_{pe}(T_e - T_s)}{h_{\text{fg}}} \right)$$

Carefully list all assumptions made concerning thermophysical properties.

2-14 For a water droplet evaporating into air at 430 K, 1 atm, and 10% RH, compare estimates of the evaporation rate given by

  i. the result of Exercise 2–13.

  ii. Eq. (2.63).

  iii. the result of a constant-property heat transfer analysis.

In all cases use the 1/3 rule for evaluating properties that were assumed constant in the analysis. Since the analyses assume adiabatic vaporization, approximate T_s as the thermodynamic wet-bulb temperature from PSYCHRO.

2-15 A small droplet of water is entrained in a high-temperature air flow at 1 atm, 430 K, and 10% RH. After a short initial transient, heat conduction into the droplet becomes negligible, and the droplet evaporates adiabatically (see Example 1.10). In order to calculate the subsequent evaporation rate, we require the droplet temperature. Compare the following three estimates:

  i. using PSYCHRO.

  ii. using Eq. (1.58).

  iii. using high mass transfer rate theory by combining the results of Exercises 2–12 and Eq. (2.63) with the 1/2 rule reference-state scheme.

Discuss the basis of each estimate.

2-16 Estimate the lifetime of a 1 mm–diameter n-pentane droplet burning in air at 1 atm pressure and 300 K, at near zero gravity. For n-pentane (n-C_5H_{12}), take $\rho_l = 605$ kg/m³, $h_{\text{fg}} = 3.57 \times 10^5$ J/kg, $\Delta h_c = 4.50 \times 10^7$ J/kg, and $T_{\text{BP}} = 309$ K. The flame temperature is $T_f = 2310$ K. At the reference temperature of $(1/2)(T_f + T_{\text{BP}}) = 1310$ K, property values for n-pentane vapor include $k = 0.179$ W/m K, $c_p = 4320$ J/kg K.

2-17 Estimate the lifetime of a 1 mm–diameter n-octane droplet burning in air at 1 atm and 300 K, at near zero gravity. For n-octane (n-C_8H_{18}), take $\rho_l = 611$ kg/m³, $h_{\text{fg}} = 3.03 \times 10^5$ J/kg, $\Delta h_c = 4.44 \times 10^7$ J/kg, and $T_{\text{BP}} = 399$ K. The flame temperature is $T_f = 2320$ K. At the reference temperature of $(1/2)(T_f + T_{\text{BP}}) = 1360$ K, property values of n-octane vapor include $k = 0.113$ W/m K, $c_p = 4280$ J/kg K.

2-18 Forced diffusion plays a key role in the operation of metal–ion lasers, such as the helium–cadmium laser or "blue laser" that is widely used in copying machines and computer printers.

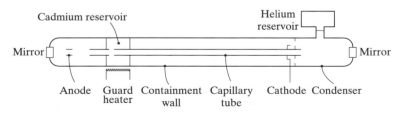

Cadmium reservoir

Helium reservoir

Mirror

Mirror

Anode   Guard   Containment   Capillary   Cathode   Condenser
        heater   wall   tube

A typical configuration is shown in the figure. The process by which the Cd ions are excited is not well understood, but the relationships between the design parameters required for satisfactory operation are known. The lasing action takes place in the positive column of a He gaseous discharge contained in a capillary tube, where the electric field strength E [V/m] is constant. The Cd vapor pressure is three orders of magnitude less than the He pressure. The electric field causes Cd ions to be transported by forced diffusion from the solid cadmium reservoir to the cathode end of the capillary tube, which is cooled to act as a condenser for the Cd ions. The anode serves to collect electrons at the other end of the tube. With the Cd ions denoted as species 1, the molar flux due to forced diffusion is

$$J_1^{*F} = \mu_1 E c x_1$$

where μ_1 [m^2/V s] is the *mobility* of the ions (and is related to the diffusion coefficient by the Nernst–Einstein equation $\mu_1 = e\mathcal{D}_{1m}/\hat{k}T$). Notice that $\mu_1 E$ has units of velocity and is called the *cataphoretic velocity*. In this situation it is of the order of 3000 m/s, which is very large indeed. The cadmium vapor is only partially ionized. If we denote the cadmium neutrals as species 2 and $x = x_1 + x_2$, then we can write $x_1 = \varepsilon x$, where ε is the *ionization fraction* and is of the order of 0.01 in this situation. Cadmium ions are continuously created and destroyed by the collision processes in the plasma of the positive column at a rate $\dot{R}_1''' = -\dot{R}_2'''$, maintaining a constant value of ε.

To analyze steady transport of Cd along the capillary tube, we can assume that radial–direction diffusion phenomena are uncoupled from axial–direction diffusion phenomena. Thus, a one-dimensional analysis is adequate, with concentrations taken as average values over the tube cross section. Proceed as follows.

1. Write down expressions for the absolute fluxes N_1 and N_2.

2. Write down appropriate species conservation equations for species 1 and 2.

3. During steady operation, the helium is stationary. By summing the two species conservation equations and neglecting convection, obtain the equation governing the distribution of total cadmium along the tube as

$$\frac{d}{dz}\left(-c\mathcal{D}\frac{dx}{dz} + \mu_1 E c \varepsilon x\right) = 0$$

where $\mathcal{D}_{1m} = \mathcal{D}_{2m} = \mathcal{D}$ has been assumed.

4. Let $z^* = z/L$, where L is the length of capillary tube under consideration. Then show that x is the solution of

$$\frac{d}{dz^*}\left(-\frac{dx}{dz^*} + \beta x\right) = 0$$

where

$$\beta = \frac{\mu_1 E \varepsilon}{\mathcal{D}/L} = \frac{\text{Cataphoretic velocity} \times \text{Ionization fraction}}{\text{Characteristic diffusion velocity}}$$

In He-Cd lasers, β is of the order of 10^2.

5. Show that the solution for the cathode end, $0 < z < L_1$, is $x = \text{constant} = x_0$, where $x_0 = x_{\text{sat}}(T_0, P)$ for a cadmium reservoir temperature T_0. Also show that $N_c = N_1 + N_2 = \mu_1 E \varepsilon c x_0$ is the cadmium transport to the cathode-end condenser.

6. If the anode region is maintained at a temperature T_a, with $x_a = x_{sat}(T_a, P)$, show that the solution for the anode end, $0 < z < -L_2$, gives the cadmium flux into the anode region as

$$N_a = \left(\frac{cD}{L}\right)\left(\frac{\beta}{1 - e^{-\beta}}\right)(x_a - x_0 e^{-\beta})$$

The results obtained in steps 5 and 6 above are important design relations for the laser (see Exercise 2–19).

2-19 A helium–cadmium laser of the type analyzed in Exercise 2–18 has a 2 mm–I.D. capillary tube to confine the discharge. Nominal values of relevant parameters include a total pressure of 467 Pa, a cadmium reservoir temperature of 500 K, a positive column-temperature of 700 K, an electric field of 3300 V/m, and an ionization fraction of 0.01.

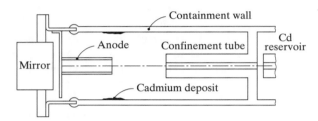

i. Determine the Cd charge required for a laser lifetime of 10,000 h.

ii. If the anode-end mirror temperature must not exceed 370 K to avoid damage to the coatings, determine the length of the anode-end confinement capillary tube required for zero Cd flow to the mirror.

iii. If the laser experiences a "hot shutdown"—for example, if the power supply is abruptly interrupted—the field is disestablished while the reservoir is still at 500 K. Make an engineering estimate of the amount of Cd deposited in the anode region for an effective one-dimensional diffusion path of 3.0 cm. Based on experimental data, the time constant for the temperature decay of the reservoir is approximately 3 min.

iv. When the laser is switched on again, Cd will be pumped out of the anode region. Determine the time required to pump out the amount deposited during the hot shutdown of part (iii) as a function of anode-region temperature T_a in the range 350–400 K. The length of the anode-end confinement capillary tube is 2.5 cm.

Required data for cadmium and the cadmium–helium mixture are:

$$\log_{10} P_{Cd} = 8.56 - (5720/T), \text{ for } P \text{ in torr, } T \text{ in kelvins}$$

$$D_{CdHe} = 5.77 \times 10^{-7} T^{1.75} \text{ m}^2/\text{s at } P = 467 \text{ Pa}$$

$$\mu_{CdHe} = 6.12 \times 10^{-3} T^{0.75} \text{ m}^2/\text{V s at } P = 467 \text{ Pa}$$

2-20

i. Show that a stagnant film analysis of high-rate mass transfer from a sphere leads to a result that is identical to the Couette–flow model result, Eq. (2.99).

ii. Repeat for a long cylinder.

2-21 Rework Example 1.8 for a bath temperature of 360 K.

2-22 A strip of wet textile passes through the central section of a dryer; in this section the temperature of the textile is substantially independent of distance and the drying rate is constant. An air–steam mixture is blown on the top surface of the textile through a perforated-plate distributor. The mixture is at 420 K and 1 atm pressure and contains 8% of water vapor by mass. When the mass velocity of the mixture ρV is 0.8 kg/m^2 s, design charts for the distributor indicate a heat transfer Stanton number of 0.08. Calculate the rate at which water is removed from the textile. (See Exercise 1–46.)

2-23 In a paper-drying unit the wet paper is on the outside of a rotating drum, the inside wall of which is heated by condensing saturated steam at 3 atm pressure. Air at 1 atm pressure, 310 K, and 40% relative humidity is blown onto the paper at a mass velocity ρV of 0.25 kg/m^2 s: design data indicate a mass transfer Stanton number of $\text{St}_m^* = 0.060$. If the overall heat transfer coefficient from the heating steam to the paper surface is estimated to be 75 W/m^2 K, calculate the paper drying rate. (See Exercise 1–47.)

2-24 A wet- and dry-thermocouple installation is to be used to measure the composition of an air–steam mixture under the following conditions: pressure 0.9–1.3 bar, temperature 330–380 K, and relative humidity 10–100%. Write down a set of relations that will allow determination of the mass fraction of water vapor. Discuss any assumptions you make.

2-25 In a turbulent flow over a surface, we can model the turbulent contribution to species diffusive transport using an eddy diffusivity model: we then write Fick's law as

$$j_i = -\rho(\mathcal{D}_{im} + \varepsilon_D)\, dm_i/dy$$

where ε_D is the eddy diffusivity of a chemical species and is a strong function of y, the distance normal to the surface (see Section 5.5 of *Heat Transfer*). Show that the Couette–flow relation

$$\frac{\mathcal{g}_m}{\mathcal{g}_m^*} = \frac{\ln(1 + \mathcal{B}_m)}{\mathcal{B}_m}; \quad \mathcal{B}_m = \frac{m_{i,e} - m_{i,s}}{m_{i,s} - n_{i,s}/\dot{m}''}$$

for transfer of an inert species i remains valid if $\int_0^\delta dy/\rho(\mathcal{D}_{im} + \varepsilon_D)$ is independent of $\dot{m}''$.

2-26 Oxyacetylene cutting of steel plate involves heating the plate with an oxygen–acetylene flame to a temperature high enough for the iron to burn: the heat release then sustains the process, and the acetylene supply can be shut off. The cutting is then a process of combustion of iron in almost pure oxygen. The oxides are molten and flow away under the action of gravity and shear forces: there are no volatile oxides to diffuse back into the oxygen flow. Also, at the high temperatures prevailing, there is chemical equilibrium at the s-surface with $m_{O_2,s} \simeq 0$. In a particular cutting process, a jet of 99% oxygen and 1% inerts by mass exits a 4 mm–diameter nozzle held 2 mm from the plate. If the jet velocity is 15 m/s, determine the rate at which iron is consumed. An appropriate correlation for the Sherwood number is [42]

$$\text{Sh}_L = 0.55\text{Re}_L^{1/2}\text{Sc}^{1/3}$$

where L is the distance of the nozzle from the plate. Assume a plate temperature of 1700 K, a Schmidt number for O_2 and inerts of 0.8, and an Fe_2O_3 oxide.

2-27 A 2.5 cm–diameter cylinder is located in a 2 m/s cross-flow of air at 1 atm and 400 K. The wall in the front stagnation region is porous and is to be maintained at 300 K by air injection at an s-surface velocity of 0.1 m/s. Determine the air reservoir temperature required to maintain a steady state. Neglect radiation heat transfer, and use the improved blowing factors of Section 2.4.2.

2-28 Nitrogen at 1 atm and 800 K flows along a porous flat plate at 100 m/s. A mixture of 50% N_2 and 50% H_2 by mass is transpired through the plate from a reservoir at 300 K. At a location 0.5 m from the leading edge determine

i. the transpiration rate required to maintain the plate at 400 K.

ii. the mass fraction of H_2 at the s-surface.

Assume a turbulent boundary layer, and ignore thermal radiation. Use the simple high mass transfer rate theory of Section 2.4.1.

2-29 In a laboratory experiment, dry air at 1350 K and 1 atm flows at 10 m/s over a 3 cm–diameter sweat-cooled cylinder. If the coolant water is supplied from a plenum chamber at 300 K, determine the water supply rate required to balance evaporation at the stagnation line, and the corresponding steady-state surface temperature. Neglect radiation heat transfer.

i. Use the simple high mass transfer rate theory of Section 2.4.1.

ii. Use the improved blowing factors given in Section 2.4.2.

2-30 Repeat Exercise 2–29 for the stagnation point of a 3 cm–diameter sphere.

2-31 In a laboratory experiment, dry air at 1350 K and 1 atm flows at 10 m/s along a sweat-cooled flat plate. If the coolant water is supplied from a plenum chamber at 300 K, determine the water supply rate required to balance evaporation at a location 10 cm from the leading edge, and the corresponding surface temperature.

i. Use the simple high mass transfer rate theory of Section 2.4.1.

ii. Use the improved blowing factors given in Section 2.4.2.

Assume a laminar boundary layer and negligible radiation heat transfer.

2-32 Repeat Exercise 2–31 for a turbulent boundary layer starting at the leading edge.

2-33 Estimate the lifetime of salt crystals suspended in pure water at 300 K. The crystals can be modeled as spheres of diameter 0.4 mm and with a density of 2170 kg/m^3.

2-34 Air at 1400 K, 1 atm flows across a 2 cm–diameter transpiration-cooled cylinder at 100 m/s. Helium is supplied from a plenum chamber at 300 K. Determine the helium injection rate required to maintain the stagnation line surface temperature at 1000 K. Allow for radiation heat transfer into black surrounds at 1400 K for a surface emittance of 0.7. The simple high mass transfer rate theory of Section 2.4.1 can be used, but comment on its applicability.

2-35 A 1 cm–radius tungsten plug for a missile nose cone is tested in a high-temperature air arc jet. Of concern is the behavior of the plug should the heat shield fail and expose the plug. Test data include a measured surface temperature of 3000 K and a surface recession rate of 0.0481 mm/s. Determine the zero-mass-transfer-limit mass transfer conductance that prevailed in the test. Assume diffusion-controlled oxidation with gaseous W_3O_9 as product and negligible oxygen at the s-surface. Take the density of solid tungsten as 19,000 kg/m^3. Use the simple high mass transfer rate theory of Section 2.4.1, but comment on its applicability.

2-36 A falling-film direct-contact steam condenser has plates spaced 3 cm apart. The steam flow is downward cocurrent with the water films, and at a particular location the following data apply: pressure = 1780 Pa, bulk steam temperature = 295 K, bulk water temperature = 282 K, water flow rate per unit film width = 6.0 kg/m s, steam velocity = 100 m/s, and mass fraction of air in the bulk steam = 0.112%. Determine the local rate of steam condensation. [*Hint:* For the liquid-side heat transfer coefficient, use the heat transfer analog to the correlation given in Exercise 1–32 for gas absorption into a turbulent falling film; and for the gas-side mass transfer conductance, use Eq. (1.47). However, comment on the applicability of these correlations.]

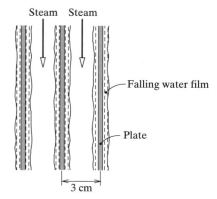

Steam   Steam

Falling water film

Plate

3 cm

2-37 Steam at 340 K and 21.69 kPa containing 0.85% air by mass flows at 3 m/s down over a horizontal brass condenser tube of 2 cm O.D. and 1 mm wall thickness. Coolant water flows through the tube at a bulk velocity of 2.5 m/s. At a location where the bulk coolant temperature is 290 K, estimate the steam condensation rate and compare it with the value if no air were present.

2-38

    i. A tungsten plug at 2600 K is exposed to a 40 m/s air stream at 1400 K, 4 atm. Calibration data suggest a mass transfer Stanton number of $St_m^* = 0.036$. The oxidation product is gaseous W_3O_9, and the concentration of oxygen at the s-surface can be assumed to be very small. Determine the tungsten recession rate in μm/s. For solid tungsten, take $\rho = 19,000$ kg/m^3.

    ii. If a porous sintered tungsten plug ($\epsilon_v = 0.2$) is tested with nitrogen blown through the surface at 3.0 kg/m^2 s, calculate the new recession rate.

    iii. Compare the N_2 mass fractions at the s-surface for the two situations.

Use the simple high mass transfer rate theory of Section 2.4.1.

2-39 Air at 1 atm and 300 K flows at 10 m/s along a 10 cm–square horizontal slab of dry ice (solid CO_2). If the back side of the slab is well insulated, determine the slab temperature and average sublimation rate. Use a transition Reynolds number of 50,000. The vapor pressure of dry ice is given by $\log_{10} P$ [mm Hg] $= 9.9082 - 1367.3/T$ [K], and the enthalpy of sublimation can be taken as 0.57×10^6 J/kg. Neglect radiation heat transfer and use simple Couette–flow model blowing factors.

2-40 Determine the adiabatic vaporization temperature for small cadmium particles evaporating into pure helium at a total pressure of 3.5 torr in the temperature range $T_e = 500$–750 K. The vapor pressure of cadmium is given by

$$\log_{10} P \text{ [torr]} = 8.56 - \frac{5720}{T \text{ [K]}}$$

Obtain $\mathcal{D}_{CdHe}$ using Eq. (1.112) with Lennard–Jones parameters for cadmium of $\sigma = 3.0$ Å and $\varepsilon/k = 3000$. Cadmium vapor can be modeled as an ideal monatomic gas for $T < 3000$ K. The enthalpy of sublimation of cadmium can be obtained from the vapor pressure relation using the Clausius–Clapeyron relation. Neglect radiation heat transfer. (*Hint:* Perform the analysis on a molar basis.)

2-41 Modify the result of Section 2.3.5 to account for a relative velocity between the fuel droplet and air stream. Hence obtain the burning-rate constant β as a function of relative velocity V for the combustion of 1 mm–diameter n-hexane droplets burning in air at 1 atm pressure and 300 K. Let $10 < V < 40$ m/s. (*Hint:* Use the same approach as was used for the Couette–flow model of Section 2.4.)

2-42 In a grinding process, iron particles of about 200 μm size are ejected and entrained in an exhaust system. The temperatures attained are sufficient for the particles to ignite and burn on leaving the workpiece. Estimate the burning times of these iron "sparks" if the ambient air is at 300 K and 1 atm. The temperature is not high enough for appreciable evaporation of the metal or the oxide formed, which can be taken to be FeO. It can be assumed that the reaction rate is controlled by diffusion of oxygen through the gas phase. Required data include:

Iron melting temperature and density: 1809 K and 7800 kg/m^3

Oxide melting temperature and density: 1650 K and 3700 kg/m^3

Heat of combustion: 4.58×10^6 J/kg of iron

Emittance of FeO: 0.5

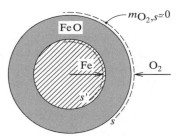

2-43 In a test rig, a steam–air mixture at 4714 Pa and 310 K flows downward at 15 m/s along a 5 cm–square vertical copper plate 3 mm thick. The rear of the plate is cooled by a jet of water at 280 K with a heat transfer coefficient of 4600 W/m^2 K. Estimate the condensation rate as a function of the air content, for $0.001 < m_{air,e} < 0.1$. (*Hint:* To simplify the problem, use average values for the film condensation heat transfer coefficient and the vapor-phase mass transfer conductance.)

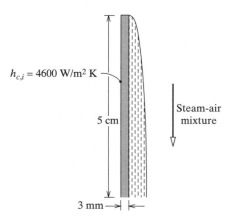

2-44 Air at 300 K, 1 atm flows at 10 m/s along a flat plate also maintained at 300 K. The plate is bare for $0 < x \le 10$ cm measured from the leading edge and is coated with a thin layer of naphthalene for $x > 10$ cm. At $x = 20$ cm, calculate Sh_x, St_{mx}, g_{mx}, and $j_{1,s}$. Take $\mathrm{Re}_{tr} = 200{,}000$.

2-45 Following the analysis in Section 5.4.5 of *Heat Transfer*, derive the integral conservation equations for a laminar natural-convection boundary layer of a gas mixture on a vertical wall in a form suitable for isothermal mass transfer, and valid for high mass transfer rates. Hence obtain the special form of the species equation for transfer of one component only. Also write down an expression for the local Sherwood number valid at low mass transfer rates.

2-46 A porous vertical plate is kept wet with water and maintained at 290 K in dry air at 290 K and 1 atm. Referring to the integral method analysis of natural convection in Section 5.4.5 of *Heat Transfer*, determine δ, U, Ra_x, Sh_x, g_{mx}, and $j_{1,s}$ at a location 10 cm from the bottom of the plate.

2-47 A flat plate exposed to intense radiative and convective heating is to be sweat cooled with water. Dry air at 840 K and 1 atm flows at 200 m/s along the plate as a laminar boundary layer.

    i. Determine the water supply rate to maintain a surface temperature of 360 K at a location 10 cm from the leading edge.

    ii. If there is a net radiation flux to the plate of $q_{\mathrm{rad}} = -102.3\ \mathrm{kW/m}^2$, determine the water supply temperature required to maintain a steady state.

Use Knuth's reference-state scheme to account for variable-property effects, in conjunction with the constant-property self-similar solution.

2-48 Air at 1500 K, 1 atm flows at 10 m/s along a porous flat plate that is cooled by injection of helium from a reservoir at 300 K. At a location 20 cm from the leading edge, determine the injection rate required to maintain a surface temperature of 500 K. Also determine the concentration of helium at the plate surface. Use Knuth's reference-state scheme and the constant-property solution for the self-similar laminar boundary layer. Neglect thermal radiation.

2-49 The good performance of the integral method demonstrated in Section 5.4.5 of *Heat Transfer* suggests that such methods might yield satisfactory results for the natural-convection noncondensable gas problem. If the molecular weight of the vapor is less than that of the gas, condensation on a vertical wall will produce a downward-flowing vapor boundary layer with motion induced by drag of the liquid film and a downward-directed buoyancy force. Perform an integral method analysis, proceeding as follows.

1. Show that the integral forms of the momentum and species equations for the vapor-phase boundary layer are

$$\frac{d}{dx} \int_0^\infty u^2\,dy - v_s u_s = -\nu \left.\frac{\partial u}{\partial y}\right|_s + g\mathcal{M} \int_0^\infty m\,dy$$

$$\frac{d}{dx} \int_0^\infty um\,dy - v_s m_s = -\mathcal{D}_{12} \left.\frac{\partial m}{\partial y}\right|_s$$

where we have written $m = m_2 - m_{2,e}$ and

$$1 - \frac{\rho_e}{\rho} = \mathcal{M}(m_2 - m_{2,e}); \quad \mathcal{M} = \frac{M_2 - M_1}{M_2 - (M_2 - M_1)m_{2,e}}$$

Subscripts 1 and 2 refer to vapor and gas, respectively. The effect of temperature on mixture density has been neglected.

2. Show that use of a boundary condition requiring the liquid surface to be impermeable to gas relates v_s to $\partial m/\partial y|_s$ as

$$v_s = \frac{\mathcal{D}_{12}}{m_{2,s}} \left.\frac{\partial m}{\partial y}\right|_s$$

3. Following Rose [43], choose the following profiles:

$$u = u_s \left(1 - \frac{y}{\Delta}\right)^2 + U\frac{y}{\Delta}\left(1 - \frac{y}{\Delta}\right)^2$$

$$\frac{m}{m_s} = \left(1 - \frac{y}{\Delta}\right)^2$$

where u_s is the surface velocity of the liquid film obtained from Nusselt's analysis [Eq. (7.4) of *Heat Transfer*], Δ is the vapor boundary layer thickness, and U is a yet-to-be-determined scaling velocity. Notice that the velocity profile is the sum of two components, the first due to drag by the liquid film and the second due to natural convection. Show that these profiles satisfy appropriate boundary conditions.

4. Substitute the profiles in the governing ordinary differential equations and, proceeding as in Section 5.4.5 with the local condensation rate taken from Nusselt's solution, show that

$$10\,\mathrm{Sc}\,\frac{\mathrm{Ja}_l}{\mathrm{Pr}_l}\left[\frac{(\rho\mu)_l}{(\rho\mu)_v}\right]\left(\frac{m_{2,e}}{m_{2,s} - m_{2,e}}\right)^2\left(\frac{20}{21} + \frac{m_{2,s}}{m_{2,e}}\,\mathrm{Sc}\right)$$

$$+ \frac{8}{\mathrm{Sc}}\left(\frac{\mathrm{Pr}_l}{\mathrm{Ja}_l}\right)^2\left[\frac{(\rho\mu)_v}{(\rho\mu)_l}\right]\left(\frac{m_{2,s} - m_{2,e}}{m_{2,s}}\right)^2\left[\frac{5}{28}\frac{\mathrm{Ja}_l}{\mathrm{Pr}_l} - \frac{\mathcal{M}(m_{2,s} - m_{2,e})}{3}\right]$$

$$= \frac{100}{21}\frac{m_{2,e}}{m_{2,s}} - 2\frac{m_{2,s} - m_{2,e}}{m_{2,s}} + 8\,\mathrm{Sc}$$

where $\mathrm{Ja}_l = c_{pl}(T_s - T_w)/h_{\mathrm{fg}}$.

2-50 Referring to Exercise 2–49, obtain the interface temperature and total condensation rate per unit width, for condensation from a steam–air mixture on a 0.05 m–high vertical surface with $P = 1$ atm, $T_e = 375$ K, $m_{2,e} = 0.02$, and $T_w = 360$ K.

2-51 Following the analysis in Section 2.5.2, derive the equation governing total energy conservation for a laminar boundary layer on a flat plate by including transport of kinetic energy, and viscous work. (Refer to Sections 5.2.1 and 5.7.3 of *Heat Transfer*.) Hence show that for unity Lewis and Prandtl numbers, the equation reduces to a form identical to Eq. (2.162) with static enthalpy h replaced by total enthalpy $H = h + (1/2)u^2$.

2-52 A solid-propellant rocket nozzle throat is lined with graphite and has an initial diameter of 5 cm. The elemental composition of the propellant is $m_{(O)} = 0.560$, $m_{(C)} = 0.269$, $m_{(N)} = 0.147$, $m_{(H)} = 0.024$. The combustion chamber pressure is $P_{cc} = 100$ atm, and the temperature is 3300 K. The characteristic velocity of the propellant is $c^* = 1200$ m/s. After ignition, the surface temperature of the graphite quickly reaches a temperature high enough to give thermodynamic equilibrium at the s-surface, with the concentrations of all oxidizing species equal to zero. Estimate the rate of enlargement of the throat if experimental data for heat transfer to regeneratively cooled liquid-propellant rocket nozzles can be correlated as

$$\text{Nu}_D = 0.020\, \text{Re}_D^{0.8}\text{Pr}^{0.33}$$

(The mass velocity in the throat of the nozzle is given by $G_{\text{th}} = P_{\text{cc}}/c^*$.) Take $\mu \simeq 7.5 \times 10^{-5}$ kg/m s and Sc $\simeq 0.7$ to evaluate the conductance.

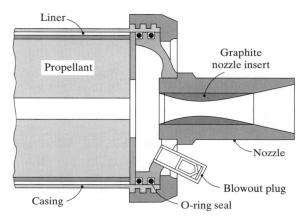

2-53 In an ammonia scrubber, a mixture of air and ammonia gas flows counter-current to a falling film of water inside 3 cm–diameter vertical tubes. The exchanger operates at 300 K and 1 atm. The zero-mass-transfer-limit gas-side and liquid-side mass transfer conductances are estimated to be 0.013 kg/m^2 s and 0.13 kg/m^2 s, respectively. At a particular location down the tubes, the bulk mass fractions of ammonia in the air and water streams are 0.62 and 0.09, respectively. Determine the local gas absorption rate. Use constant-property Couette–flow blowing factors.

2-54 In a direct-contact condenser for a pilot open-cycle ocean thermal energy conversion plant, cold seawater films fall down 3 mm–thick parallel plates at a pitch of 4.0 cm. Low-pressure steam flows cocurrently with the seawater. At a particular location down the plates, the following data apply.

Bulk gas temperature $T_e = 283$ K

Bulk gas air mass fraction $m_{2,e} = 9 \times 10^{-4}$

Bulk liquid temperature $T_b = 278$ K

Bulk liquid air mass fraction $m_{2,b} = 4 \times 10^{-5}$

Total pressure $= 1215$ Pa

Gas-side mass transfer conductance $g_{mG}^* = 6.0 \times 10^{-3}$ kg/m^2 s

Liquid-side heat transfer conductance $g_{hL}^* = 6.3$ kg/m^2 s

Liquid-side mass transfer conductance $g_{mL}^* = 1.58$ kg/m^2 s

Determine the local condensation rate and air desorption rate. Use constant-property Couette–flow blowing factors.

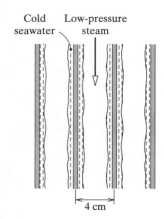

Cold   Low-pressure
seawater   steam

4 cm

2-55 Evaporation of a multicomponent droplet is analyzed in Section 2.6.3. If good mixing within the droplet is assumed, the governing equations are Eqs. (2.208), (2.211), and (2.212). Write a computer program to calculate the lifetime of a binary hydrocarbon fuel droplet evaporating into stagnant air. Use your program to obtain the lifetime of a 1 mm–diameter pentane–hexane droplet, initially at 300 K with 50% by mass of each component, in stagnant air at 2000 K and 1 atm. Also plot the bulk composition, bulk temperature, and evaporation rates of each component as a function of time. Property data that can be used include $\rho_l = 610$ kg/m^3; $c_{pl} = 2400$ J/kg K; and for pentane and hexane designated as species 1 and 2, respectively, $M_1 = 72.0$, $M_2 = 86.2$, $h_{fg1} = 3.57 \times 10^5$ J/kg, $h_{fg2} = 3.35 \times 10^5$ J/kg. For the vapor pressures,

$$P = \exp\left[-A\left(\frac{1}{T} - \frac{1}{B}\right)\right] \text{atm}$$

with $A_1 = 3330$, $A_2 = 3810$, $B_1 = 309$, $B_2 = 314$. Use Raoult's law, namely, $P_{i,s} = x_{i,u} P_{i,s} (x_{i,u} = 1)$. The gas-phase properties can be evaluated using Hubbard's 1/3 rule extended to two evaporating components.

2-56 Iron and ferrous alloys are soluble in molten aluminum. Dissolution rates must be estimated for applications including casting of aluminum, high-temperature containers, and safety analysis of advanced nuclear reactors. For example, in a postulated core meltdown accident in a heavy-water reactor, a pool of molten aluminum may form in the bottom of the reactor vessel. Decay heating of the radioactive materials dissolved in the aluminum could ensure that the pool remains molten ($T_{MP} = 933$ K) for a long time, and thus erosion of the carbon steel vessel may be a serious problem. To obtain a preliminary estimate of the expected erosion rates, consider a pool of aluminum at 1000 K in a vessel with a 5 m–square base, and calculate the erosion rate of the steel base. The solubility of iron in aluminum is approximately

$$m_{1,s} = 0.92 \exp\left[-\left(\frac{T_s \text{ K} - 1820}{450}\right)^2\right]$$

Also, at 1000 K, take $\rho_s = 2537$ kg/m^3, $\rho_{Al} = 2360$ kg/m^2 s, $\nu_{Al} = 0.47 \times 10^{-6}$ m^2/s, and Sc$_{FeAl} = 260$.

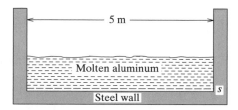

2-57 A large tank contains carbon dioxide and connects to ambient air through a 5 cm–I.D., 1 m-long tube. Motion of the CO_2 and air at each end of the tube maintains essentially pure CO_2 at one end and ambient air (0.03% CO_2 by volume) at the other.

    i. If the pressure and temperature are 1 atm and 300 K, respectively, calculate the rate at which CO_2 is lost from the tank by diffusion along the tube (in grams per day). Perform your analysis on a molar basis to take advantage of the fact that both temperature and pressure are very nearly constant along the tube.

    ii. In practice in such situations, the pressure inside the tank will seldom be the same as ambient pressure because of ambient pressure and temperature changes. The diffusion flow rate calculated in part (i) is very small: to gain an appreciation of how small it is, calculate the pressure difference across the ends of the tube required to give a bulk flow rate of pure CO_2 equal to the diffusion flow.

    iii. Calculate the mass average velocity and corresponding pressure gradient at the tank end of the tube for part (i).

2-58 A laboratory experiment requires a supply of ammonia gas. In order to maintain the supply pressure close to 1 atm, it is proposed to provide a 3 mm–I.D., 10 m–long vent line to connect the supply pipe with ambient air outside the laboratory. To evaluate the performance of the venting process, your supervisor has asked you to determine the rate at which ammonia gas diffuses through the vent line.

    i. If the temperature and pressure are 300 K and 1 atm, respectively, calculate the rate at which NH_3 is lost by diffusion in g/day. Perform your analysis on a molar basis to take advantage of the almost constant temperature and pressure along the tube.

    ii. The diffusion flow rate calculated in part (i) is very small. Calculate the pressure difference required to give a bulk flow rate of NH_3 equal to the diffusion flow. Since this pressure difference is extremely small, discuss the relevance of the diffusion calculation to the practical venting problem.

2-59 A large tank contains helium and connects to ambient air through a 5 cm–diameter, 1 cm–thick perforated disk. Motion of the helium and air at each end of the disk maintains essentially pure He on one side of the disk and ambient air at the other. The disk has 50 μm–diameter perforations with an open area ratio of 0.7.

    i. If the pressure and temperature are 1 atm and 300 K, respectively, calculate the rate at which helium is lost from the tank by diffusion through the disk (in grams/day). Perform your analysis on a molar basis.

    ii. Calculate the pressure difference across the disk required to give a bulk flow of pure helium equal to that calculated in part (i).

    iii. Calculate the mass average velocity and corresponding pressure gradient at each end of the disk for part (i). Is a one-dimensional model meaningful in this context?

2-60 Water is supplied to a porous plug at the top of a 3 cm–long, 1 cm–diameter tube. The lower end is open, and dry air at 1 bar flows gently over the mouth (see Exercise 2–5). The water supply rate is controlled to keep the plug surface wet. The complete system is temperature-controlled and maintained at 350 K. Calculate the rate of evaporation using the following approaches:

    i. A molar-based analysis.

   ii. A mass-based analysis assuming the mixture density to be constant at the average of the end values.

  iii. A mass-based analysis without assuming a constant mixture density (a numerical integration may be required).

2-61 A large tank contains hydrogen and connects to ambient air through a 5 cm–diameter, 1 cm–thick porous plug. Motion of the hydrogen and air on each side of the plug maintains essentially pure hydrogen on one side and pure air on the other. The plug has a volume void fraction of 0.8 and a tortuosity factor of 1.8.

    i. If the pressure and temperature are 1 atm and 300 K, respectively, calculate the rate at which hydrogen is lost from the tank by diffusion through the plug (in grams/day). Perform your analysis on a molar basis.

   ii. Calculate the pressure difference across the plug required to give a bulk flow of hydrogen equal to that calculated in part (i). The Darcy permeability of the plug is $3 \times 10^{-10} \, \text{m}^2$.

  iii. Calculate the mass average velocity and pressure gradient at each end of the plug for part (i). Is a one-dimensional model meaningful in this context?

Mass Exchangers

3.1 INTRODUCTION

Heat exchangers are devices that effect a change in the temperature of one or more flowing streams. Analogously, mass exchangers are devices that effect a change in the composition of one or more flowing streams. In a **catalytic reactor,** the composition of a single stream is changed as it flows over a catalyst that promotes a reaction between components of the stream: the catalytic reactor is a *single-stream* exchanger and is analogous to the single-stream heat exchangers considered in Sections 1.6 and 8.4 of *Heat Transfer*. A catalytic converter for an automobile is shown in Fig. 3.1. **Filters** and **electrostatic precipitators** are single-stream exchangers in which particles are removed from an air stream. In a **gas scrubber,** an unwanted component of a gas stream is removed by absorption into a liquid stream. In general, the compositions of the two streams vary through a scrubber, so that this is a *two-stream* exchanger, analogous to the two-stream heat exchangers considered in Section 8.5. However, some scrubbers can be analyzed as single-stream exchangers, as will be seen in Section 3.3.2. As was the case for heat exchangers, a large transfer area per unit volume is desirable for mass exchangers. Large surface areas are obtained by spraying liquid into a gas as very small droplets, by bubbling gas through a liquid, or by flowing one or both phases through a packed bed.

Figure 3.1 A catalytic converter for a diesel engine automobile. (Photograph courtesy Mercedes–Benz AG.)

Simultaneous heat and mass exchangers are devices that effect changes in both the temperature and the composition of one or more flowing streams. The **adiabatic humidifier** considered in Section 3.5.1 is an example of a single-stream simultaneous heat and mass exchanger; it is also particularly simple, since the heat and mass transfer processes can be *uncoupled* and the performance determined by analyzing the mass transfer process only. **Cooling towers** are examples of two-stream simultaneous heat and mass exchangers. In a power plant, the condenser coolant water rejects heat to air in the cooling tower and is then recycled to the condenser. The water temperature decreases through the tower, and the air temperature and moisture content increase. Counterflow cooling towers are analyzed in Section 3.5.2. Cross-flow cooling towers are analyzed in Section 3.5.3. Hand calculations of cooling tower performance can be long and tedious and are impractical for tower design. The computer program CTOWER performs these calculations efficiently and reliably.

3.2 SYSTEM BALANCES

In Section 8.3 of *Heat Transfer*, exchanger energy balances were made on heat exchangers in order to relate inlet and outlet temperatures. Similarly, exchanger, or system, mass balances are made on mass exchangers to relate inlet and outlet compositions. Such balances can be made on a mass or molar basis. In addition, if chemical reactions occur within the system, it is often useful to base such balances on chemical elements, because chemical elements are conserved in chemical reactions. Owing to the greater complexity of mass transfer processes, system balances on mass exchangers play a more important role than energy balances on heat exchangers. System balances on mass exchangers can yield much useful information, and should always be made before considering the design of a mass exchanger to effect the desired mass transfer processes.

3.2.1 Mass, Mole, and Element Balances

At steady state, a *system* balance for a selected chemical species i is made by equating the production rate due to reactions within the system to the rate of net outflow of the species in the various streams entering or leaving the system. For species i we can write

$$\text{Rate of internal production} = \text{Rate of outflow} - \text{Rate of inflow}$$

The mass flow rate of the species i in a stream is the total flow in the stream $\dot{m}$ times the mass fraction m_i:

$$\dot{m}_i = m_i \dot{m} \tag{3.1a}$$

Similarly, the molar flow rate of species i, $\dot{M}_i$, is the total molar flow rate $\dot{M}$ times the mole fraction x_i:

$$\dot{M}_i = x_i \dot{M} \tag{3.1b}$$

Consider now two or more inlet streams that mix to form a single outlet stream, as shown in Fig. 3.2. The mass balance for a system with k inlet streams is

$$\sum_k (\dot{m})_k = \dot{m}_{\text{out}} \tag{3.2}$$

If the chemical species of interest is inert, there is no internal production, and the species balance for the system takes a simple form,

$$\sum_k (m_i \dot{m})_k = m_{i,\text{out}} \dot{m}_{\text{out}} \tag{3.3a}$$

on a mass basis. On a molar basis,

$$\sum_k (x_i \dot{M})_k = x_{i,\text{out}} \dot{M}_{\text{out}} \tag{3.3b}$$

Often it is useful to perform a system balance on a chemical element, denoted (α), irrespective of which chemical species it is contained in. If nuclear reactions are

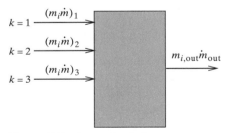

Figure 3.2   A species balance on a system with k inlet streams and one outlet stream.

absent, there cannot be production of chemical elements within the system; thus

$$\text{Rate of inflow of element } (\alpha) \; = \; \text{Rate of outflow of element } (\alpha)$$

The mass flow of chemical element (α) in a stream is obtained by multiplying each species flow rate $m_i \dot{m}$ by the mass fraction of chemical element (α) in species i, $m_{(\alpha)i}$, and summing over all species containing (α). Recall that an element mass fraction $m_{(\alpha)i}$ is obtained from the chemical formula of species i and atomic weights. For example, if species i is water, H_2O, and element (α) is hydrogen, (H), $m_{(\alpha)i} = m_{(H)H_2O} = 2/18 = 1/9$. Again, with k inlet streams forming a single outlet stream,

$$\sum_k \left(\left[\sum_i m_{(\alpha)i} m_i \right] \dot{m} \right)_k = \left[\sum_i m_{(\alpha)i} m_{i,\text{out}} \right] \dot{m}_{\text{out}} \qquad (3.4)$$

In this text the balances described above will be applied to systems as diverse as power plants, automobile engines, catalytic reactors, and cooling towers. Provided that the flows and compositions of all the streams but one are known, one can find the flow and elemental composition of the unknown stream. One need not know what happens in the system. Also, with minimal additional knowledge of what happens in the system—for example, knowledge that no reactions occur or that a particular reaction goes to completion—the molecular composition of the outlet stream can be determined.

3.2.2 A Balance on an Automobile Engine

In most urban areas of the United States, a principal source of nitric oxide in the atmosphere is automobile engine exhaust. Nitric oxide, NO, is itself harmful to humans; additionally, in the presence of sunlight it reacts with oxygen, O_2, to produce nitrogen dioxide, NO_2, which is also harmful. Ozone, O_3, is produced as an intermediate in this reaction, but it reacts very rapidly with any excess NO to produce more NO_2. However, when most of the available NO has been oxidized, which usually occurs some miles downwind of the source, ozone builds up in concentration and produces the characteristic irritating effects of photochemical smog. The precise mechanisms of subsequent reactions are not well understood. One view is that as ozone builds up, it oxidizes unburned hydrocarbons from automobile and industrial emissions. The ozone is replenished by a photochemical reaction involving NO_2,

$$NO_2 + \text{photon} \rightarrow NO + O$$

$$O + O_2 \rightarrow O_3$$

Some of the resulting NO then further reacts with the oxidized hydrocarbons to form particularly dangerous chemical species such as peroxyacetylnitrate (PAN). These species damage plants and are highly irritating (and presumably injurious) to humans. The remaining NO is reoxidized by O_3 to NO_2, and the cycle repeats itself. In strong sunlight, O_3 concentrations in a polluted atmosphere tend to be approximately one-half of the NO_2 concentrations.

Consider a typical automobile that gets 7 km per liter of gasoline (14.3 liters/100 km). An oxygen sensor in the exhaust allows the fuel injection system to maintain an approximately stoichiometric fuel–air mixture. If the gasoline is approximated as octane (C_8H_{18}), one can write

$$C_8H_{18} + 12.5O_2 \rightarrow 8CO_2 + 9H_2O$$

assuming complete combustion. The specific gravity of gasoline can be taken as 0.70, and thus we have the following.

Fuel consumption: (1/7 liter/km)(0.70 kg/liter) $= 0.10$ kg/km; or, on a molar basis, $(0.10/114) = 8.8 \times 10^{-4}$ kmol/km, since the molecular weight of octane is ($8 \times 12) + 18 = 114$

Oxygen consumption: $(8.8 \times 10^{-4})(12.5) = 1.10 \times 10^{-2}$ kmol/km $= 0.35$ kg/km

Nitrogen throughput: $(79/21)(1.10 \times 10^{-2}) = 4.14 \times 10^{-2}$ kmol/kg $= 1.16$ kg/km

Total exhaust flow: $(0.10 + 0.35 + 1.16) = 1.61$ kg/km

Table 3.1 gives the allowable emissions of pollutants according to 1993–94 State of California standards for passenger cars. The unburnt hydrocarbons and carbon monoxide (CO) result from incomplete combustion; the nitrogen oxides (NO_x) result from side reactions between N_2 and O_2 at the high temperatures of the combustion process. The pollutant NO_x is actually emitted largely as NO, with a little NO_2. Because the NO oxidizes after entering the atmosphere, the amount of NO allowed is multiplied by 46/30, the ratio of molecular weights of NO_2 and NO, and added to the amount of NO_2. This sum is then referred to as the allowable emission rate of pollutant NO_x. Accordingly, the allowable mass fraction of NO_x in the automobile exhaust is

$$m_{NO_x,\text{out}} = (0.25 \times 10^{-3}/1.61) = 1.55 \times 10^{-4}$$

Catalytic converters for automobiles must be sized to reduce the concentration of NO_x in the engine exhaust to this value: the required mass exchanger effectiveness proves to be typically of the order of 99%.

To estimate the total amount of pollutants entering the atmosphere above an urban area such as the Los Angeles basin, we will assume 4 million automobiles in the basin, each traveling, on average, 50 km/day. Then the total exhaust flow into the atmosphere is

$$(1.61 \text{ kg/km})(50 \text{ km/day})(4 \times 10^6 \text{ automobiles}) = 3.22 \times 10^8 \text{ kg/day}$$

Table 3.1 California mobile source emission standards for new passenger cars, 1993–94.

Pollutant	Allowable Emission	
	g/mi	g/km
Hydrocarbons[a]	0.25	0.16
Carbon monoxide	3.4	2.1
NO_x	0.4	0.25
Diesel particulates	0.08	0.05

[a] Excluding methane.

and the NO_x flow is

$$(m_{NO_x})(\dot{m}_{exhaust}) = (1.55 \times 10^{-4})(3.22 \times 10^8) = 5.0 \times 10^4 \text{ kg/day}$$

Similarly, the flow rates of hydrocarbons and carbon monoxide can be calculated to be 3.2×10^4 and 4.2×10^5 kg/day, respectively.

3.2.3 A Balance on an Oil-Fired Power Plant

Oil- and coal-fired power plants are the principal sources of the pollutant sulfur dioxide (SO_2) in our atmosphere. The SO_2 primarily results from the oxidation of free sulfur in the fuel. Typically, fuel oils have free sulfur contents ranging from 0.2% to 3% by mass. Consider an 800-MW power plant that burns a fuel oil containing 1% free sulfur by mass. If the fuel oil is approximated as $C_{21}H_{44}$, the main combustion reaction is

$$C_{21}H_{44} + 32O_2 \rightarrow 21CO_2 + 22H_2O$$

The overall efficiency of such a plant is typically 40%, and the heat of combustion of the fuel is approximately 4.2×10^7 J/kg. Thus the fuel flow rate can be estimated as follows.

$$\text{Power output} = (\text{Fuel flow rate})(\text{Heat of combustion})(\text{Efficiency})$$

$$(800 \times 10^6 \text{ J/s}) = (\dot{m}_f \text{ kg/s})(4.2 \times 10^7 \text{ J/kg})(0.40)$$

Solving, $\dot{m}_f = 47.6$ kg/s.

The molecular weight of the fuel is $(21 \times 12) + 44 = 296$; thus, the molar flow rate of fuel is

$$\dot{M}_f = \dot{m}_f/M_f = (47.6 \text{ kg/s})/(296 \text{ kg/kmol}) = 0.161 \text{ kmol/s}$$

From the fuel flow rate we can immediately find the sulfur flow rate:

$$(\dot{m}_S)_f = m_{(S)f}\dot{m}_f = (0.01 \times 47.6 \text{ kg/s}) = 0.476 \text{ kg/s}$$

$$(\dot{M}_S)_f = (\dot{m}_S)_f/M_S = (0.476 \text{ kg/s})/(32 \text{ kg/kmol}) = 0.0149 \text{ kmol/s}$$

The chemical reaction states that for each kmol of fuel, 32 kmol of oxygen are required for stoichiometric combustion:

$$\dot{M}_{O_2} = (32)(0.161 \text{ kmol/s}) = 5.15 \text{ kmol/s}$$

Typically, 10% excess oxygen is used to reduce CO emissions from the stack. The required flow of oxygen is then

$$\dot{M}_{O_2} = (1.1)(5.15) = 5.67 \text{ kmol/s}$$

The nitrogen flow rate is obtained easily since air is approximately 21% O_2 and 79% N_2 by volume: $\dot{M}_{N_2} = (79/21)\dot{M}_{O_2} = (3.76)(5.67) = 21.3$ kmol/s. The stack gas flow rates of the major constituents in the effluent are summarized below.

CO_2	$(0.161)(21)$	$= 3.38$ kmol/s	$= 149$ kg/s
H_2O	$(0.161)(22)$	$= 3.54$ kmol/s	$= 63.8$ kg/s
SO_2		$= 0.0149$ kmol/s	$= 0.954$ kg/s
O_2	$5.67 - 5.15 - 0.0149$	$= 0.505$ kmol/s	$= 16.2$ kg/s
N_2		$= 21.3$ kmol/s	$= 596$ kg/s
Totals		28.7 kmol/s	826 kg/s

The mean molecular weight of the effluent is $(826/28.7) = 28.8$ kg/kmol. The mass fraction of SO_2 in the effluent is $(0.954/826) = 1.15 \times 10^{-3}$, and the mole fraction is $(0.0149/28.7) = 5.2 \times 10^{-4}$. Small concentrations of pollutants are often expressed as parts per million on a volume (molar) basis: thus, the concentration of SO_2 is 520 ppm. If the effluent is untreated, the flow rate of SO_2 from the plant into the atmosphere is $(0.954)(3600)(24) = 8.2 \times 10^4$ kg/day.

3.2.4 A Balance on an Urban Air Volume

The nature of the air pollution problem in an urban area can be explored by performing a system balance on the air volume above the city. Consider the Los Angeles basin with a surface area of 3200 km^2, for which air pollution is a serious problem owing to the frequent occurrence of an *inversion layer* over the basin. Figure 3.3 shows the temperature distribution in the atmosphere that causes an inversion layer. We will assume that the height of the layer is 500 m and that the average residence time for the air in the layer is one day (that is, a mass of air equal to that contained in the layer flows into, mixes with, and flows out of the layer each day). The volume of system under consideration is $(3200 \times 10^6)(500) = 1.6 \times 10^{12}$ m^3. Air at 1 atm pressure and 300 K has a density of 1.18 kg/m^3, so that the corresponding mass is $(1.6 \times 10^{12})(1.18) = 1.9 \times 10^{12}$ kg. Thus, the mass flow rate $\dot{m}$ through the system is on the order of 1.9×10^{12} kg/day.

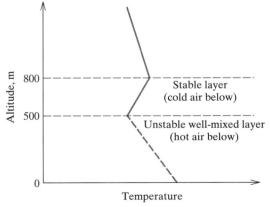

Figure 3.3 Temperature versus altitude variation characteristic of an inversion layer.

The effects of pollutants on plant life and human health are not too well defined as yet. Plant damage has been observed with chronic exposure to as little as 0.03 ppm SO_2. There is evidence that exposure to modest concentrations of such chemical species as CO, NO_x, SO_2, and O_3 is unhealthful. Indeed, it is well established that episodes of high concentrations can kill susceptible segments of a population. For this reason the Los Angeles air quality management authority has established what are termed *first-stage alert* pollution levels for CO (40 ppm, 1 h; 20 ppm, 12 h), SO_x (0.5 ppm), and O_3 (0.2 ppm). The *third-stage alert* levels, at which a serious emergency is considered to exist, are 100, 50, 2, and 0.5 ppm, respectively. Species balances on the urban air volume can be used to determine the corresponding allowable flow rates of these pollutants into the volume. For example, to prevent a third-stage alert, the flow rate of SO_x (taken as SO_2) into the air volume must not exceed

$$\dot{m}_{SO_2} = m_{SO_2}\dot{m} = (2 \times 10^{-6})(64/29)(1.9 \times 10^{12}) = 8.4 \times 10^6 \text{ kg/day}$$

Notice that for these dilute mixtures the ppm by volume value is converted to mass fraction by multiplying by $10^{-6}(M_i/M_{air})$, since $M \simeq M_{air}$.

To put the preceding result in perspective, the total power-generating capacity in the Los Angeles basin is about 10,000 MW. If all these plants burned fuel with 1% free sulfur, then the results of Section 3.2.3 indicate that the flow of SO_2 into the atmosphere would be $(8.2 \times 10^4)(10,000/800) = 1.03 \times 10^6$ kg/day—that is, about 12% of the allowable amount.

3.3 SINGLE-STREAM MASS EXCHANGERS

A single-stream mass exchanger is an exchanger in which there is only one fluid stream, or, if there are two streams, the composition of only one stream varies significantly along the exchanger. Examples of exchangers with only one stream are catalytic reactors (Section 3.3.1), electrostatic precipitators (Section 3.3.3), and filters (Section 3.3.4). Examples of exchangers with two streams, of which the composition of only one varies significantly, include some gas absorption or desorption processes, as described in Section 3.3.2. The analysis of single-stream mass exchangers is particularly simple. In fact, the result is always the familiar effectiveness–number of transfer units relation first derived for single-stream heat exchangers as Eq. (1.59) of *Heat Transfer*.

3.3.1 Catalytic Reactors

A catalytic reactor is a single-stream exchanger by virtue of the fact that there is no second stream, or adjacent phase, where the reaction products accumulate. Figure 3.4 depicts a model catalytic reactor involving the flow of a single stream with continuous removal of chemical species 1 along its flow path. Since the applications considered in this section all involve a gas stream, the mass flow rate of the stream will be denoted $\dot{m}_G$; m_1 will denote the bulk mass fraction of the transferred species 1 (as for heat exchangers, we do not use a subscript to denote "bulk" in order to keep the notation simple); $j_{1,s}$ is the rate of transfer of species 1 across the s-surface; and $\mathcal{P}\Delta x$ is the area of s-surface between locations x and $x + \Delta x$; that is, $\mathcal{P}$ is the *perimeter* of the

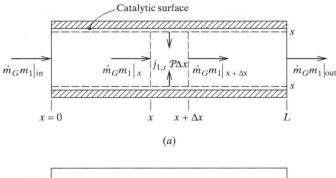

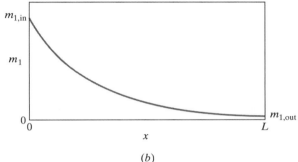

Figure 3.4 A catalytic reactor. (*a*) Schematic showing a species balance on an elemental volume Δx long.
(*b*) Reactant concentration variation along the reactor.

transfer surface. By using the perimeter, the result of our analysis will be of general applicability, irrespective of whether the catalyst is in the form of pellets, plates, or a honeycomb matrix. The principle of conservation of species applied to the elemental control volume bounded by the s-surface and located between x and $x + \Delta x$ requires that, at steady state,

Rate of inflow of species 1 $=$ Rate of outflow of species 1

since species 1 is inert in the gas phase. It is assumed that there are no *homogeneous* reactions involving species 1. If x-direction diffusion is neglected, then

$$\dot{m}_G m_1|_x + j_{1,s} \mathcal{P} \Delta x = \dot{m}_G m_1|_{x+\Delta x} \tag{3.5}$$

Also, the principle of mass conservation applied to the elemental control volume requires that

$$\dot{m}_G|_x = \dot{m}_G|_{x+\Delta x} = \text{Constant} \tag{3.6}$$

since there is no net mass transfer across the s-surface. The transfer rate of the reactants exactly balances the transfer rate of the products since the catalyst does not enter into the reaction. Then, dividing Eq. (3.5) by $\dot{m}_G \Delta x$ and letting $\Delta x \to 0$,

$$\frac{dm_1}{dx} - \frac{j_{1,s} \mathcal{P}}{\dot{m}_G} = 0 \tag{3.7}$$

Further progress requires an expression for $j_{1,s}$; for a first-order reaction and a catalyst in the form of plates or foil, we can use the result of the analyses in Sections 1.3.3 and 1.4.5. Replacing $m_{1,e}$ in Eq. (1.53) by the bulk value m_1 gives

$$j_{1,s} = -\frac{\dot{m}_1}{A} = \frac{-m_1}{\dfrac{1}{\rho k''} + \dfrac{1}{g_{m1}}} = -g_{m1}^{\text{oa}} m_1 \tag{3.8}$$

where g_{m1}^{oa} is an *overall mass transfer conductance* for species 1. Substituting in Eq. (3.7) gives

$$\frac{dm_1}{dx} + \frac{g_{m1}^{\text{oa}} \mathcal{P}}{\dot{m}_G} m_1 = 0 \tag{3.9}$$

which is a first-order ordinary differential equation for $m_1(x)$ The required initial condition is obtained from the composition of the inlet gas mixture as

$$x = 0; \quad m_1 = m_{1,\text{in}}$$

If $\mathcal{P}$ and g_{m1}^{oa} can be assumed constant along the exchanger, the solution is

$$\frac{m_1}{m_{1,\text{in}}} = e^{-g_{m1}^{\text{oa}} \mathcal{P} x / \dot{m}_G}$$

In particular, the outlet concentration of species 1 is obtained by setting $x = L$,

$$\frac{m_{1,\text{out}}}{m_{1,\text{in}}} = e^{-g_{m1}^{\text{oa}} \mathcal{P} L / \dot{m}_G}$$

and subtracting each side from unity gives

$$\frac{m_{1,\text{in}} - m_{1,\text{out}}}{m_{1,\text{in}}} = 1 - e^{-g_{m1}^{\text{oa}} \mathcal{P} L / \dot{m}_G} \tag{3.10}$$

The concentration variation is shown in Fig. 3.4. The numerator of the left-hand side of Eq. (3.10) is the actual change in the concentration of species 1, and the denominator $(m_{1,\text{in}} - 0)$ is the maximum possible change that could be achieved in an infinitely long exchanger: thus, the left-hand side is recognized to be the *effectiveness* ε of the mass exchanger, analogous to the effectiveness of a heat exchanger. Similarly, the dimensionless exponent on the right-hand side of Eq. (3.10) is recognized to be the number of transfer units N_{tu} of the mass exchanger. Thus, Eq. (3.10) can be written as

$$\varepsilon = 1 - e^{-N_{\text{tu}}} \tag{3.11}$$

which is identical to the result for a single-stream heat exchanger.

Example 3.1 involves the preliminary design of an automobile catalytic converter, in which the catalyst is in the form of oxidized copper foil wrapped in a spiral or packed in wafer form. However, current practice for automobiles is to use either a packed bed of porous catalyst pellets or a ceramic matrix with walls coated with a porous catalyst layer. Mass transfer in porous catalysts was analyzed in Section 1.6.

For a pellet in a packed bed, Eq. (1.85) gives the molar flux of species 1 across the s-surface surrounding the pellet as

$$J_{1,s} = -\frac{\dot{M}_1}{S_p} = \frac{-x_{1,e}}{\dfrac{1}{(V_p a_p/S_p)\eta_p k''c} + \dfrac{1}{G_{m1}}}$$

where V_p and S_p are the pellet volume and s-surface area, a_p is the area of catalyst per unit volume of pellet, and η_p is the pellet effectiveness obtained from Eq. (1.82). Thus, an overall mole transfer conductance can be defined as

$$\frac{1}{G_{m1}^{\text{oa}}} = \frac{1}{(V_p a_p/S_p)\eta_p k''c} + \frac{1}{G_{m1}} \tag{3.12}$$

and is used in Eq. (3.11), written on a molar basis, as

$$\varepsilon = \frac{x_{1,\text{in}} - x_{1,\text{out}}}{x_{1,\text{in}}} = 1 - e^{-G_{m1}^{\text{oa}} \mathcal{P} L / \dot{M}_G} \tag{3.13}$$

Use of Eq. (3.12) is required in Exercise 3–11, for example. Analysis of reactors containing a porous wall matrix is required in Exercises 3–10, 3–12, and 3–13.

As discussed in Section 1.3, CO oxidation on a platinum or cupric oxide catalyst can be approximately modeled as a first-order reaction. Thus, these results can be used for the design of CO oxidation converters. Such converters are used in automobiles in countries where NO emissions are not controlled. Also, oxidation converters are used on two-stroke engines, for which NO emissions are inherently low, and on diesel engines that run on lean fuel–air mixtures. So called "three-way" catalytic converters that promote the reduction of CO and NO in automobile exhaust cannot be properly designed using Eq. (3.13), since the reduction reaction is not first-order. Indeed, the kinetics of this reaction are poorly understood, and current practice is to develop such converters by extensive testing rather than by detailed design.

Example 3.1   Removal of Carbon Monoxide from Automobile Exhaust

A 99% effective catalytic converter is to be designed for removing CO from automobile engine exhaust. A suitable catalyst is cupric oxide, and we wish to examine the feasibility of using a packed bed of thin oxidized copper foil wrapped in a spiral or stacked in wafer form, with a passage width of 1 mm. The desired operating temperature is 1200 K, and the rate constant for CO oxidation by CuO is $k'' = 50.6 \exp(-9361/T)$ m/s, for T in kelvins. Suggest dimensions of a reactor suitable for an automobile that obtains 8 km/liter at 80 km/h. Assume an air/fuel ratio of 14/1 and exhaust air injection at 10% of the exhaust flow rate.

Solution

Given: Copper foil packed-bed catalytic converter for an automobile that obtains 8 km/liter at 80 km/h.

Required: Dimensions for 99% effective removal of CO from an automobile exhaust.

Assumptions:

1. Air properties can be used for the gas stream.
2. The air/fuel ratio is 14/1, and air is injected into the exhaust at a rate equal to 10% of the exhaust flow rate.
3. A gasoline specific gravity of 0.705.

The first step is to determine the number of transfer units required. Solving Eq. (3.11) for N_{tu} gives

$$N_{tu} = \frac{\mathcal{g}_{m1}^{oa} \mathcal{P} L}{\dot{m}_G} = \ln \frac{1}{1-\varepsilon} = \ln \frac{1}{1-0.99} = 4.61$$

Next we evaluate $\mathcal{g}_{m1}^{oa}$; from Eq. (3.8),

$$\frac{1}{\mathcal{g}_{m1}^{oa}} = \frac{1}{\rho k''} + \frac{1}{\mathcal{g}_{m1}}$$

We approximate ρ by the value for air at 1200 K and 1 atm; $\rho = 0.294 \text{ kg/m}^3$, and $k'' = 50.6 \exp(-9361/1200) = 2.07 \times 10^{-2}$ m/s.

$$\mathcal{g}_{m1} = \text{Sh}\rho \mathcal{D}_{1m}/D_h; \quad D_h = 2b, \quad \text{where } b \text{ is the passage width}$$

From Table 4.5 in *Heat Transfer*, the Sherwood number for laminar flow between parallel plates with a uniform wall concentration and $L \gg D_h$ is 7.54: there is no Reynolds- or Schmidt-number dependence. The assumption of laminar flow will be checked later. For $\mathcal{D}_{1m}$ an approximate value is $\mathcal{D}_{CO,air} = 209 \times 10^{-6} \text{ m}^2/\text{s}$ at 1200 K from Table A.17a; then

$$\mathcal{g}_{m1} = \frac{7.54\rho \mathcal{D}_{1m}}{2b} = \frac{(7.54)(0.294 \text{ kg/m}^3)(209 \times 10^{-6} \text{ m}^2/\text{s})}{(2)(1.0 \times 10^{-3} \text{ m})} = 0.232 \text{ kg/m}^2 \text{ s}$$

$$\frac{1}{\mathcal{g}_{m1}^{oa}} = \frac{1}{(0.294)(2.07 \times 10^{-2})} + \frac{1}{0.232} = 164 + 4.3 \quad \text{(rate-limited reaction)}$$

$$\mathcal{g}_{m1}^{oa} = 5.94 \times 10^{-3} \text{ kg/m}^2 \text{ s}$$

For a gasoline specific gravity of 0.705, the exhaust flow rate is

$$\dot{m}_{exhaust} = (80/3600 \text{ km/s})(1/8 \text{ liter/km})(10^{-3} \text{ m}^3/\text{liter})(705 \text{ kg/m}^3)$$

$$\times (15 \text{ kg fuel and air/kg fuel})$$

$$= 2.94 \times 10^{-2} \text{ kg/s}$$

and, with the additional air subsequently added to ensure proper operation of the reactor,

$$\dot{m}_G = (1.1)(2.94 \times 10^{-2}) = 3.23 \times 10^{-2} \text{ kg/s}$$

Then

$$N_{tu} = 4.61 = \frac{\mathcal{g}_{m1}^{oa} \mathcal{P} L}{\dot{m}_G} = \frac{(5.94 \times 10^{-3})\mathcal{P} L}{3.23 \times 10^{-2}}$$

Solving, $\mathcal{P} L = 25.1 \text{ m}^2$.

If a cylindrical can of 20 cm diameter is chosen, the perimeter is the number of turns times the average circumference times 2 for both sides of the foil: assuming a 0.1 mm foil thickness, $\mathcal{P} = [0.10/(0.001 + 0.0001)](0.10\pi)(2) = 57.1$ m. Hence, $L = 25.1/57.1 = 0.44$ m (44 cm).

Comments

1. A reactor of this size may not satisfy size limitations: since the reaction is seen to be rate-limited, means for increasing the effective area of catalyst should be explored (e.g., by using a porous catalyst).

2. Check the assumption of laminar flow:

$$\dot{m}_G = \rho V A_c; \quad V = \frac{\dot{m}_G}{\rho A_c} = \frac{(3.23 \times 10^{-2})}{(0.294)(\pi)(0.1)^2(1.0/1.1)} = 3.85 \text{ m/s}$$

$$\text{Re} = V(2b)/v = (3.85)(2)(0.001)/(159 \times 10^{-6}) = 48 < 2800; \text{ laminar}$$

3. A spark-ignition engine obtains maximum power on a slightly *rich* mixture, that is, an air/fuel ratio less than the stoichiometric value ($\sim$15:1 for gasoline). An automobile fuel-injection system can be designed to take advantage of this characteristic for acceleration. Since there is insufficient O_2 for complete combustion, CO is produced—hence the need for a catalytic converter to reduce CO concentrations in the exhaust gases. The table below shows the relation between the air/fuel ratio and exhaust CO content for typical gasoline.

Percent CO by volume	0.2	0.5	1.0	2.0	3.0	4.0	5.0	6.0
Air/fuel ratio		14.53	14.27	14.10	13.76	13.37	12.99	12.63	12.24

4. The oxidation of CO to CO_2 in the converter is exothermic and releases 10.1×10^6 J/kg CO of heat. An energy balance on the converter will show that the resulting adiabatic temperature rise of the exhaust is approximately 100 K/percent CO by volume. Thus, proper thermal design of the converter is required to ensure its survival. Local burnup of matrix-type converters is a significant problem: when a hot spot develops, the reaction rate increases, and the local temperature can continue to increase. ▲

3.3.2 Gas Absorption

Mass exchangers, in which a chemical species is absorbed into a liquid, are usually two-stream exchangers and will be dealt with in Section 3.4. Under certain circumstances, however, the performance of such exchangers can be described by the simple single-stream mass exchanger equations. Such situations are best illustrated by numerical examples. One such example follows as Example 3.2 and concerns the absorption of a gaseous species into a liquid, in which it subsequently enters into a chemical reaction. The reaction is very fast and maintains a near-zero concentration of the solute species at the s-surface adjacent to the liquid surface. Thus, the liquid-side mass transfer resistance is negligible, and the absorption is *gas-side controlled*. The s-surface concentration can be taken to be zero, and hence does not vary through the exchanger. In an analogous heat exchanger, it is the wall temperature T_s that does not vary through the exchanger, as, for example, in a simple condenser or evaporator.

Exercises 3–16 through 3–19 concern absorption of sparingly soluble gases such as air, CO, and CO_2 into water. Since mole fractions in the liquid phase are very small, usually so too are mole-fraction gradients. Thus, except in unusual circumstances, such gas absorption processes are *liquid-side controlled:* the gas-side mass transfer resistance can be ignored. In addition, the situations in these exercises are such that the u-surface composition varies negligibly along the exchanger so that the single-stream exchanger equations apply. These single-stream mass exchanger problems are, of course, special cases of the more general two-stream mass exchanger, which will be analyzed in Section 3.4. In Sections 3.4.1 and 3.4.2 the concept of gas-side or liquid-side mass transfer control will be carefully developed. However, it is not premature to deal with these special cases here. There are many important applications of gas absorption processes similar to those described above, both in technology and in the environment. Examples include aeration of sewage, carbonation of soft drinks, and aeration of lakes and rivers; the literature on these topics usually assumes liquid-side mass transfer control without discussion, since the validity of such a model is well established.

Example 3.2 Scrubbing of Sulfur Dioxide from a Steam Generator Exhaust

The 50,000 kg/h exhaust stream from a steam generator contains 3% sulfur dioxide by volume. To reduce SO_2 air pollution, the gas is passed through a bed of rocks, which can be approximated as 2 cm–diameter spheres, with a volume void fraction of 58%. The bed is 50 m^2 in cross-sectional area, 2 m deep, and at approximately 320 K temperature. Forty percent of the rock surface area is kept wet with an alkaline solution such that the SO_2 concentration adjacent to the wet surface is approximately zero. Estimate the effectiveness of the bed, the mole fraction of SO_2 at the exit, and the kilograms per hour of SO_2 that escape to the atmosphere.

Solution

Given: Packed-bed scrubber for SO_2 absorption.
Required: Bed effectiveness, outlet gas SO_2 mole fraction, and outlet SO_2 flow rate.
Assumptions:

1. 40% of packing wetted.
2. An s-surface concentration of SO_2 equal to zero, giving a gas side–controlled mass transfer process.
3. Gas properties can be approximated by those for air.

The bed can be modeled as a single-stream mass exchanger because $x_{SO_2,s} \simeq 0$, independent of depth; only the composition change of the exhaust stream is of concern. The number of transfer units for the bed can be calculated from $N_{tu} = \mathcal{G}_m \mathcal{P} H / \dot{M}_G$, where H is the height of the bed. Since the bed is only partially wet, the correlations for a dry packed bed given in Section 4.5.2 of *Heat Transfer* will be used to obtain $\mathcal{G}_m$; by analogy, Eq. (4.131) becomes

$$\text{Sh} = (0.5\text{Re}^{1/2} + 0.2\text{Re}^{2/3})\text{Sc}^{1/3}$$

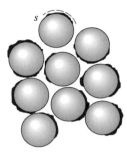

The characteristic length is

$$\mathcal{L} = d_p \left(\frac{\varepsilon_v}{1 - \varepsilon_v} \right) = 0.02 \left(\frac{0.58}{1 - 0.58} \right) = 0.0276 \text{ m}$$

and the characteristic velocity is

$$\mathcal{V} = \frac{\dot{m}}{\rho \varepsilon_v A_c} = \frac{(50{,}000/3600)}{(1.106)(0.58)(50)} = 0.433 \text{ m/s}$$

where A_c is the cross-sectional (frontal) area of the bed, and the properties of the exhaust gases have been approximated by those for air at 320 K and 1 atm.

$$\text{Re} = \frac{\mathcal{V}\mathcal{L}\rho}{\mu} = \frac{(0.433)(0.0276)(1.106)}{1.929 \times 10^{-5}} = 685$$

From Table A.17b, Sc = 1.24,

$$\text{Sh} = [0.5(685)^{1/2} + 0.2(685)^{2/3}](1.24)^{1/3} = 30.8$$

$$\mathcal{G}_m = c\mathcal{D}\text{Sh}/\mathcal{L} = \mu\text{Sh}/M\text{Sc}\mathcal{L}$$

Thus, $\mathcal{G}_m = (1.929 \times 10^{-5})(30.8)/(29)(1.24)(0.0276) = 5.98 \times 10^{-4} \text{ kmol/m}^2 \text{ s}$.
The molar flow rate of the exhaust gas stream is

$$\dot{M}_G = (50{,}000)/(3600)(29) = 0.479 \text{ kmol/s}$$

Next we must find the perimeter of the transfer area: the specific surface area of the bed is

$$a = \frac{A_p}{V_p}(1 - \varepsilon_v) = \frac{6}{d_p}(1 - \varepsilon_v) \quad \text{for spheres}$$

$$a = (6/0.02)(1 - 0.58) = 126 \text{ m}^{-1}$$

$\mathcal{P} = aA_c = (126)(50) = 6300 \text{ m}$; but only 40% of the rocks are wet, so that the effective perimeter is $(0.4)(6300) = 2520 \text{ m}$.

$$N_{tu} = \mathcal{G}_m \mathcal{P} H/\dot{M}_G = (5.98 \times 10^{-4})(2520)(2.0)/(0.479) = 6.29$$

Equation (3.11) gives the effectiveness as

$$\varepsilon = 1 - e^{-N_{tu}} = 1 - e^{-6.29} = 1 - 0.00185 = 0.99815$$

The effectiveness of this single-stream exchanger is the actual change in concentration of the gas stream, divided by the maximum possible change. Since the concentration of SO_2 in the outlet stream will be zero in an infinitely long exchanger,

$$\varepsilon = \frac{x_{SO_2,in} - x_{SO_2,out}}{x_{SO_2,in} - 0}$$

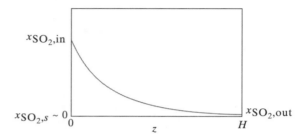

Solving for $x_{SO_2,out}$,

$$x_{SO_2,out} = x_{SO_2,in}(1 - \varepsilon) = (0.03)(0.00185) = 5.55 \times 10^{-5}$$

$$\dot{M}_{SO_2,out} = x_{SO_2,out}\dot{M}_G = (5.55 \times 10^{-5})(0.459) = 2.55 \times 10^{-5} \text{ kmol/s}$$

$$\dot{m}_{SO_2,out} = (\dot{M}_{SO_2,out})(M_{SO_2}) = (2.55 \times 10^{-5})(64)(3600) = 5.87 \text{ kg/h}$$

Comments

1. Notice that we multiply by the molecular weight of SO_2, not of the exhaust mixture, to obtain the mass flow of SO_2.
2. The effectiveness of this rock bed is very high; it is more convenient to quote its *ineffectiveness,* $1 - \varepsilon = 1 - 0.99815 = 0.00185 = 0.185\%$ (i.e., only 0.185% of the entering SO_2 escapes from the bed). ▲

3.3.3 Electrostatic Precipitators

Electrostatic precipitation is widely used to remove particles from the exhausts of coal-fired power plants and solid waste incinerators. The process is particularly effective in removing particles in the 0.1–10 μm size range. Energy requirements are lower than for other particulate removal systems, such as venturi scrubbers and fabric filters. Although particles in industrial emissions may possess an electric charge, commercial precipitators are based on particle charging in coronas generated upstream or within the precipitator, in order to obtain a maximum charge on the particles. Figure 3.5*a* shows a dry precipitator, and Fig. 3.5*b* shows a weir-type wet-wall precipitator. An electric field set up in the precipitator causes the charged particles to migrate to the collector plates by the phenomenon of *forced diffusion*. The electric field produces a

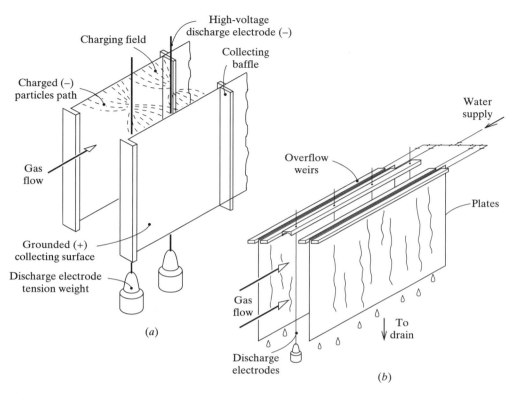

Figure 3.5 Electrostatic precipitators: (*a*) dry wall, (*b*) weir-type wet wall.

corona of gas molecules ionized by high-energy electrons. Excess electrons combine with electronegative gas molecules such as O_2 and SO_2, whereas the negative ions are adsorbed onto particles, which migrate to the grounded plates. Once collected, the particles begin to lose their charge to the plates; this charge transfer completes the electrical circuit to maintain the current flow and voltage drop. In a wet-wall precipitator, the water film serves to prevent dust buildup on the electrodes that would increase resistivity, and also minimizes reentrainment of particles into the gas stream. In dry precipitators, dust is removed by periodic *rapping* of the electrodes using mechanical hammers: reentrainment of particles is then a problem, but, on the other hand, dry units do not have the corrosion and scaling problems of wet units. Dry electrostatic precipitators are used for coal-fired power plants and cement kilns; wet units can be found on paint finishing lines, sewage sludge incinerators, and for mist removal downstream of SO_2 scrubbers.

Migration Velocity

The steady migration velocity V^E of a particle due to forced diffusion is obtained from a force balance, in which the electrical force on the particle equals the viscous drag on the particle. For a particle diameter d_p, a particle charge Q [A s], and an electric field E [V/m], the electric force is QE and

$$QE = 3\pi d_p \mu V^E$$

where it has been assumed that the Stokes' drag law, Eq. (4.73), applies. For very small particles it is necessary to introduce a slip correction to Stokes' law to account for departure from continuum flow conditions [1]. (See also Section 1.7.3.) Solving for V^E,

$$V^E = \frac{QE}{3\pi d_p \mu} \tag{3.14}$$

The particle charge Q can be calculated from well-known principles of electrostatics, but the theory is complicated [2, 3]. Figure 3.6 shows a graph of V^E as a function of particle size for typical precipitator operating conditions.

Precipitator Performance

Figure 3.7 shows an elemental control volume Δx long, used to derive the equation governing particle concentration through a precipitator. The concentration of particles is expressed in terms of a number density $\mathcal{N}$ that has units $[\text{m}^{-3}]$ but as a conceptual aid can be assigned units $[\text{particles/m}^3]$. The flow of gas in a commercial precipitator is always turbulent, and the characteristic turbulent mixing velocity is quite large compared with the electrical migration velocity, except very close to the electrodes. Thus, we will assume that the particle concentration $\mathcal{N}$ is uniform across

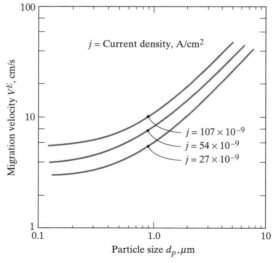

Figure 3.6   Particle migration velocity V^E as a function of particle size and current density at 30 kV [2].

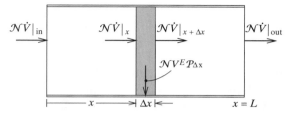

Figure 3.7 Schematic of an electrostatic precipitator showing a particle balance on an elemental volume Δx long.

the channel. The principle of conservation of particles applied to the elemental control volume requires that, at steady state,

$$
\begin{array}{ccc}
\text{Rate of inflow} & = & \text{Rate of deposition} \\
\text{of particles} & & \text{of particles}
\end{array} + \begin{array}{c} \text{Rate of outflow} \\ \text{of particles} \end{array}
$$

$$
\mathcal{N}\dot{V}|_x = \mathcal{N}V^E\mathcal{P}\Delta x + \mathcal{N}\dot{V}|_{x+\Delta x}
$$

where $\dot{V}$ is the volume flow rate of gas, $\mathcal{P}$ is the perimeter of the collecting plates, and V^E is taken as positive toward the plates. Rearranging, dividing by Δx, and letting $\Delta x \to 0$,

$$
\frac{d\mathcal{N}}{dx} + \frac{V^E\mathcal{P}}{\dot{V}}\mathcal{N} = 0 \tag{3.15}
$$

which is the differential equation governing the particle concentration distribution through the precipitator, $\mathcal{N}(x)$. The required boundary condition is the inlet condition,

$$
x = 0: \ \mathcal{N} = \mathcal{N}_{\text{in}} \tag{3.16}
$$

The solution gives $\mathcal{N}(x)$,

$$
\mathcal{N} = \mathcal{N}_{\text{in}}e^{-V^E\mathcal{P}x/\dot{V}} \tag{3.17}
$$

In particular, at $x = L$, $\mathcal{N} = \mathcal{N}_{\text{out}}$,

$$
\mathcal{N}_{\text{out}} = \mathcal{N}_{\text{in}}e^{-V^E\mathcal{P}L/\dot{V}}
$$

or

$$
\frac{\mathcal{N}_{\text{in}} - \mathcal{N}_{\text{out}}}{\mathcal{N}_{\text{in}}} = 1 - e^{-V^E\mathcal{P}L/\dot{V}} \tag{3.18}
$$

In the absence of reentrainment, the particle concentration can be reduced to zero in an infinitely long precipitator. Thus, we recognize that the left-hand side of Eq. (3.18) is the effectiveness[1] of the precipitator as a mass exchanger, and write

$$
\varepsilon = 1 - e^{-N_{\text{tu}}}; \quad N_{\text{tu}} = \frac{V^E\mathcal{P}L}{\dot{V}} \tag{3.19}
$$

Notice that the electrical migration velocity can be viewed as a particle transfer conductance; the product $\mathcal{N}V^E$ [particles/m² s] gives the local deposition rate.

---

[1] The effectiveness is called the *collection efficiency* by aerosol technologists.

Equation (3.19) is called the Deutsch–Anderson equation by particle (aerosol) technologists, after E. Anderson, who obtained it experimentally in 1919, and W. Deutsch, who supplied a theoretical derivation in 1922. Equation (3.19) gives the effectiveness for a particular particle size, since the migration velocity is size-dependent. In practice, precipitators collect particles with a wide range of sizes. Design calculations can be made by dividing the expected particle size distribution into finite size increments to obtain an overall collection effectiveness.

The assumption of a uniform particle concentration across the channel is, in fact, rather poor. Figure 3.8 shows concentration profiles calculated from more exact theory, where it is seen that the particle concentration increases as the channel wall is approached. Thus, Eq. (3.19) underestimates the effectiveness. Interestingly enough, although the more exact theory is confirmed by laboratory experiments [5], the measured performance of industrial precipitators is always less than that given by Eq. (3.19). The discrepancy is attributed to such problems as particle reentrainment and gas leakage between precipitator sections. Thus, current design practice is to use the Deutsch–Anderson equation with an empirical value of the migration velocity V^E that is less than the theoretical value. Typical design parameters are:

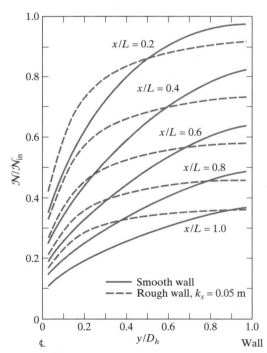

Figure 3.8 Particle concentration profiles in a plate-plate electrostatic precipitator showing the effect of plate roughness [4]. For $V^E \mathcal{P} L / \dot{V} = 10$; coordinate y is measured from the centerplane.

plate spacing 0.2–0.28 m, gas velocity 0.8–2.5 m/s, plate height 4–13 m, plate length 0.5 to 2.0 times height, effective migration velocity 3–20 cm/s, and pressure drop 25–125 Pa.

Particle Resistivity

The electrical resistivity of the particles plays a very important role. If the resistance of the dust layer is too low, the electrostatic charge is lost too quickly and collected particles can be reentrained into the gas flow. On the other hand, if the resistance of the layer is too high, the charge is retained at the collecting plates, producing a *back corona* that reduces the degree of ionization and hence lowers the effective migration velocity. Also, since the particles remain strongly attracted to the plate, removal by rapping is difficult. The resistivity of fly ash in the flue gas of coal-fired boilers depends on temperature, and the chemical composition of the ash and gases. Resistivity tends to have a maximum value in the range of about 120–180°C, which coincides with typical flue gas temperatures (below 120°C there is a risk of condensation of sulfuric acid on duct walls; above 180°C the energy loss out of the stack is excessive). Resistivity also decreases with increased sulfur content of the coal. The practical problem is usually one of ensuring that the fly ash resistivity is not too high. For example, if a power plant changes from using a high-sulfur coal to a low-sulfur coal, the performance of the precipitator is adversely affected. One remedy is to chemically condition the flue gases by the addition of a small amount of ammonium salts to reduce the fly ash resistivity. Empirical data for the effect of particle resistivity on effective migration velocity are available [2] and are used by design engineers.

Example 3.3   An Electrostatic Precipitator for a Toxic Waste Incinerator

An electrostatic precipitator for a toxic waste incinerator has a plate spacing of 0.2 m and is 20 m long. Gas at 1 atm and 410 K flows at a bulk velocity of 1.5 m/s through the unit. The current density is 27×10^{-9} A/cm^2 at 30 kV. Determine the collection efficiency for particles of sizes 0.5 μm, 0.9 μm, 2.0 μm, and 8.0 μm.

Solution

Given: A 20 m–long electrostatic precipitator operating at 30 kV, giving a current density of 27×10^{-9} A/cm^2.

Required: Collection efficiency for particles of sizes 0.5, 0.9, 2.0, and 8.0 μm.

Assumptions:
  1. A uniform particle concentration across the channel (the Deutsch model).
  2. The electrical migration velocity is given by the data in Fig. 3.6.

From Eq. (3.19) the effectiveness (collection efficiency) is

$$\varepsilon = 1 - e^{-N_{tu}}; \quad N_{tu} = V^E \mathcal{P} L / \dot{V}$$

For a plate spacing d and precipitator height H, the volume flow rate is $\dot{V} = Hdu_b$, and the perimeter is $\mathcal{P} = 2H$. Hence,

$$N_{tu} = \frac{V^E \mathcal{P} L}{\dot{V}} = \frac{V^E 2HL}{Hdu_b} = \frac{2V^E L}{du_b}$$

For a particle size of 0.5 μm and current density $j = 27 \times 10^{-9}$ A/cm^2, Fig. 3.6 gives $V^E = 4.0$ cm/s $= 0.040$ m/s. Thus, the number of transfer units is

$$N_{tu} = \frac{(2)(0.040)(20)}{(0.2)(1.5)} = 5.33$$

giving an effectiveness of

$$\varepsilon = 1 - e^{-5.33} = 1 - 0.00483 = 0.9952$$

The results for all particle sizes are tabulated below.

Particle Size μm	V^E m/s	N_{tu}	$1 - \epsilon$	ϵ
0.5	0.040	5.33	4.83×10^{-3}	0.99517
0.9	0.055	7.33	6.53×10^{-4}	0.999347
2.0	0.105	14.0	8.32×10^{-7}	0.999999
8.0	0.42	56.0	4.78×10^{-25}	1.000000

Comments

1. The collection efficiency is seen to be very dependent on particle size.
2. For this high-effectiveness exchanger, it is more appropriate to look at $(1-\varepsilon)$, which is the fraction of particles *not* collected.
3. For the larger particles, the very low values of $(1 - \varepsilon)$ suggest that, in practice, secondary factors might determine the true collection efficiency (for example, gas leakage or reentrainment while rapping).   ▲

3.3.4 Fibrous Filters

A simple and effective method of removing solid particles from a gas stream is to use a filter of some porous material. Commonly used are fabrics of fine fibers woven from cotton or wool, as well as fiberglass and other synthetic materials. Familiar examples include vacuum cleaner bags and air filters on automobiles. Large-scale industrial fabric filters are often called "bag houses" and consist of several large fabric bags. When operating properly, the *collection efficiency* (mass exchanger effectiveness) can exceed 99.99%, with pressure drops ranging from 500 to 1500 Pa. The bags are cleaned by periodic mechanical shaking or using a reverse air flow to dislodge the collected dust layer. Selection of suitable fiber diameters allows particles of a wide range of sizes to be collected.

The process of particle collection by a filter is primarily a combination of diffusion to, interception by, and inertial impaction on the fibers making up the filter. Diffusion is important only for very small particles (see Section 1.7.3), whereas inertial impaction is important only for the larger particles that cannot follow flow streamlines around a fiber. Particles are intercepted by the fiber when their centers come within a particle radius of the fiber, as shown in Fig. 3.9. Analyses of these processes are presented in the aerosol science literature [6, 7, 8]. The collection efficiency of a single fiber in a bed, γ, is defined as the ratio of particles that are collected to the

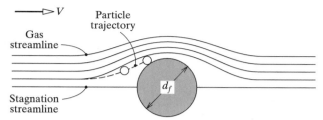

Figure 3.9 Schematic of particle deposition on a filter fiber by inertial impaction.

number that would strike the fiber if the particle path lines were not affected by the presence of the fiber.

The parameters that affect the single-fiber collection efficiency are as follows:

1. The Reynolds number $\mathrm{Re} = \mathcal{V}d_f/\nu$, where $\mathcal{V}$ is a characteristic gas velocity, d_f is the average fiber diameter, and ν is the gas kinematic viscosity. The Reynolds number characterizes the flow around the fiber.

2. The Peclet number $\mathrm{Pe} = \mathcal{V}d_f/\mathcal{D}$, where $\mathcal{D}$ is the particle diffusion coefficient that was introduced in Section 1.7.3. The Peclet number characterizes the particle concentration boundary layer for diffusional deposition of very small particles.

3. The interception number $R = d_p/d_f$, where d_p is the particle diameter. The interception number characterizes particle deposition by interception.

4. The Stokes number $\mathrm{Stk} = \rho_p d_p^2 \mathcal{V}/18\mu d_f$ where ρ_p is the particle density and μ the gas dynamic viscosity. The Stokes number characterizes inertial effects and hence inertial impaction of large particles that do not follow streamlines.

5. The filter porosity, or volume void fraction, ε_v. The flow field around a fiber depends on the porosity of the filter (and the already defined Reynolds number).

In general, we can write the single-fiber collection efficiency as

$$\gamma = \gamma(\mathrm{Re},\ \mathrm{Pe},\ R,\ \mathrm{Stk},\ \varepsilon_v)$$

Engineering correlations for γ recognize two Reynolds number regimes: creeping flow for $\mathrm{Re} < 1$, and transitional flow for $1 < \mathrm{Re} < 300$. Also, two regimes of Stokes numbers are appropriate: for $\mathrm{Stk} < 0.5$ impaction is ignored, and for $\mathrm{Stk} > 0.5$ diffusion is ignored. For $\mathrm{Stk} < 0.5$ it is sufficient to consider only the limit $\mathrm{Pe} \gg 1$, which corresponds to a small particle diffusion coefficient and hence a thin particle concentration boundary layer. Section 1.7.3 gives formulas and data for the particle diffusion coefficient that can be used to estimate the Peclet number. Table 3.2 lists recommended formulas for γ.

Figure 3.10 shows a schematic of a filter. The superficial velocity (based on the frontal area, A_{fr}) of the gas flowing through the bed is V, and the free-space velocity (based on flow cross-sectional area) is $\mathcal{V} = V/\varepsilon_v$. The average diameter of the fiber

Table 3.2 Recommended formulas for single-fiber collection efficiency in fibrous filters.

	Viscous Flow, Re < 1	Transitional Flow, $1 < \text{Re} < 300$
Stk < 0.5	$\gamma = 1.6(\varepsilon_v/H)^{1/3}\text{Pe}^{-2/3}$	$\gamma = \gamma_D + \gamma_R + 1.24R^{2/3}/(H\text{Pe})^{1/2}$
Pe $\gg 1$	$\quad + 0.6(\varepsilon_v/H)R^3/(1+R),$	$\gamma_D = (1 + 0.55\pi\text{Pe}^{1/3})/\text{Pe}$
Impaction	where H is the hydrodynamic factor	$\gamma_R = [(1+R)^{-1} - (1+R) + 2(1+R)$
negligible	for Kuwabara flow,	$\ln(1+R)]/(2H)$
	$H = -0.5\ln(1-\varepsilon_v) - 0.75 + (1-\varepsilon_v)$	
	$\quad - 0.25(1-\varepsilon_v)^2$	
Stk > 0.5	$\gamma = F_1(R,\text{Stk})F_2(\varepsilon_v)$	$\gamma = 1 - (1-\gamma_R)(1-\gamma_I)$
Diffusion	$F_1 = R + (0.5 + 0.8R)\text{Stk} - 0.105R\,\text{Stk}^2$	γ_R as above
negligible	$F_2 = 0.16 + 10.9(1-\varepsilon_v) - 17(1-\varepsilon_v)^2$	$\gamma_I = F_3(\text{Re},\text{Stk})F_4(\varepsilon_v)$
		$F_3 = \{1 + [1.53 - 0.23\ln\text{Re}$
		$\quad + 0.0167(\ln\text{Re})^2]/\text{Stk}\}^{-2}$
		$F_4 = F_2/0.16$

is d_f, and the length of fiber per unit volume of bed is ℓ m/m^3. At steady state, a particle balance on the gas flow in a length of bed Δx long requires that

$$\text{Particle inflow} - \text{Particle outflow} = \text{Rate of particle collection}$$

$$\mathcal{N}VA_{\text{fr}}|_x - \mathcal{N}VA_{\text{fr}}|_{x+\Delta x} = \gamma(V/\varepsilon_v)\mathcal{N}d_f\ell A_{\text{fr}}\Delta x$$

where $\mathcal{N}[m^{-3}]$ is the particle concentration. Dividing by $VA_{\text{fr}}\Delta x$ and letting $\Delta x \to 0$,

$$\frac{d\mathcal{N}}{dx} + \frac{d_f\ell\gamma}{\varepsilon_v}\mathcal{N} = 0 \qquad (3.20)$$

The length of fiber per unit volume ℓ is related to the void fraction as

$$1 - \varepsilon_v = (\pi/4)d_f^2\ell; \quad \text{hence} \quad \ell = 4(1-\varepsilon_v)/\pi d_f^2$$

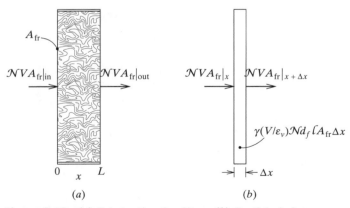

(a)   (b)

Figure 3.10 (a) Schematic of a filter. (b) Particle balance on an elemental volume Δx long.

Substituting for ℓ in Eq. (3.20),

$$\frac{d\mathcal{N}}{dx} + \frac{4(1 - \varepsilon_v)\gamma}{\varepsilon_v \pi d_f} \mathcal{N} = 0 \tag{3.21}$$

which must be solved subject to the boundary condition

$$x = 0: \quad \mathcal{N} = \mathcal{N}_{in} \tag{3.22}$$

The solution gives $\mathcal{N}(x)$ as

$$\mathcal{N} = \mathcal{N}_{in} e^{-4(1-\varepsilon_v)\gamma x/\varepsilon_v \pi d_f} \tag{3.23}$$

Substituting $\mathcal{N} = \mathcal{N}_{out}$ at $x = L$ and rearranging gives

$$1 - \frac{\mathcal{N}_{out}}{\mathcal{N}_{in}} = \frac{\mathcal{N}_{in} - \mathcal{N}_{out}}{\mathcal{N}_{in}} = 1 - e^{-4(1-\varepsilon_v)\gamma L/\varepsilon_v \pi d_f} \tag{3.24}$$

If there is no particle reentrainment, $\mathcal{N}_{out} \to 0$ in an infinitely thick filter; thus, again we recognize that the left-hand side of Eq. (3.24) is the effectiveness (or collection efficiency) of the filter as a mass exchanger, and we write

$$\varepsilon = 1 - e^{-N_{tu}}; \quad N_{tu} = \frac{4(1 - \varepsilon_v)\gamma L}{\epsilon_v \pi d_f} \tag{3.25}$$

When filters are used to remove radioactive particles, the quantity $\mathcal{N}_{in}/\mathcal{N}_{out} = 1/(1 - \varepsilon)$ is called the **decontamination factor (DF)**. For example, an effectiveness or collection efficiency of 99.99% corresponds to a decontamination factor of 10^4. When the effectiveness is very high, fewer significant figures are required to specify the decontamination factor to the same accuracy as the effectiveness.

The Computer Program FILTER

FILTER calculates the effectiveness and decontamination factor of fibrous filters according to the theory presented in this section. Spherical particles and air properties are assumed. The formulas for single-fiber collection efficiency are given in Table 3.2.

Example 3.4 A Fiberglass Filter for Decontaminating Radioactive Waste Gas

The effluent gas from a process vessel used to treat a radioactive ore is passed through a fiberglass filter of cross-sectional area 1.30 m^2 and thickness 0.5 m. The fibers have a diameter of 30 μm and are packed at a density of 390 kg/m^3. The gas flow is 25 m^3/min at 300 K and 1 atm. Determine the decontamination factor for 0.5 μm particles of average density 5000 kg/m^3.

Solution

Given: Fiberglass filter of thickness 0.5 m.
Required: Decontamination factor for 0.5 μm particles.
Assumptions:

1. No reentrainment of particles.
2. A clean filter.
3. The gas properties can be approximated by those of air.

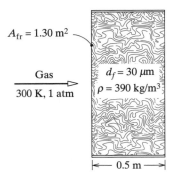

$A_{fr} = 1.30$ m^2

Gas

300 K, 1 atm

$d_f = 30$ μm
$\rho = 390$ kg/m^3

|← 0.5 m →|

To obtain the decontamination factor, we must first calculate the number of transfer units for the filter; from Eq. (3.25),

$$N_{tu} = \frac{4(1 - \varepsilon_v)\gamma L}{\varepsilon_v \pi d_f}$$

With the glass density taken as 2300 kg/m^3, the volume void fraction of the filter is

$$\varepsilon_v = 1 - \frac{390}{2300} = 0.830$$

To determine the single-fiber collection efficiency γ, we first calculate the particle Stokes number and the Peclet and Reynolds numbers, using the free-space velocity as the characteristic velocity:

$$\mathcal{V} = \frac{25}{(60)(1.3)(0.830)} = 0.386 \text{ m/s}$$

$$\text{Stk} = \frac{\rho_p d_p^2 \mathcal{V}}{18\mu d_f} = \frac{(5000)(0.5 \times 10^{-6})^2(0.386)}{(18)(18.43 \times 10^{-6})(30 \times 10^{-6})} = 0.0485$$

$$\text{Pe} = \frac{\mathcal{V} d_f}{\mathcal{D}} = \frac{(0.386)(30 \times 10^{-6})}{6.38 \times 10^{-11}} = 1.815 \times 10^5$$

where $\mathcal{D}$ has been obtained from Table A.30;

$$\text{Re} = \frac{\mathcal{V} d_f}{\nu} = \frac{(0.386)(30 \times 10^{-6})}{15.66 \times 10^{-6}} = 0.739$$

Also, the interception number is

$$R = d_p/d_f = 0.5/30 = 0.0167$$

Since Re < 1.0 and Stk < 0.5, the flow is creeping and impaction is negligible; also Pe ≫ 1, and thus from Table 3.2, the collection efficiency of a single fiber is

$$\gamma = 1.6(\varepsilon_v/H)^{1/3}\text{Pe}^{-2/3} + 0.6(\varepsilon_v/H)R^3/(1 + R)$$

where

$$H = -0.5\ln(1 - \varepsilon_v) - 0.75 + (1 - \varepsilon_v) - 0.25(1 - \varepsilon_v)^2$$
$$= -0.5\ln(1 - 0.830) - 0.75 + (1 - 0.830) - 0.25(1 - 0.830)^2 = 0.30$$

Hence, we calculate γ as

$$\gamma = 1.6(0.830/0.30)^{1/3}(1.815 \times 10^5)^{-2/3} + \frac{0.6(0.830/0.30)(0.0167)^3}{(1 + 0.0167)}$$
$$= 7.01 \times 10^{-4} + 7.74 \times 10^{-6} = 7.09 \times 10^{-4}$$

Then the number of transfer units of the filter is

$$N_{\text{tu}} = \frac{4(1 - \varepsilon_v)\gamma L}{\varepsilon_v \pi d_f} = \frac{4(1 - 0.83)(7.09 \times 10^{-4})(0.5)}{(0.83)(\pi)(30 \times 10^{-6})} = 3.08$$

giving an effectiveness of

$$\varepsilon = 1 - e^{-N_{\text{tu}}} = 1 - e^{-3.08} = 95.4\%$$

or a decontamination factor of

$$\text{DF} = \frac{1}{1 - \varepsilon} = 21.8$$

Comments

1. In practice, a number of filter elements are often arranged in series; the first is designed to filter out the largest particles, and succeeding elements filter out smaller particles.
2. Use FILTER to check this result.   ▲

3.4 TWO-STREAM MASS EXCHANGERS

Two-stream mass exchangers are widely used in the chemical and process industries and thus are given considerable attention in standard chemical engineering texts. Here we will restrict our attention to countercurrent (counterflow) gas scrubbers, in which a component of a gas stream is removed by absorption into a liquid stream. No chemical reactions will be considered. Thus, the transfer process involves convection from the bulk gas to the gas–liquid interface, dissolution at the interface, and convection into the bulk liquid stream. The solubility of the gas will be described by Henry's law, Eq. (1.13). In the case of systems for which Henry's law is invalid, for example, SO_2-H_2O, an *effective* Henry number will be defined as an average value over the concentration range of interest. Use of Henry's law gives a linear system of equations to be solved, and an analytical result can be obtained that is analogous to Eq. (8.41) of *Heat Transfer* for a counterflow heat exchanger. This relatively simple analytical result conveniently displays the essential features of gas scrubber operation.

Gas scrubbers are usually in the form of vertical packed columns (towers), with liquid flowing down over the packing and gas flowing upward (see Fig. 1.1). The design of such columns also requires pressure drop calculations and the establishment of *flooding limits*. A column is said to flood when the gas flow is sufficient to blow the liquid upward against gravity. Correlations for pressure drop and flooding are presented to complete this section.

3.4.1 Overall Mole Transfer Conductances

Figure 3.11 shows a chemical species in a gaseous mixture being absorbed into a nonvolatile liquid, for example, the NH_3-N_2-H_2O system at low temperatures with NH_3 the solute and H_2O the solvent. The concentration profiles of the species in question are shown as solid lines, and its bulk concentration in each stream is shown as a dotted line. To avoid excessive subscripting, liquid- and gas-phase mole fractions are denoted x and y, respectively. This notation is commonly used by chemical engineers. Bulk mole fractions are not subscripted, consistent with the conventional notation for two-stream heat exchangers; values at the interface are subscripted s and u as usual. From Henry's law, Eq. (1.13),

$$y_s = \text{He}\, x_u \tag{3.26}$$

The Henry number, He, is obtained from the Henry constant $C_{\text{He}}(T)$ using Eq. (1.14):

$$\text{He} = \frac{C_{\text{He}}(T)}{P} \tag{3.27}$$

where P is total pressure. It is understood that C_{He} and He are properties of the solute-solvent combination.

Actually, the Henry constant is found to vary somewhat with pressure and liquid-phase concentration; however, for sparingly soluble gases the variation is small and can be ignored. Henry constant data for aqueous solutions of sparingly soluble solutes are given in Table A.21. In concentration ranges of practical interest for

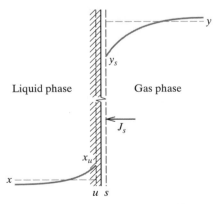

Figure 3.11 Concentration profiles for absorption of a gaseous species into a liquid.

the NH_3-water and SO_2-water systems, the Henry constant varies appreciably with liquid-phase concentration. Table A.22 gives equilibrium compositions for these systems; often a suitable average value for the Henry constant is sufficient.

Low mass transfer rates will be assumed; thus, from Eq. (1.37), the molar flux across the interface can be written as

$$J_s = \mathcal{G}_G(y - y_s) = \mathcal{G}_L(x_u - x) \tag{3.28}$$

Our sign convention here makes J_s positive for absorption into the liquid (and negative for desorption from the liquid). The subscripts G and L have been introduced to distinguish mole transfer conductances in the gas and liquid phases, and the subscript m has been dropped for convenience. Also, since only one species is transferred, there is no ambiguity in denoting the flux of that species J_s. Substituting Eq. (3.26) into Eq. (3.28) and rearranging,

$$J_s = \mathcal{G}_G(y - \text{He}\, x_u) = \mathcal{G}_G\left(y - \frac{\text{He}\, J_s}{\mathcal{G}_L} - \text{He}\, x\right)$$

Solving for J_s,

$$J_s = \frac{1}{1/\mathcal{G}_G + \text{He}/\mathcal{G}_L}(y - \text{He}\, x) = \mathcal{G}_G^{\text{oa}}(y - \text{He}\, x) \tag{3.29}$$

where

$$\frac{1}{\mathcal{G}_G^{\text{oa}}} = \frac{1}{\mathcal{G}_G} + \frac{\text{He}}{\mathcal{G}_L} \tag{3.30}$$

and $\mathcal{G}_G^{\text{oa}}$ is called the overall mole transfer conductance for a gas-side driving force $(y - \text{He}\, x)$. The product $(\text{He}\, x)$ may be viewed as a pseudo gas-phase mole fraction: it is the gas-phase mole fraction that would exist at the interface if the liquid phase were at a uniform concentration x.

Equations (3.29) and (3.30) remain valid even when Henry's law is not obeyed: the Henry number is then an *effective* one, as will be illustrated in Example 3.5. Figure 3.12 shows a mass transfer circuit interpretation, which is particularly useful when one has to know y_s in order to obtain the effective Henry number. From the circuit,

$$y_s = y - J_s(1/\mathcal{G}_G) \tag{3.31}$$

or

$$y_s = y - \frac{\mathcal{G}_G^{\text{oa}}}{\mathcal{G}_G}(y - \text{He}\, x) \tag{3.32}$$

and the partial pressure is then $y_s P$.

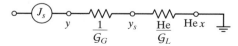

Figure 3.12 Mass transfer circuit for gas absorption (per unit area of interface).

Equation (3.29) is not a simple analogy to the corresponding heat transfer result: the heat flux from the gas to the liquid is

$$q = \frac{1}{1/h_{cG} + 1/h_{cL}}(T_G - T_L) \tag{3.33}$$

Notice that $q = 0$ for $T_G = T_L$, that is, when the two phases are in thermal equilibrium. For the two phases to be in chemical equilibrium $y = \text{He}\,x$, and, appropriately, Eq. (3.29) gives $J_s = 0$ for $y = \text{He}\,x$. Also, Eq. (3.29) shows that the liquid-side contribution to the overall mass transfer resistance is $\text{He}/\mathcal{G}_L$, not simply $1/\mathcal{G}_L$. For sparingly soluble gases He is very large, and as a result the liquid-side resistance dominates. The reason for this feature is that, since mole fractions are much lower in the liquid phase, so too are mole-fraction gradients and hence diffusion rates (the $c\mathcal{D}$ products and conductances are of the same order in both phases). An analogous situation does not occur for heat transfer because, whereas concentration is discontinuous across the interface, temperature is continuous. Example 3.5 will illustrate the concept of a *controlling* resistance for gas absorption. (See also the discussion of the mass transfer Biot number in Section 1.4.4.)

3.4.2 Mole Transfer Conductances in Packed Columns

Since the number of transfer units of a gas scrubber is proportional to interfacial area, it is common practice to break up the flow of the stream by packing the column with, for example, irregularly shaped solid particles. Typically, these particles range from about 5 to 75 mm in size, are usually made from a ceramic, and are loaded in the column in a random fashion. Some common types are shown in Fig. 3.13. Breaking

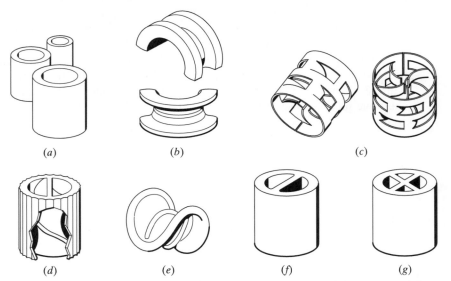

Figure 3.13 Common random packings: (*a*) Raschig rings, (*b*) Intalox saddle, (*c*) Pall ring, (*d*) Cyclohelix spiral ring, (*e*) Berl saddle, (*f*) Lessing ring, (*g*) cross-partition ring.

up the flow also serves to increase the mass transfer conductances. Alternatively, structured packings are used. These packings give lower pressure drops for a given throughput of gas, usually at the expense of a higher installation cost. The column is usually circular in cross section and contains an open space below the packing support to ensure good distribution of the incoming gas stream, and a liquid feed system to spray liquid on the top of the packing (see Fig. 1.1). Typical columns range from 0.3 to 15 m in diameter and from 2 to 50 m in height. From a fluid flow standpoint, problems arise with respect to flooding (gas stream flow rate too high, causing reversal of the liquid downflow), channeling (nonuniform downflow of liquid), entrainment (droplets of liquid carried off at the top as mist), and calculation of pressure drop. Some of these topics will be discussed in Section 3.4.4; here our interest is in calculating mole transfer conductances and the interfacial area.

Random Packings

We first consider random packings such as Raschig rings and Berl saddles. Table 3.3 gives relevant data for some dry packings, including the nominal size D_p, void fraction ε_v, specific surface area a (surface area per unit volume), and drag factor Φ for pressure drop calculations. As can be imagined, the convective mass transfer processes in a random packing are very complicated. Not only is the geometry of the dry packing, and hence the resulting flow patterns, complex, but also as the liquid flow rate is increased, so the liquid *holdup* in the tower increases, reducing the area of interface for transfer and the cross-sectional area available for gas flow. Holdup depends not only on liquid and gas flow rates, but also on wettability of the packing and hence on surface tensions. One can find many different correlation schemes for random packing conductances in the chemical engineering literature. The correlations given here were developed by Onda and co-workers [9, 10, 11] and were chosen as a compromise between simplicity, generality, and accuracy. If more accurate correlations are required, the text by Treybal is recommended [12].

Table 3.3 Characteristics of commercial packings.

	Nominal Size, D_p [m]	Void Fraction, ε_v	Specific Surface Area, a [m^{-1}]	Drag Factor, Φ
Ceramic Raschig rings	0.006	0.73	787	1600
	0.013	0.63	364	580
	0.025	0.73	190	155
	0.038	0.71	125	95
	0.050	0.74	92	65
Metal Raschig rings (1.6 mm wall)	0.013	0.73	387	410
	0.025	0.85	186	137
	0.050	0.92	103	57
Ceramic Berl saddles	0.013	0.63	466	240
	0.025	0.69	249	110
	0.038	0.75	144	65
Interlox saddles (plastic)	0.025	0.91	207	33
	0.050	0.93	108	21

Reynolds numbers are defined in terms of superficial mass velocities $\dot{m}/A_{\mathrm{fr}}$, where A_{fr} is the column cross-sectional area, and the reciprocal of the specific surface area, a^{-1}, as characteristic length:

$$\mathrm{Re}_G = \frac{\dot{m}_G}{A_{\mathrm{fr}}a\mu}; \quad \mathrm{Re}_L = \frac{\dot{m}_L}{A_{\mathrm{fr}}a\mu_l} \tag{3.34a, b}$$

The liquid-side Stanton number is then

$$\mathrm{St}_L = \frac{G_L}{c_l(\nu_l g)^{1/3}} = C_1 \mathrm{Re}_L^{2/3}\mathrm{Sc}_l^{-1/2} \tag{3.35}$$

where

$$C_1 = 0.0051(aD_p)^{0.4}$$

The subscript l is used to distinguish liquid-phase properties. The quantity $(\nu_l g)^{1/3}$ has the dimensions of velocity, as is required to form a Stanton number.

The gas-side Sherwood number is

$$\mathrm{Sh}_G = \frac{G_G}{cDa} = C_2 \mathrm{Re}_G^{0.7}\mathrm{Sc}^{1/3} \tag{3.36}$$

where

$$C_2 = 2.0(aD_p)^{-2.0}, \quad D_p < 0.013 \text{ m}$$
$$= 5.23(aD_p)^{-2.0}, \quad D_p > 0.013 \text{ m}$$

A correlation is also needed for the transfer area perimeter,

$$\frac{\mathcal{P}/A_{\mathrm{fr}}}{a} = 1 - \exp\left[-1.45\mathrm{Re}_L^{0.1}\mathrm{We}_L^{0.2}\mathrm{Fr}_L^{-0.05}(\sigma_c/\sigma)^{0.75}\right] \tag{3.37}$$

where the Weber and Froude numbers are defined as

$$\mathrm{We}_L = \frac{(\dot{m}_L/A_{\mathrm{fr}})^2}{\rho_l \sigma a}; \quad \mathrm{Fr}_L = \frac{(\dot{m}_L/A_{\mathrm{fr}})^2 a}{\rho_l^2 g} \tag{3.38}$$

and σ_c is a critical surface tension for the packing material; values of σ_c are given in Table 3.4. The above correlations are valid for the Reynolds number ranges

$$5 < \mathrm{Re}_G < 1000; \quad 1 < \mathrm{Re}_L < 40, \quad \text{organic liquids}$$
$$4 < \mathrm{Re}_L < 400, \quad \text{water}$$

and may be used for packings listed in Table 3.3 (but with caution for the Intalox saddles). Most of the experimental data used to develop the correlations fall within 20% of the values given by the correlation, though some data deviate by as much as 50%.

Structured Packings

Structured packings are used when it is important to have a low gas-flow pressure drop in order to reduce fan power requirements or for vacuum service. Figure 3.14a shows a typical commercial structured packing. As a benchmark, it is useful to be able to design a column with a simple vertical-plate packing that would give the lowest possible pressure drop. Figure 3.14b shows a packing that has a pitch p, plate

Table 3.4 Critical surface tension for various packing materials.

Packing Material	σ_c [N/m $\times 10^3$]
Carbon	56
Ceramic	61
Glass	73
Polyethylene	33
Polyvinylchloride	40
Steel	75

thickness t, and falling-film thickness δ. The gas-side Reynolds number is defined in terms of the velocity $\dot{m}_G/\rho A_c$, where A_c is the flow cross-sectional area, and the hydraulic diameter D_h of the flow:

$$\mathrm{Re}_G = \frac{(\dot{m}_G/A_c)D_h}{\mu}$$

$$A_c = A_{\mathrm{fr}}\left(1 - \frac{t + 2\delta}{p}\right); \quad D_h = 2[p - (t + 2\delta)] \tag{3.39}$$

The gas-side Sherwood number is given by Eq. (1.47):

$$\mathrm{Sh}_G = \frac{\mathcal{G}_G D_h}{c\mathcal{D}} = 0.00814\,\mathrm{Re}_G^{0.83}\mathrm{Sc}^{0.44}\mathrm{Re}_L^{0.15} \tag{3.40}$$

which is valid for $2000 < \mathrm{Re}_G < 10{,}000$; $\mathrm{Re}_L < 1200$. The liquid-side Reynolds number is the film Reynolds number introduced for falling films in Chapter 7 of *Heat Transfer*:

$$\mathrm{Re}_L = \frac{4\Gamma}{\mu_l} = \frac{2(\dot{m}_L/A_{\mathrm{fr}})p}{\mu_l} \tag{3.41}$$

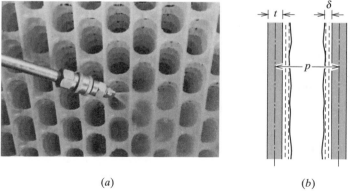

(a)   (b)

Figure 3.14 Structured packings. (a) Munters PN Fill structured packing (showing how it is cleaned). (b) Parallel-plate packing.

where Γ is the mass flow rate per unit perimeter. The characteristic velocity $(\nu_l g)^{1/3}$ is used to define the liquid-side Stanton number, and separate correlations are required for the laminar, wavy laminar, and turbulent regimes, as follows.

Laminar regime, $\mathrm{Re}_L < 30$:

$$\mathrm{St}_L = \frac{\mathcal{G}_L}{c_l (\nu_l g)^{1/3}} = 0.233\,\mathrm{Re}_L^{2/3}\mathrm{Pe}_m^{-1/2} \tag{3.42}$$

where the mass transfer Peclet number is $\mathrm{Pe}_m = u_\delta H / \mathcal{D}_l$; H is the packing height, and the film surface velocity is

$$u_\delta / (\nu_l g)^{1/3} = 0.413\,\mathrm{Re}_L^{2/3} \tag{3.43}$$

Equation (3.42) can be derived from the results obtained in Exercise 1–30.

Wavy laminar regime [13], $30 < \mathrm{Re}_L < 1100$:

$$\mathrm{St}_L = 0.04\,\mathrm{Re}_L^{0.2}\mathrm{Sc}_l^{-0.53} \tag{3.44}$$

Turbulent regime [14], $1100 < \mathrm{Re}_L < 10{,}000$:

$$\mathrm{St}_L = 6.79 \times 10^{-9}\,\mathrm{Re}_L^n \mathrm{Sc}_l^{-\alpha}\mathrm{Ka}_l^{-1/2} \tag{3.45}$$

where

$$n = 3.49\,\mathrm{Ka}_l^{0.068}$$

$$\alpha = 0.36 + 2.43(\sigma\ \mathrm{N/m})$$

and the Kapitza number is $\mathrm{Ka}_l = \nu_l^4 \rho_l^3 g / \sigma^3$. Notice the transition Reynolds number is ~1100, rather than the value of ~1800 used for film condensation. As explained in Section 7.2.2 of *Heat Transfer*, the lower value corresponds to transition in the outer region of the film, and it is this region that is relevant to gas absorption into the film.

The film thickness is required to calculate the flow cross-sectional area A_c and hydraulic diameter D_h; from reference [15],

$$\frac{\delta}{(\nu_l^2 / g)^{1/3}} = 0.909\,\mathrm{Re}_L^{1/3}\quad \mathrm{Re}_L < 1600 \tag{3.46a}$$

$$= 0.0672\,\mathrm{Re}_L^{2/3}\quad \mathrm{Re}_L > 1600 \tag{3.46b}$$

Note that these liquid film correlations, Eqs. (3.42) through (3.46), are valid only if the shear stress exerted by the gas flow on the film surface is relatively small. The transfer area perimeter and specific area are obtained directly from the geometry as

$$\frac{\mathcal{P}}{A_{\mathrm{fr}}} = \frac{2}{p} = a \tag{3.47}$$

A great variety of commercial structured packings are used in practice. Manufacturers usually supply performance data in the form of correlations of the conductance times the specific area, rather than separate correlations of the conductance and transfer perimeter.

Example 3.5 Controlling Resistances for Gas Absorption

A mixture of air and a trace amount of an inert species flows countercurrent to a falling film of water between parallel plates of 2 mm thickness at a pitch of 5 cm. The air velocity between the plates is 1.8 m/s, and the water flow rate is 0.20 kg/s per unit perimeter. The exchanger operates at 300 K and 1 atm. The water enters saturated with air but contains no species 1. Determine the relative resistances to mass transfer in each phase ($1/G_G$ and He/G_L) and the overall mole transfer conductance G_G^{oa} when the trace gas is (i) ammonia, (ii) sulfur dioxide, (iii) hydrogen sulfide, and (iv) carbon dioxide.

Solution

Given: Falling water film with a countercurrent flow of air containing an inert species.

Required: Relative resistances to mass transfer in each phase for NH_3, SO_2, H_2S, and CO_2.

Assumptions:

1. Properties of pure air and pure water for the gas and liquid phases, respectively.

2. An *effective* Henry number can be used for the highly soluble SO_2 and NH_3.

The overall mole transfer conductance is given by Eq. (3.30) as

$$\frac{1}{G_G^{oa}} = \frac{1}{G_G} + \frac{He}{G_L}$$

First consider the liquid phase. Assuming pure water at 300 K and using Table A.8, $\rho_l = 996$ kg/m³, $v_l = 0.87 \times 10^{-6}$ m²/s, $\mu_l = 8.67 \times 10^{-4}$ kg/m s, $c_l = \rho_l/M = (996 \text{ kg/m}^3)/(18 \text{ kg/kmol}) = 55.3$ kmol/m³. The film Reynolds number is

$$Re_L = \frac{4\Gamma}{\mu_l} = \frac{(4)(0.20)}{8.67 \times 10^{-4}} = 923$$

Hence, using Eq. (3.44), the liquid-side mole transfer conductance is obtained from

$$G_L Sc_l^{0.53} = 0.04 c_l (v_l g)^{1/3} Re_L^{0.2}$$
$$= (0.04)(55.3)(0.87 \times 10^{-6} \times 9.81)^{1/3}(923)^{0.2} = 0.177 \text{ kmol/m}^2 \text{ s}$$

The film thickness is calculated from Eq. (3.46a):

$$\delta = 0.909 \left(\frac{v_l^2}{g}\right)^{1/3} Re_L^{1/3} = (0.909)\left[\frac{(0.87 \times 10^{-6})^2}{9.81}\right]^{1/3}(923)^{1/3}$$
$$= 3.77 \times 10^{-4} \text{ m } (0.377 \text{ mm})$$

Now consider the gas phase and assume its properties to be those of pure air at 300 K and 1 atm; from Table A.7, $\rho = 1.177 \text{ kg/m}^3$, $\mu = 18.43 \times 10^{-6} \text{ kg/m s}$, $c = \rho/M = (1.177 \text{ kg/m}^3)/(29 \text{ kg/kmol}) = 0.0406 \text{ kmol/m}^3$. From Eq. (3.39) the hydraulic diameter is

$$D_h = 2[p - (t + 2\delta)] = 2\{0.05 - [0.002 + (2)(0.000377)]\} = 0.0945 \text{ m}$$

Then the gas-phase Reynolds number is

$$\text{Re}_G = \frac{\rho V D_h}{\mu} = \frac{(1.177)(1.8)(0.0945)}{18.43 \times 10^{-6}} = 10{,}870$$

Using Eq. (3.40), with $\mathcal{D} = \mu/\rho\text{Sc}$, the gas-side mole transfer conductance is obtained from

$$\mathcal{G}_G \text{Sc}^{0.56} = 0.00814 \left(\frac{c\mu}{\rho D_h} \right) \text{Re}_G^{0.83} \text{Re}_L^{0.15}$$

$$= 0.00814 \left[\frac{(0.0406)(18.43 \times 10^{-6})}{(1.177)(0.0945)} \right] (10{,}870)^{0.83} (923)^{0.15}$$

$$= 3.42 \times 10^{-4} \text{ kmol/m}^2 \text{ s}$$

At 1 atm total pressure, the Henry number equals Henry's constant. From Table A.21, at 300 K the Henry numbers for H_2S and CO_2 are found to be 560 and 1710, respectively. For NH_3 in dilute solution, Henry's law is approximately valid; at 300 K Table A.22a gives $P_{NH_3,s}/x_{NH_3,u} \simeq 1.1$. For SO_2, Table A.22b shows that the *effective* Henry number at 300 K and $P_{SO_2,s} = 0.003$ atm is $0.003/0.18 \times 10^{-3} = 17$. [The value of $P_{SO_2,s}$ must be initially guessed; subsequently, Eq. (3.32) can be used to find a better value.]

Gas- and liquid-phase Schmidt numbers are obtained from Tables A.17b and A.18, respectively. The results are summarized below.

Derived Quantities	Absorbed Species			
	NH_3	SO_2	H_2S	CO_2
Sc (gas)	0.61	1.24	0.94	1.00
$\mathcal{G}_G = 3.42 \times 10^{-4}/\text{Sc}^{0.56}$ ($\times 10^3$)	0.451	0.303	0.354	0.342
Sc_l (liquid)	410	520	430	420
$\mathcal{G}_L = 0.177/\text{Sc}_l^{0.53}$ ($\times 10^3$)	7.30	6.43	7.12	7.21
He	1.1	17.0	560	1710
Gas-side resistance, $1/\mathcal{G}_G$ ($\times 10^{-3}$)	2.22	3.30	2.82	2.92
Liquid-side resistance, $\text{He}/\mathcal{G}_L$ ($\times 10^{-3}$)	0.15	2.64	78.7	237
Total resistance, $1/\mathcal{G}_G + \text{He}/\mathcal{G}_L$ ($\times 10^{-3}$)	2.37	5.94	81.5	240
$\mathcal{G}_G^{\text{oa}}$, kmol/m^{2}s ($\times 10^3$)	0.422	0.168	0.0123	0.00417

Comments

1. The liquid-side resistance controls the absorption of CO_2 and H_2S, neither resistance controls for SO_2, and the gas-side resistance controls for NH_3.

2. When the liquid-side resistance is very large, common practice is to add a reactant to the solvent (see Example 3.2).

3. The value of Re_G (10,870) is a little higher than the specified upper limit for Eq. (3.40) to be valid (10,000). This modest extrapolation should be satisfactory, since there is no change in flow patterns at these Reynolds numbers.

4. SCRUB (see Section 3.4.3) can be used to check these results.   ▲

Example 3.6   Overall Mole Transfer Conductance for a Packed Column

Air containing a trace amount of NH_3 flows countercurrent to water in a 1 m²–cross-sectional-area column packed with 0.038 m–nominal-size ceramic Berl saddles. The flow rates of gas and liquid are 0.35 kg/s and 1.90 kg/s, respectively, and the tower operates at 300 K and 1 atm. Determine the overall mole transfer conductance and transfer perimeter.

Solution

Given: Absorption column packed with Berl saddles.
Required: Overall mole transfer conductance and transfer perimeter for NH_3 absorption into water.
Assumptions: Properties for pure air and pure water can be used for the gas and liquid phases, respectively.
The overall mole transfer conductance is given by Eq. (3.30) as

$$\frac{1}{\mathcal{G}_G^{oa}} = \frac{1}{\mathcal{G}_G} + \frac{He}{\mathcal{G}_L}$$

For the liquid phase, assuming pure water at 300 K, $\rho_l = 996$ kg/m³, $\mu_l = 8.67 \times 10^{-4}$ kg/m s, $\nu_l = 0.87 \times 10^{-6}$ m²/s, $c_l = 996/18 = 55.3$ kmol/m³, and $Sc_l = 410$. From Table 3.3, the specific area of the packing is $a = 144$ m⁻¹. Thus, from Eq. (3.34b), the liquid-phase Reynolds number is

$$\mathrm{Re}_L = \frac{\dot{m}_L}{A_{fr}a\mu_l} = \frac{1.90}{(1)(144)(8.67 \times 10^{-4})} = 15.2$$

From Eq. (3.35), the liquid-side mole transfer conductance is

$$\mathcal{G}_L = 0.0051(aD_p)^{0.4}c_l(\nu_l g)^{1/3}\mathrm{Re}_L^{2/3}Sc_l^{-1/2}$$

$$= 0.0051(144 \times 0.038)^{0.4}(55.3)(0.87 \times 10^{-6} \times 9.81)^{1/3}(15.2)^{2/3}(410)^{-1/2}$$

$$= 3.45 \times 10^{-3} \text{ kmol/m}^2 \text{ s}$$

For the gas phase we use properties of pure air at 300 K and 1 atm: $\rho = 1.177$ kg/m³, $\mu = 18.43 \times 10^{-6}$ kg/m s, $c = 1.177/29 = 0.0406$ kmol/m³, $Sc = 0.61$. From Eq. (3.34a), the gas-phase Reynolds number is

$$\mathrm{Re}_G = \frac{\dot{m}_G}{A_{fr}a\mu} = \frac{0.35}{(1)(144)(18.43 \times 10^{-6})} = 132$$

and from Eq. (3.36), the gas-side mole transfer conductance is

$$G_G = 5.23(aD_p)^{-2}c\left(\frac{\mu}{\rho Sc}\right) a Re_G^{0.7} Sc^{1/3}$$

$$= 5.23(144 \times 0.038)^{-2}(0.0406)\frac{18.43 \times 10^{-6}}{1.177 \times 0.61}(144)(132)^{0.7}(0.61)^{1/3}$$

$$= 6.78 \times 10^{-4} \text{ kmol/m}^2 \text{ s}$$

Then, using Eq. (3.30), we obtain the overall mole transfer conductance as

$$\frac{1}{G_G^{oa}} = \frac{1}{G_G} + \frac{He}{G_L}; \quad He \simeq 1.1 \text{ from Table A.22a   (SCRUB uses } He = 1.042)$$

$$\frac{1}{G_G^{oa}} = \frac{1}{6.78 \times 10^{-4}} + \frac{1.1}{3.45 \times 10^{-3}}$$

$$= 1475 + 319 = 1794; \quad G_G^{oa} = 5.57 \times 10^{-4} \text{ kmol/m}^2 \text{ s}$$

Equation (3.37) gives the transfer perimeter as

$$P = aA_{fr}\left\{1 - \exp[-1.45Re_L^{0.1}We_L^{0.2}Fr_L^{-0.05}(\sigma_c/\sigma)^{0.75}]\right\}$$

For water at 300 K, Table A.11 gives $\sigma = 71.7 \times 10^{-3}$ N/m, and Table 3.4 gives $\sigma_c = 61 \times 10^{-3}$ N/m. The Weber and Froude numbers are

$$We_L = \frac{(\dot{m}_L/A_{fr})^2}{\rho_l \sigma a} = \frac{(1.90/1)^2}{(996)(71.7 \times 10^{-3})(144)} = 3.51 \times 10^{-4}$$

$$Fr_L = \frac{(\dot{m}_L/A_{fr})^2 a}{\rho_l^2 g} = \frac{(1.90/1)^2(144)}{(996)^2(9.81)} = 5.34 \times 10^{-5}$$

The perimeter is then

$$P = (144)(1)\left\{1 - \exp[-1.45(15.2)^{0.1}(3.51 \times 10^{-4})^{0.2}\right.$$

$$\left.(5.34 \times 10^{-5})^{-0.05}(61/71.7)^{0.75}]\right\}$$

$$= 61.7\text{m}$$

Comments

1. SCRUB (see Section 3.4.3) can be used to check these results.

2. The gas-side resistance dominates, but the liquid-side resistance is not negligible. ▲

3.4.3 Countercurrent Mass Exchangers

As depicted in Fig. 3.15, a countercurrent[2] mass exchanger has streams flowing in opposite directions. In many situations the liquid stream flows down under the influence of gravity, with countercurrent upflow of the gas stream. For convenience, we separate the gas and liquid into two distinct streams. However, it should be kept in mind that the usual column has multiple paths for the flow of each phase, in order to increase the contact area $\mathcal{P}\Delta z$ and hence the mass transfer rate $J_s \mathcal{P}\Delta z$. The liquid and gas flow rates, $\dot{M}_L$ and $\dot{M}_G$, are both taken as positive. In the usual situation, the species to be absorbed is in small concentration in a carrier gas stream: the amount of gas absorbed is small compared with the stream flow rates, and $\dot{M}_L$ and $\dot{M}_G$ can be taken to be constant.

The exchanger species balance is

$$\dot{M}_G y|_{\text{in}} - \dot{M}_L x|_{\text{out}} = \dot{M}_G y|_{\text{out}} - \dot{M}_L x|_{\text{in}} \tag{3.48}$$

which is evident upon inspection of Fig. 3.15. A species balance on the gas stream for an element of length Δz between z and $z + \Delta z$ gives

$$\dot{M}_G y|_z = \dot{M}_G y|_{z+\Delta z} + J_s \mathcal{P}\Delta z$$

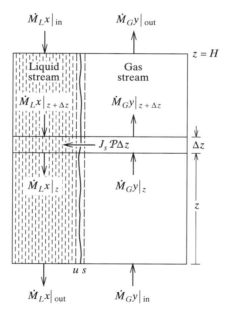

Figure 3.15 Schematic of a countercurrent mass exchanger, showing species balances on an element of packing Δz high.

---

[2] Following chemical engineering practice, we will refer to counterflow mass exchangers as *countercurrent* units.

where J_s is positive for absorption. Taking $\dot{M}_G$ constant, introducing the overall mole transfer conductance from Eq. (3.29), and allowing $\Delta z \to 0$ gives

$$\frac{dy}{dz} = -\frac{\mathcal{G}_G^{oa}\mathcal{P}}{\dot{M}_G}(y - \text{He }x) \tag{3.49}$$

For gas absorption, $y > \text{He }x$, and y is seen to decrease with increasing z. A similar balance for the liquid stream gives

$$\frac{dx}{dz} = -\frac{\mathcal{G}_G^{oa}\mathcal{P}}{\dot{M}_L}(y - \text{He }x) \tag{3.50}$$

Likewise, x decreases with increasing z. This equation is now multiplied by He, and the quantity $\dot{M}_L/\text{He}$ is written $\dot{M}_L^\dagger$. As a result,

$$\frac{d(\text{He }x)}{dz} = -\frac{\mathcal{G}_G^{oa}\mathcal{P}}{\dot{M}_L^\dagger}(y - \text{He }x) \tag{3.51}$$

Subtracting Eq. (3.51) from Eq. (3.49) gives

$$\frac{d}{dz}(y - \text{He }x) = -\mathcal{G}_G^{oa}\mathcal{P}\left(\frac{1}{\dot{M}_G} - \frac{1}{\dot{M}_L^\dagger}\right)(y - \text{He }x) \tag{3.52}$$

Equations (3.49), (3.51), and (3.52) are equivalent forms of the governing differential equation for mass transfer in the column. There is an exact parallelism between these equations and those for a two-stream countercurrent heat exchanger. The boundary conditions required to solve the governing equations are

$$z = 0: \ y = y_{\text{in}}, x = x_{\text{out}} \tag{3.53a}$$

$$z = H: \ y = y_{\text{out}}, x = x_{\text{in}} \tag{3.53b}$$

Equations (3.48) and (3.52) can be viewed as two equations in essentially eight parameters:

$$\dot{M}_L, \dot{M}_G, H, \mathcal{P}, x_{\text{in}}, x_{\text{out}}, y_{\text{in}}, y_{\text{out}}$$

(The quantities $\mathcal{G}_G^{oa}$ and He are not regarded as independent parameters, since they are functions of the other parameters as well as temperature and pressure, which are assumed to be known.) Thus, a total of six parameters must be considered as known, leaving two unknowns to be solved for. As was the case for two-stream heat exchangers, the choice of unknowns dictates the method of solution. Recall that in the design of a heat exchanger to meet a specified performance, the inlet and outlet temperatures were specified, and the unknowns were H and $\mathcal{P}$ (the *design* problem). Alternatively, for the determination of the performance of an existing exchanger, H and $\mathcal{P}$ were known and the outlet temperatures were unknown (the *rating* problem). The former analysis led to the concept of log mean temperature difference, and the latter led to the concepts of effectiveness and number of transfer units. However, it was also seen that once an $\varepsilon - N_{\text{tu}}$ relation had been obtained, it could be inverted to give an $N_{\text{tu}} - \varepsilon$ relation (either analytically or numerically). This feature, together with the useful physical significance of effectiveness and number of transfer units, suggests that it will be sufficient to solve the rating problem for mass exchangers. Thus, we will assume that the unknowns are x_{out} and y_{out}, and proceed as follows. Equation (3.52) is integrated as

$$\int_{y_{in} - He \, x_{out}}^{y_{out} - He \, x_{in}} d \ln(y - He \, x) = -\left(\frac{1}{\dot{M}_G} - \frac{1}{\dot{M}_L^{\dagger}} \right) (G_G^{oa} \mathcal{P} H)$$

$$y_{out} = He \, x_{in} + (y_{in} - He \, x_{out}) \exp \left[-\left(\frac{1}{\dot{M}_G} - \frac{1}{\dot{M}_L^{\dagger}} \right) (G_G^{oa} \mathcal{P} H) \right] \quad (3.54)$$

To obtain the effectiveness, we must find an expression for $\dot{M}_G(y_{in} - y_{out})$, or equivalently, $\dot{M}_L(x_{out} - x_{in})$. For convenience, define γ as

$$\gamma = \left(\frac{1}{\dot{M}_G} - \frac{1}{\dot{M}_L^{\dagger}} \right) (G_G^{oa} \mathcal{P} H) \quad (3.55)$$

then

$$
\begin{aligned}
y_{in} - y_{out} &= y_{in} - [He \, x_{in} + (y_{in} - He \, x_{out})e^{-\gamma}] \\
&= (y_{in} - He \, x_{in}) - (y_{in} - He \, x_{out})e^{-\gamma} \\
&= (y_{in} - He \, x_{in}) - \left[y_{in} - He \, x_{in} - \frac{\dot{M}_G}{\dot{M}_L^{\dagger}}(y_{in} - y_{out}) \right] e^{-\gamma} \\
&= (y_{in} - He \, x_{in})(1 - e^{-\gamma}) + (y_{in} - y_{out}) \frac{\dot{M}_G}{\dot{M}_L^{\dagger}} e^{-\gamma}
\end{aligned}
$$

where x_{out} has been eliminated by means of the exchanger balance, Eq. (3.48). Rearranging,

$$\frac{y_{in} - y_{out}}{y_{in} - He \, x_{in}} = \frac{1 - e^{-\gamma}}{1 - (\dot{M}_G/\dot{M}_L^{\dagger})e^{-\gamma}} \quad (3.56)$$

Suppose $\dot{M}_G < \dot{M}_L^{\dagger}$ so that $\gamma > 0$. Then as the column height becomes large, $\gamma \to \infty$, the right-hand side of Eq. (3.56) approaches 1 and $y_{out} \to He \, x_{in}$. That is, referring to Fig. 3.16a, the streams approach equilibrium at the top of the column. Thus, the left-hand side of Eq. (3.56) is the ratio of the actual absorption from the gas stream to the maximum possible in an infinitely high column, which we call the effectiveness ε for a countercurrent exchanger,

$$\varepsilon = \frac{\dot{M}_G(y_{in} - y_{out})}{\dot{M}_G(y_{in} - y_{out})_{max}} = \frac{\dot{M}_G(y_{in} - y_{out})}{\dot{M}_G(y_{in} - He \, x_{in})} \quad (3.57)$$

and from Eq. (3.56),

$$\varepsilon = \frac{1 - e^{-\gamma}}{1 - (\dot{M}_G/\dot{M}_L^{\dagger})e^{-\gamma}} \quad (3.58)$$

For this case with $\dot{M}_G < \dot{M}_L^\dagger$, the performance of the exchanger can be viewed as being limited by the amount of the solute gas carried by the gas stream rather than the capacity of the liquid stream to absorb the gas: y_{out} approaches He x_{in} but He x_{out} does not approach y_{in} as closely, as shown in Fig. 3.16a.

Now suppose $\dot{M}_L^\dagger < \dot{M}_G$, so that $\gamma < 0$. Then Eq. (3.54) requires that He $x_{out} \to y_{in}$ as $|\gamma| \to \infty$. Referring to Fig. 3.16b, the streams approach equilibrium at the bottom of the column. The performance of the exchanger is now limited by the capacity of the liquid stream to absorb gas, which indicates that the effectiveness is more usefully expressed as

$$\varepsilon = \frac{\dot{M}_G(y_{in} - y_{out})}{\dot{M}_L(x_{out} - x_{in})_{max}} = \frac{\dot{M}_G(y_{in} - y_{out})}{\dot{M}_L^\dagger(y_{in} - \text{He } x_{in})} = \frac{\dot{M}_L(x_{out} - x_{in})}{\dot{M}_L^\dagger(y_{in} - \text{He } x_{in})} \tag{3.59}$$

Using Eq. (3.56), it is easily shown that

$$\varepsilon = \frac{1 - e^\gamma}{1 - (\dot{M}_L^\dagger/\dot{M}_G)e^\gamma} \tag{3.60}$$

Equations (3.58) and (3.60) can be written in the more compact forms

$$\varepsilon = \frac{\dot{M}_G}{C_{min}}\left(\frac{y_{in} - y_{out}}{y_{in} - \text{He } x_{in}}\right) = \frac{\dot{M}_L}{C_{min}}\left(\frac{x_{out} - x_{in}}{y_{in} - \text{He } x_{in}}\right) \tag{3.61}$$

$$\varepsilon = \frac{1 - e^{-N_{tu}(1-R_C)}}{1 - R_C e^{-N_{tu}(1-R_C)}} \tag{3.62}$$

where

$$N_{tu} = \frac{\mathcal{G}_G^{oa}\mathcal{P}H}{C_{min}}; \quad R_C = \frac{C_{min}}{C_{max}} = \frac{\min(\dot{M}_G, \dot{M}_L^\dagger)}{\max(\dot{M}_G, \dot{M}_L^\dagger)}$$

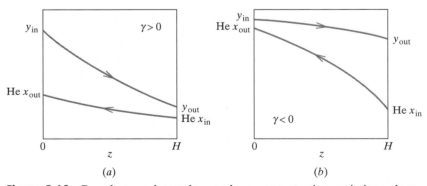

Figure 3.16 Gas-phase and pseudo-gas-phase concentration variations along a countercurrent exchanger: (a) $\gamma > 0$ ($\dot{M}_G < \dot{M}_L^\dagger$), (b) $\gamma < 0$ ($\dot{M}_G > \dot{M}_L^\dagger$).

Equation (3.62) is used for determining the performance of a given mass exchanger. In designing an exchanger to obtain a desired effectiveness, the inverse relation is required, which is

$$N_{tu} = \frac{1}{1 - R_C} \ln \frac{1 - R_C \varepsilon}{1 - \varepsilon} \tag{3.63}$$

Because of their higher effectiveness, countercurrent exchangers are usually preferred over cocurrent exchangers. However, the engineer may well choose a cocurrent exchanger if floor space is limited and a high gas throughput is required: a cocurrent unit has a lower pressure drop and superior liquid distribution, and does not have the flooding limitation of a countercurrent unit. The required $\varepsilon - N_{tu}$ and $N_{tu} - \varepsilon$ relations are given in Table 8.3 of *Heat Transfer*. Similarly, equipment configuration constraints may suggest use of a cross-flow unit; again, the required relations are given in Table 8.3.

The Computer Program SCRUB

SCRUB calculates the performance of countercurrent scrubbers for absorption into water. Dilute concentrations and low mass transfer rates are assumed. Packings include Raschig rings, Berl saddles, Interlox saddles, and parallel plates. The "rating" problem is solved; that is, the stream flow rates, inlet compositions, and packing dimensions are specified, and the outlet compositions are calculated. Also, the pressure drop is calculated and the flooding limit is checked, following the recommendations in Section 3.4.4.

Example 3.7   Effectiveness of a Packed-Column Ammonia Scrubber

Suppose that ammonia is to be removed from an air stream in the packed column described in Example 3.6. Given that the inlet water is pure and that the inlet air stream contains 3% ammonia by volume, calculate the effectiveness of the column and the outlet air composition if the column is 2 m high.

Solution

Given: A 2 m–high column packed with 0.038 m Berl saddles using 1.90 kg/s of water to scrub NH_3 from a 0.35-kg/s air stream.

Required: Column effectiveness, and y_{out} for $y_{in} = 0.03$.

Assumptions: Properties of pure air and pure water may be used for gas and liquid phases, respectively.

The effectiveness of a countercurrent column is given by Eq. (3.62) as

$$\varepsilon = \frac{1 - e^{-N_{tu}(1 - R_C)}}{1 - R_C e^{-N_{tu}(1 - R_C)}}; \quad N_{tu} = \frac{G_G^{oa} \mathcal{P} H}{C_{min}}; \quad R_C = \frac{C_{min}}{C_{max}}$$

The molar flow rates and capacity rate ratio are

$$\dot{M}_G = \frac{\dot{m}_G}{M_G} = \frac{0.35 \text{ kg/s}}{29 \text{ kg/kmol}} = 0.0121 \text{ kmol/s}$$

$$\dot{M}_L = \frac{\dot{m}_L}{M_L} = \frac{1.90 \text{ kg/s}}{18 \text{ kg/kmol}} = 0.1056 \text{ kmol/s}$$

$$\dot{M}_L^\dagger = \frac{\dot{M}_L}{\text{He}} = \frac{0.1056}{1.1} = 0.0960 \text{ kmol/s}$$

$$C_{\min} = \min(\dot{M}_G, \dot{M}_L^\dagger) = \min(0.0121, \ 0.0960)$$

$$= 0.0121 \text{ kmol/s}$$

$$R_C = \frac{C_{\min}}{C_{\max}} = \frac{0.0121}{0.0960} = 0.126$$

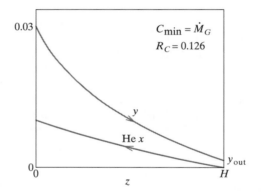

From Example 3.6, the overall mole transfer conductance and transfer perimeter are $\mathcal{G}_G^{\text{oa}} = 5.57 \times 10^{-4} \text{ kmol/m}^2 \text{ s}$, $\mathcal{P} = 61.7 \text{ m}$. Thus, the number of transfer units is

$$N_{\text{tu}} = \frac{\mathcal{G}_G^{\text{oa}} \mathcal{P} H}{C_{\min}} = \frac{(5.57 \times 10^{-4})(61.7)(2)}{0.0121} = 5.68$$

and the effectiveness is

$$\varepsilon = \frac{1 - e^{-N_{\text{tu}}(1-R_C)}}{1 - R_C e^{-N_{\text{tu}}(1-R_C)}} = \frac{1 - e^{-(5.68)(1-0.126)}}{1 - 0.126 e^{-5.68(1-0.126)}} = 99.39\%$$

From Eq. (3.61), noting that $x_{\text{in}} = 0$, the mole fraction of ammonia in the outlet air stream is

$$y_{\text{out}} = y_{\text{in}}(1 - \varepsilon) = 0.03(1 - 0.9939) = 1.83 \times 10^{-4}$$

Comments

1. Use SCRUB to check N_{tu}, ε, and y_{out}. Note that SCRUB uses He = 1.042.
2. Since the effectiveness is very high, it is more convenient to quote the ineffectiveness as $1 - \varepsilon = 0.611\%$.   ▲

3.4.4 Pressure Drop and Flooding in Packed Columns

Random Packings

Figure 3.17 shows typical pressure drop data for random packings. At low gas flow rates ΔP is proportional to velocity squared, even at high liquid flow rates. But as liquid holdup increases, the *loading point* is reached, where ΔP begins to increase more rapidly; finally, in a countercurrent unit, the *flooding limit* is reached, where flow reversal of the liquid occurs. Industrial practice is such that columns usually operate with gas flow rates between the loading point and the flooding limit. Empirical data for pressure drop in this range is usually supplied by the manufacturer of the particular packing. However, most such data are well approximated by the curves in Fig. 3.17, which also shows an approximate flooding limit.

The flooding phenomenon is obviously a complex one. In addition to depending on gas and liquid flow rates, the flooding limit is also found to depend on a packing factor ε_v^3/a, gravity, the inverse cube of surface tension, and other fluid properties. Figure 3.17 shows one widely used correlation. The strong effect of surface tension gives a low flooding limit when water is contaminated by a surfactant.

Structured Packings

Figure 3.18 shows typical pressure drop data for a patented structured packing. For parallel-plate packings, a simple correlation based on work by Bharathan and Wallis [17] is recommended for rough estimates. The idea behind this correlation is that the effect of surface waves on the liquid film of thickness δ is similar to that of wall

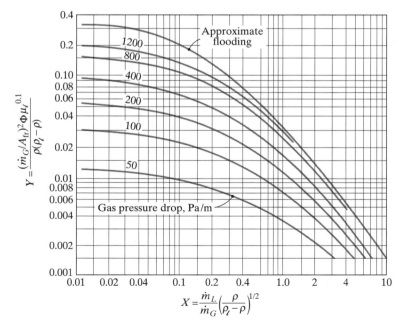

Figure 3.17 Pressure drop and flooding limit in towers with random packings [16]. The drag factor Φ is given in Table 3.3.

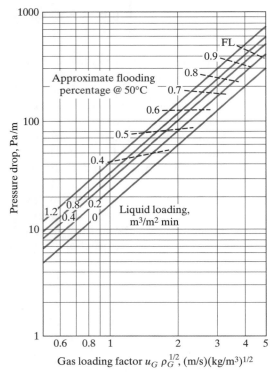

Figure 3.18 Pressure drop and flooding limit for Munters PN Fill structured packing.

roughness, and the friction factor is approximately

$$f \simeq 4(0.005 + C\delta^{*n}) \tag{3.64a}$$

$$\log_{10} C = -0.56 + \frac{9.07}{D_h^*}; \quad n = 1.63 + \frac{4.74}{D_h^*}$$

$$\delta^* = \delta/L_c; \quad D_h^* = D_h/L_c; \quad L_c = [\sigma/(\rho_l - \rho)g]^{1/2}$$

Notice that the film thickness and hydraulic diameter are made dimensionless with the characteristic length L_c introduced for nucleate boiling in Section 7.4.3 of *Heat Transfer*. Also, the friction factor f is defined in terms of a velocity based on the gas-flow cross-sectional area A_c [see Eq. (3.39)]. For large passages ($D_h^* > 160$), Eq. (3.64a) can be approximated as

$$f = 0.02(1 + 55\delta^{*1.63}) \tag{3.64b}$$

Equation (3.64) is based on data for tubes of circular cross section and may be less reliable for flow between parallel plates. Notice that Fig. 3.18 and Eq. (3.64) apply to downward flow of the liquid phase and upward counterflow of the gas phase.

Flooding in circular tubes has been extensively studied, though there remains considerable controversy as to which of the many available correlations is preferred.

Part of the confusion is related to the associated phenomenon of entrainment: at relatively low liquid flow rates, droplets can be stripped from wave crests and be entrained at gas flow rates lower than those required to cause bridging of the gas flow passage by liquid, and consequent flow reversal. Some widely used flooding correlations are, in fact, entrainment correlations. The true flooding limit can be estimated from the correlation developed by Tien and co-workers [18, 19]:

$$K_G^{1/2} + 0.8K_L^{1/2} = 2.1 \tanh(0.8D_h^{*1/4}) \tag{3.65}$$

where

$$K_G = \alpha V_G \rho^{1/2}/[g\sigma(\rho_l - \rho)]^{1/4}$$

$$K_L = (1 - \alpha)V_L \rho_l^{1/2}/[g\sigma(\rho_l - \rho)]^{1/4}$$

V_G and V_L are the velocities based on passage cross-sectional area, and α is the fraction of the cross section occupied by the gas flow. In a circular tube, $\alpha \simeq (1 - 2\delta/D)^2$, where δ is the film thickness. The quantities K_G and K_L are dimensionless gas and liquid flow rates: when $(K_G^{1/2} + 0.8K_L^{1/2})$ exceeds the critical value specified by Eq. (3.65), flooding occurs.

Example 3.8 Pressure Drop in a Column Packed with Berl Saddles

Determine the pressure drop in a 1 m²–cross-section, 3 m–high column packed with 0.038 m ceramic Berl saddles, operating at 300 K and 1 atm, when 1.4 kg/s of air flows countercurrent to 4.2 kg/s water.

Solution

Given: Column packed with Berl saddles.
Required: Pressure drop for countercurrent flow if column is 3 m high.
Assumptions: Figure 3.17 is sufficiently accurate.
To use Fig. 3.17, we must calculate the abscissa and ordinate values, X and Y, from the given data. From Tables A.7 and A.8 for 300 K and 1 atm,

$$\rho = 1.177 \text{ kg/m}^3, \quad \rho_l = 996 \text{ kg/m}^3, \quad \mu_l = 8.67 \times 10^{-4} \text{ kg/m s}$$

Table 3.3 gives the drag factor for 0.038 m ceramic Berl saddles as $\Phi = 65$. Referring to Fig. 3.17, the abscissa and ordinate values are

$$X = \frac{\dot{m}_L}{\dot{m}_G}\left(\frac{\rho}{\rho_l - \rho}\right)^{1/2} = \frac{4.2}{1.4}\left(\frac{1.177}{996 - 1.177}\right)^{1/2} = 0.103$$

$$Y = \frac{(\dot{m}_G/A_{\text{fr}})^2 \Phi \mu_l^{0.1}}{\rho(\rho_l - \rho)} = \frac{(1.4/1)^2(65)(8.67 \times 10^{-4})^{0.1}}{(1.177)(996 - 1.177)} = 5.38 \times 10^{-2}$$

giving a pressure gradient of $\Delta P/H \simeq 310$ Pa/m. Hence, the column pressure drop is $\Delta P = (310)(3) = 930$ Pa.

Comments

1. Check ΔP using SCRUB. The discrepancy is due to the approximate curve fits of Figure 3.17 used in SCRUB.

2. Figure 3.17 is of very general applicability and thus is not too accurate. If at all possible, the engineer should obtain pressure drop data for the specific packing under consideration. Usually the manufacturer of the packing can supply such data.   ▲

3.5 SIMULTANEOUS HEAT AND MASS EXCHANGERS

In general, simultaneous heat and mass exchangers are more difficult to analyze than simple heat or mass exchangers, since coupled differential equations governing conservation of mass species and energy must be solved simultaneously. Direct numerical solution of the differential equations is usually necessary. However, for some exchangers it is possible to simplify the analysis and yet obtain satisfactory results. In the case of a humidifier, the assumption of an adiabatic system reduces the problem to that of a simple single-stream mass exchanger, similar to those of Section 3.3. For counterflow cooling towers, it will be shown that *Merkel's approximation* allows the governing equations to be reduced to a single differential equation in terms of enthalpy, which can be solved easily by numerical integration.

3.5.1 Adiabatic Humidifiers

Figure 3.19 shows examples of adiabatic humidifiers used in air-conditioning systems. The purpose of the humidifier could be to increase the humidity of the air stream; alternatively, the purpose could be to cool the air stream. Since the water is continuously recirculated, with only a small fraction evaporating, the enthalpy of vaporization is supplied by heat transfer from the air stream to the water surface. Various types of packing are used with the objective of obtaining a large water surface area exposed to the air stream without an excessive pressure drop. If the design requires a very low pressure drop for the air stream, a spray chamber may be used in place of the packing. However, such units tend to be very large since the air velocity must be low to prevent water droplets from being blown out of the chamber.

If the amount of makeup water required to replace the evaporated water is relatively small, and if heat losses or gains between the unit and its surroundings are small, the process can be assumed to be adiabatic; that is, all the enthalpy of vaporization is supplied from the air itself. At steady state, the recirculated water will approach the wet-bulb temperature of the incoming air, and T_s and $m_{1,s}$ will be unchanging along the exchanger. For engineering purposes it is adequate to assume that the water is at the thermodynamic wet-bulb temperature. Thus, such humidifiers can be classified as single-stream exchangers since only the condition of the air changes along the exchanger. Both heat and mass are transferred in a humidifier, but, in the case of the adiabatic humidifier, the mass transfer and heat transfer can be analyzed separately. In exchangers such as cooling towers, the mass and heat transfer are coupled and must be considered simultaneously; the resulting analysis is more complicated, as will

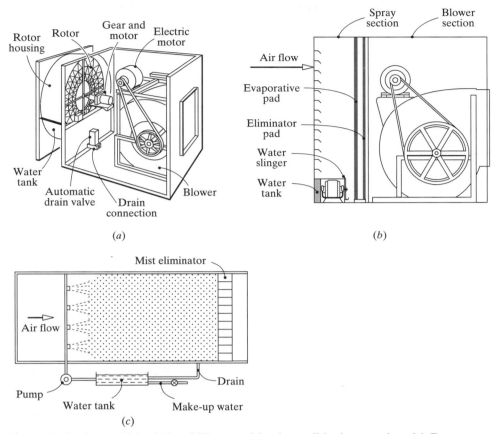

Figure 3.19 Some abiabatic humidifiers used in air conditioning practice. (*a*) Rotary type. (*b*) Slinger type. (*c*) Spray chamber.

be seen in Section 3.5.2. Low mass transfer rate theory is applicable to the adiabatic humidifiers used in air-conditioning practice. The analysis of heat and mass transfer follows the procedures described in Sections 1.4 and 1.5.

Figure 3.20 depicts an adiabatic humidifier with water, species 1, evaporating into an air stream. The flow rate of the mixture of water vapor and air will be denoted $\dot{m}_G$, m_1 denotes the bulk mass fraction of water vapor, $j_{1,s}$ is the rate of evaporation of water, and $\mathcal{P}$ is the perimeter of transfer surface (*s*-surface). Application of the principle of conservation of species to the elemental control volume between x and Δx gives an equation identical to Eq. (3.5):

$$\dot{m}_G m_1|_x + j_{1,s} \mathcal{P} \Delta x = \dot{m}_G m_1|_{x+\Delta x}$$

In addition, the principle of mass conservation gives

$$\dot{m}_G|_x + j_{1,s} \mathcal{P} \Delta x = \dot{m}_G|_{x+\Delta x} \tag{3.66}$$

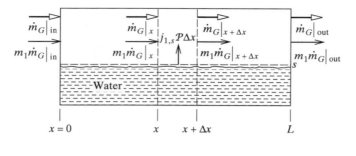

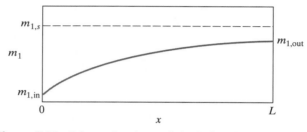

Figure 3.20 Schematic of an adiabatic humidifier showing mass and species balances, and the expected variation of the water vapor mass fraction along the humidifier.

However, for humidifier operation at close to ambient temperatures, the evaporation rate is small compared with the air flow rate, and $\dot{m}_G$ can be *assumed* to be constant: then Eq. (3.66) can be discarded and Eq. (3.7) is obtained as before:

$$\frac{dm_1}{dx} - \frac{j_{1,s}\,\mathcal{P}}{\dot{m}_G} = 0 \tag{3.67}$$

For the humidifier, the required expression for the evaporation rate $j_{1,s}$ is simple:

$$j_{1,s} = g_{m1}(m_{1,s} - m_1)$$

where, as stated before, $m_{1,s} = m_{1,\text{sat}}(T_{\text{WB}})$ is constant along the exchanger, and m_1 is the bulk value. Substituting in Eq. (3.67),

$$\frac{dm_1}{dx} + \frac{g_{m1}\,\mathcal{P}}{\dot{m}_G}(m_1 - m_{1,s}) = 0 \tag{3.68}$$

Integrating with $m_1 = m_{1,\text{in}}$ at $x = 0$ and $m_1 = m_{1,\text{out}}$ at $x = L$, and rearranging gives

$$\frac{m_{1,\text{out}} - m_{1,\text{in}}}{m_{1,s} - m_{1,\text{in}}} = 1 - e^{-g_{m1}\mathcal{P}L/\dot{m}_G} \tag{3.69}$$

Again, the left-hand side of this equation is recognized as the exchanger effectiveness since it is the actual increase in moisture content of the air stream divided by the maximum possible increase that could be obtained in an infinitely long exchanger: when m_1 equals $m_{1,s}$, no further evaporation can occur (see Fig. 3.20). The dimensionless exponent on the right-hand side is the number of transfer units of the humidifier. Thus, Eq. (3.11) again applies, with appropriate expressions for ε and N_{tu}.

To obtain the amount of makeup water, $\dot{m}_{add}$, that must be added, we perform a species balance on the exchanger as a whole:

$$(\dot{m}_G m_1)_{in} + \dot{m}_{add} = (\dot{m}_G m_1)_{out} \tag{3.70}$$

For an approximate result we can take $\dot{m}_G$ constant and write

$$\dot{m}_{add} \simeq \dot{m}_G (m_{1,out} - m_{1,in}) = \dot{m}_G \varepsilon (m_{1,s} - m_{1,in}) \tag{3.71}$$

Often the purpose of a humidifier is to cool the air rather than to humidify it: it is then called an evaporative or "swamp" cooler. An easy way to determine the outlet air temperature is to recognize that, in an adiabatic humidifier, the wet-bulb temperature of the air does not change as it flows through the exchanger. After calculation of $m_{1,out}$, the humidity ratio $\omega = m_{1,out}/(1 - m_{1,out})$ kg H_2O per kilogram of dry air can be calculated, and a psychrometric chart used to determine the dry-bulb temperature, which is T_{out}. Alternatively, the computer program PSYCHRO can be used with specification of P, T_{WB}, and m_1 in the option that holds P and water vapor concentration constant.

Example 3.9   A Laminar-Flow Evaporative Cooler

A solar air-conditioning process under development involves evaporative cooling of the air followed by dehumidification in a silica gel desiccant bed: hot air from a solar air heater is used to regenerate (dry) the silica gel. The only power requirements of the cycle are those of the fans, pumps, and controls. An existing evaporative cooler installation has a cross section 1.1 m high and 0.5 m wide and contains a patented packing. Since fan power and hence pressure drop are key factors in the system design, a laminar-flow device is to be evaluated, in which there is cross-flow of air between vertical plastic plates with falling water films. Find the required length of an 80% effective exchanger if the volumetric air flow rate is 50 m³/min at 1 atm pressure. The inlet air is at 300 K and 40% RH. Take the plate center-to-center spacing as 9.5 mm, and the combined width of one plate plus two liquid films as 2.5 mm. Also calculate the outlet air temperature.

Solution

Given: Parallel-plate evaporative cooler.

Required: (i) Length for 80% effectiveness when the cooler is required to cool 50 m³/min of air. (ii) Outlet air temperature.

Assumptions:

1. Negligible heat gain from the surroundings.

2. Laminar flow (check).

3. Properties of pure air may be used.

Geometrical data for the packing are

$$\text{Number of plates} = (0.5)/(9.5 \times 10^{-3}) = 53$$

$$\text{Perimeter } \mathcal{P} = (2)(53)(1.1) = 116 \text{ m}$$

$$\text{Flow area } A_c = (1.1)(0.5) - (53)(1.1)(2.5 \times 10^{-3})$$
$$= 0.404 \text{ m}^2$$

$$D_h = 2 \times \text{passage width} = (2)(7 \times 10^{-3}) = 1.4 \times 10^{-2} \text{ m}$$

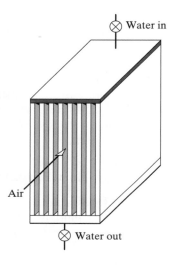

⊗ Water in

Air

⊗ Water out

The exchanger is required to be 80% effective; thus, the required number of transfer units is

$$N_{\text{tu}} = \ln \frac{1}{1-\varepsilon} = \ln \frac{1}{1-0.8} = 1.61 = \frac{g_{m1} \mathcal{P} L}{\dot{m}_G}$$

$$L = \frac{1.61 \dot{m}_G}{g_{m1} \mathcal{P}}$$

To evaluate g_{m1}, assume laminar flow and no entrance effects; then $\text{Sh} = 7.54$ from Table 4.5 in *Heat Transfer*. Take properties for pure air at 300 K and 1 atm; then $\rho = 1.177 \text{ kg/m}^3$, $\nu = 15.66 \times 10^{-6} \text{ m}^2/\text{s}$. Also, $\text{Sc}_{12} = 0.61$ for the H_2O-air mixture.

$$g_{m1} = 7.54 \frac{\rho \mathcal{D}_{12}}{D_h} = 7.54 \frac{\rho \nu}{\text{Sc}_{12} D_h} = \frac{(7.54)(1.177)(15.66 \times 10^{-6})}{(0.61)(1.4 \times 10^{-2})}$$
$$= 1.63 \times 10^{-2} \text{ kg/m}^2 \text{ s}$$

$$L = \frac{1.61 \dot{m}_G}{g_{m1} \mathcal{P}} = \frac{(1.61)(50/60)(1.177)}{(1.63 \times 10^{-2})(116)} = 0.835 \text{ m}$$

To check if the flow is laminar, the Reynolds number is calculated:

$$V = \frac{50/60 \text{ m}^3/\text{s}}{0.404 \text{ m}^2} = 2.06 \text{ m/s}$$

$$\text{Re}_{D_h} = \frac{V D_h}{\nu} = \frac{(2.06)(1.4 \times 10^{-2})}{15.66 \times 10^{-6}} = 1842 < 2800, \quad \text{that is, laminar.}$$

To check the entrance effect we use the mass transfer analog to Eq. (4.51):

$$\overline{\text{Sh}}_{D_h} = 7.54 + \frac{0.03(D_h/L)\text{Re}_{D_h}\text{Sc}}{1 + 0.016[(D_h/L)\text{Re}_{D_h}\text{Sc}]^{2/3}}$$

$$= 7.54 + \frac{(0.03)(18.8)}{1 + 0.016(18.8)^{2/3}} = 8.05$$

New estimates are:

$$g_{m1} = (8.05/7.54)(1.63 \times 10^{-2}) = 1.74 \times 10^{-2} \text{ kg/m}^2 \text{ s}$$

$$L = (1.63/1.74)(0.835) = 0.782 \text{ m}$$

To obtain T_{out}, we first need to calculate the wet-bulb temperature of the inlet air, as well as $m_{1,\text{in}}$ and $m_{1,s}$. PSYCHRO can be used as follows.

1. Option 1.
   Input: $P = 1.013 \times 10^5$ Pa, $T_{\text{DB}} = 300$ K, RH = 40%
   Output: $T_{\text{WB}} = 290.7$ K, $m_{1,\text{in}} = 8.713 \times 10^{-3}$

2. Option 1.
   Input: $P = 1.013 \times 10^5$ Pa, $T_{\text{DB}} = 290.7$ K, RH = 100%
   Output: $m_{1,s} = 1.239 \times 10^{-2}$
   Then, from Eq. (3.69), we obtain the outlet mass fraction as

   $$m_{1,\text{out}} = m_{1,\text{in}} + \varepsilon(m_{1,s} - m_{1,\text{in}})$$

   $$= 0.00871 + 0.8(0.01239 - 0.00871) = 0.01165$$

3. Option 2.
   Input: $P = 1.013 \times 10^5$ Pa, $m_1 = 0.01165$, $T_{\text{WB}} = 290.7$ K
   Output: $T_{\text{DB}} = 292.8$ K $= T_{\text{out}}$

Comments

1. Air-conditioning engineers sometimes use a "temperature effectiveness" to characterize the performance of an evaporative cooler,

   $$\varepsilon = \frac{T_{\text{in}} - T_{\text{out}}}{T_{\text{in}} - T_s} = \frac{300 - 292.8}{300 - 290.7} = 0.77 = 77\%$$

   which is a little lower than the mass transfer effectiveness. However, more advanced texts will show that it is more appropriate to use an effectiveness based on enthalpy change to characterize the thermal performance of mass exchangers (see Exercise 3–57).

2. Use Eq. (1.58) to determine T_{out} in step 3 above.

3. Use a psychrometric chart to determine T_{out} and compare the result.

4. Notice that accounting for entrance effects reduces the required length by 6%.

5. Use PSYCHRO to check that the enthalpy of the air is constant through the exchanger, consistent with the adiabatic assumption. Of course, assuming a constant T_{WB} through the exchanger ensures this result.

6. If the effect of ripples on the water film surfaces is ignored, the friction factor can be obtained from Table 4.5 as $f = 96/\text{Re}_{D_h} = 96/1842 = 0.0521$. The pressure drop is then

$$\Delta P = f(L/D_h)(1/2)\rho V^2 = (0.0521)(0.782/1.4 \times 10^{-2})(0.5)(1.177)(2.06)^2$$
$$= 7.3 \text{ Pa}$$

which is very small. The sum of the inlet and outlet pressure drops is likely to be considerably larger, unless the air ducts have the same cross-section as the exchanger (see Section 8.7.1 of *Heat Transfer*).

7. See Exercise 3–42 for a performance comparison with a patented packing.▲

3.5.2 Counterflow Cooling Towers

In a wet cooling tower, water is evaporated into air with the objective of cooling the water stream. Since both the air and the water streams change state along the exchanger, the wet cooling tower is a *two-stream exchanger*. Both natural- and mechanical-draft cooling towers are popular, and examples are shown in Fig. 3.21. Large natural-draft cooling towers are used in power plants for cooling the water supply to the condenser. Smaller mechanical-draft towers are preferred for oil refineries and other process industries, as well as for central air-conditioning systems. Figure 3.21 shows *counterflow* units in which the water flows as thin films down over a suitable packing, and air flows upward. In a natural-draft tower, the air flows upward due to the buoyancy of the warm, moist air leaving the top of the packing. In a mechanical-draft tower, the flow is forced or induced by a fan. Since the air inlet temperature is usually lower than the water inlet temperature, the water is cooled both by evaporation and by sensible heat loss. For usual operating conditions, q_{evap} is considerably larger than q_{conv}.

Exchanger Balances

Figure 3.22 shows the schematic of a counterflow cooling tower packing (or *fill*). The subscript G is used to denote the gas stream, which is a water vapor and air mixture; the subscript L denotes the liquid stream, which is taken to be pure water. Species 1 and 2 are H_2O and air, respectively. The exchanger balances for mass, mass species (H_2O), and energy are

$$\dot{m}_{G,\text{in}} - \dot{m}_{L,\text{out}} = \dot{m}_{G,\text{out}} - \dot{m}_{L,\text{in}} \tag{3.72}$$
$$(\dot{m}_G m_1)_{\text{in}} - \dot{m}_{L,\text{out}} = (\dot{m}_G m_1)_{\text{out}} - \dot{m}_{L,\text{in}} \tag{3.73}$$
$$(\dot{m}_G h_G)_{\text{in}} - (\dot{m}_L h_L)_{\text{out}} = (\dot{m}_G h_G)_{\text{out}} - (\dot{m}_L h_L)_{\text{in}} \tag{3.74}$$

where the water flow rate, $\dot{m}_L$, is *positive*.[3]

---

[3] This convention is identical to that used for counterflow heat exchangers in Chapter 8 of *Heat Transfer* and for countercurrent mass exchangers in Section 3.4.3. In some texts the convention of a negative water flow rate is used.

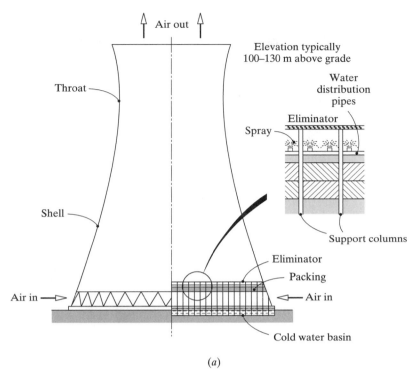

(a)

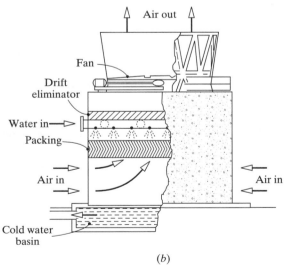

(b)

Figure 3.21 Counterflow cooling towers. (a) A natural-draft tower for a power plant. (b) A mechanical-draft tower for an air-conditioning system.

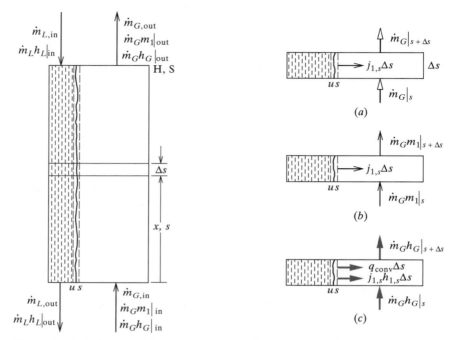

Figure 3.22 Schematic of a counterflow cooling tower packing showing an elemental volume containing a transfer area Δs, and balances on the gas stream: (a) mass, (b) species, (c) energy.

Mixture Enthalpy and Specific Heat

The gas stream is a mixture of water vapor and air. The enthalpy of the mixture is

$$h = \sum_{i=1}^{2} m_i h_i = m_1 h_1 + m_2 h_2 \qquad (3.75)$$

Typical enthalpy tables for water vapor and air mixtures are based on a datum temperature of $0°C$, at which the enthalpy of dry air and liquid water are set equal to zero. Hence,

$$h_1 = h_{fg}(0°C) + \int_0^t c_{p1}\, dt \qquad (3.76a)$$

$$h_2 = \int_0^t c_{p2}\, dt \qquad (3.76b)$$

where t is used to denote temperature in degrees Celsius. Table A.31 and the computer program PSYCHRO are based on these datum states.

The range of temperature encountered in humidifier and cooling tower operation is relatively small, so that the species specific heats can be assumed constant for most purposes. Then

$$h_1 = h_{fg} + c_{p1}t \qquad (3.77a)$$

$$h_2 = c_{p2}t \qquad (3.77b)$$

Substituting in Eq. (3.75), the mixture enthalpy is

$$h = m_1 h_{fg} + (m_1 c_{p1} + m_2 c_{p2})t$$

or

$$h = m_1 h_{fg} + c_p t \qquad (3.78)$$

where $c_p = \sum_{i=1}^{2} m_i c_{pi}$ is the mixture specific heat. For water vapor and air mixtures, the mixture specific heat depends strongly on composition because $c_{p\,H_2O}$ is almost twice $c_{p\,air}$.

The Governing Differential Equations

For convenience we define the *contact area* in a length of exchanger Δx as $\Delta s = \mathcal{P}\Delta x$, where $\mathcal{P}$ is the transfer area perimeter. Then the total contact area in a packing of height H is

$$S = \int_0^H \mathcal{P} dx = \mathcal{P}H, \quad \text{if } \mathcal{P} \text{ is constant} \qquad (3.79)$$

Mass, species, and energy balances on each stream for an element of packing containing contact area Δs gives the differential conservation equations. The mass and species balances are straightforward. For example, the mass balance on an element of the gas stream shown in Fig. 3.22 gives

$$\dot{m}_G|_s + j_{1,s}\Delta s = \dot{m}_G|_{s+\Delta s}$$

Dividing by Δs and letting $\Delta s \to 0$,

$$\frac{d\dot{m}_G}{ds} = j_{1,s} \qquad (3.80)$$

Similarly,

$$\frac{d\dot{m}_L}{ds} = j_{1,s} \qquad (3.81)$$

$$\frac{d}{ds}(\dot{m}_G m_1) = j_{1,s} \qquad (3.82)$$

Low mass transfer rate theory has been used to approximate the rate of evaporation of the water as $j_{1,s}$ [see Eq. (1.38)]. The energy balances require special consideration. The steady-flow energy equation cannot be used in precisely the form of Eq. (1.4) of *Heat Transfer* because the stream flow rates $\dot{m}_G$ and $\dot{m}_L$ are not constant. Referring to Fig. 3.22, the energy balance on the gas stream is

$$q_{conv}\Delta s = \dot{m}_G h_G|_{s+\Delta s} - \dot{m}_G h_G|_s - j_{1,s} h_{1,s}\Delta s$$

where the last term on the right-hand side accounts for the enthalpy added to the control volume by the evaporated water. Dividing by Δs and letting $\Delta s \to 0$,

$$\frac{d}{ds}(\dot{m}_G h_G) = q_{conv} + j_{1,s} h_{1,s} \qquad (3.83)$$

Similarly, for the liquid phase,

$$\frac{d}{ds}(\dot{m}_L h_L) = q_{conv} + j_{1,s}h_{1,s} \tag{3.84}$$

because energy is conserved between the s- and u-surfaces. It will prove unnecessary in this analysis to write the liquid stream energy balance in terms of u-surface fluxes: Eq. (3.84) will be sufficient. Since $q_{conv} = h_c(T_s - T_G)$ the energy equation, Eq. (3.83), is in terms of two dependent variables, namely, h and T. Unless we are going to solve the equation directly by numerical means, one variable should be eliminated in favor of the other. Enthalpies are easily written in terms of mass fraction and temperature, but if we go to T as the dependent variable, the mass fraction m_1 will remain as an additional dependent variable. Instead we will eliminate T in favor of h, and after we make some reasonable assumptions, the equation set will reduce to a single equation in terms of only one dependent variable, the enthalpy h.

Simplification and Solution of the Governing Equations

Our first step in a rather involved manipulation of Eq. (3.83) is to expand its left-hand side and substitute Eq. (3.80):

$$\frac{d}{ds}(\dot{m}_G h_G) = \dot{m}_G \frac{dh_G}{ds} + h_G \frac{d\dot{m}_G}{ds} = \dot{m}_G \frac{dh_G}{ds} + j_{1,s}h_G \tag{3.85}$$

Thus, Eq. (3.83) becomes

$$\dot{m}_G \frac{dh_G}{ds} = q_{conv} + j_{1,s}(h_{1,s} - h_G) \tag{3.86}$$

Next the mass and heat transfer conductances are introduced. The evaporation rate is

$$j_{1,s} = g_m(m_{1,s} - m_{1,G}) \tag{3.87}$$

and we will write the convective heat transfer as

$$q_{conv} = g_h c_{pG}(T_s - T_G) \tag{3.88}$$

where $g_h = h_c/c_{pG}$ is the *heat transfer conductance* introduced in Section 9.4.4. Advanced mass transfer theory (see Section 2.5.2) shows that it is appropriate to use the bulk gas specific heat in the definition of heat transfer conductance, but this point is of no practical significance in the present analysis. Substituting Eqs. (3.87) and (3.88) in Eq. (3.86) gives

$$\dot{m}_G \frac{dh_G}{ds} = g_h c_{pG}(T_s - T_G) + g_m(m_{1,s} - m_{1,G})(h_{1,s} - h_G) \tag{3.89}$$

The Lewis number for dilute water vapor and air mixtures is Le $= $ Pr/Sc $\simeq$ $0.69/0.61 = 1.13$, and thus the ratio $g_m/g_h \simeq (1.13)^{2/3} = 1.08$. This result was used in the analysis of the wet- and dry-bulb psychrometer in Section 1.5. Here we will be more bold and take $g_h = g_m$, which is equivalent to assuming a Lewis number of unity.[4] Equation (3.89) then becomes

$$\dot{m}_G \frac{dh_G}{ds} = g_m \left\{ c_{pG}(T_s - T_G) + (m_{1,s} - m_{1,G})(h_{1,s} - h_G) \right\} \qquad (3.90)$$

where we choose to use g_m rather than g_h because q_{evap} is usually considerably larger than q_{conv}. We now come to the key step in the analysis, and assert that Eq. (3.90) can be approximated as

$$\dot{m}_G \frac{dh_G}{ds} \simeq g_m(h_s - h_G) \qquad (3.91)$$

for typical cooling tower operating conditions, *provided that the usual enthalpy datum states are used.* The validity of the approximation is best demonstrated by a numerical calculation. For example, at the bottom of the tower, possible conditions are $P = 1\text{atm}$, $T_s = 30°C$, $T_G = 15°C$, $m_{1,G} = 0$, where dry air has been chosen for convenience. Using PSYCHRO or Tables A.7, A.12a, and A.31, $m_{1,s} = 0.0266$ and

$$h_s - h_G = 97.36 - 15.09 = 82.27 \text{ kJ/kg}$$

$$c_{pG}(T_s - T_G) + (m_{1,s} - m_{1,G})(h_{1,s} - h_G)$$
$$= 1.007(30 - 15) + (0.0266 - 0)(2556 - 15) = 82.70 \text{ kJ/kg}$$

The discrepancy is very small. In general, a discrepancy of up to 5% may be expected for usual cooling tower operating conditions. Notice that it is **essential** to use the usual enthalpy datum states, that is, such that enthalpies of dry air and liquid water are zero at 0°C. The magnitude of a term such as $(h_s - h_G)$ depends on the choice of enthalpy datum states since the compositions of the bulk and s-surface mixtures are different. It is possible to arbitrarily choose datum states to obtain $(h_s - h_G) = 0$. On the other hand, a term such as $c_{pG}(T_s - T_G)$ is independent of the choice of datum states.

A similar manipulation for the liquid stream gives

$$\dot{m}_L \frac{dh_L}{ds} = g_m(h_s - h_G) \qquad (3.92)$$

Dividing Eq. (3.91) by Eq. (3.92),

$$\frac{dh_G}{dh_L} = \frac{\dot{m}_L}{\dot{m}_G} \qquad (3.93)$$

which, if $\dot{m}_L/\dot{m}_G$ is assumed constant, can be integrated from the bottom of the tower where $h_G = h_{G,\text{in}}$ and $h_L = h_{L,\text{out}}$, to give

$$h_G = h_{G,\text{in}} + \frac{\dot{m}_L}{\dot{m}_G}(h_L - h_{L,\text{out}}) \qquad (3.94)$$

---

[4] The equality $g_h = g_m$ is often called the *Lewis relation*.

The definition of the number of transfer units is $N_{tu} = g_m S/\dot{m}_L$, where $S = \mathcal{P}H$ is the contact area.[5] Introducing $dN_{tu} = g_m ds/\dot{m}_L$ into Eq. (3.91) gives

$$\frac{\dot{m}_G}{\dot{m}_L} \frac{dh_G}{dN_{tu}} = h_s - h_G \qquad (3.95)$$

which can be integrated using Eqs. (3.93) and (3.94):

$$dN_{tu} = \frac{\dot{m}_G}{\dot{m}_L} \frac{dh_G}{h_s - h_G} = \frac{dh_L}{h_s - h_G}$$

$$N_{tu} = \int_{h_{L,out}}^{h_{L,in}} \frac{dh_L}{h_s - h_G} \qquad (3.96)$$

where $h_G = h_{G,in} + (\dot{m}_L/\dot{m}_G)(h_L - h_{L,out})$ from Eq. (3.94). Our final assumption (in a long list of assumptions!) is that the liquid-side heat transfer resistance is negligible; then $T_s = T_L$ or

$$h_s(P, T_s) = h_s(P, T_L) \qquad (3.97)$$

Equation (3.96) can be integrated numerically using Eq. (3.97), as will be demonstrated in Example 3.10.

This method of calculating the number of transfer units was originally developed by Merkel in 1925 [20]. When the conventional enthalpy datum states are used, the method is accurate up to temperatures of about $60°C$. Comparisons with more exact solution procedures seldom show errors greater than 10%. Notice that the method does not give the outlet state of the air; however, in situations encountered in practice, the outlet air can be assumed saturated for the purpose of calculating its density. Before computers became widely used, graphical methods, based on the enthalpy-versus-composition chart for the water–water vapor–air system, were popular for analyzing simultaneous heat and mass exchangers. Spalding [21] gives a comprehensive treatment of such methods. Merkel's method is based on the observation that the mixed-phase isotherms on an enthalpy-composition chart are nearly horizontal below $60°C$; assuming these isotherms to be horizontal is equivalent to the approximation made in going from Eq. (3.90) to Eq. (3.91). Merkel's method can be used for higher temperature ranges, or different systems, by defining appropriate enthalpy datum states, that is, states that yield nearly horizontal mixed-phase isotherms in the temperature range under consideration. It is also possible to extend Merkel's method to include a finite liquid-side heat transfer resistance (see Section 3.5.4 and Exercise 3–48), but such refinement is seldom warranted. For a laminar film, the liquid-side resistance is approximately $(\delta/2)/k$, where k is the liquid conductivity and δ the film thickness: for typical operating conditions, the bulk liquid temperature is seldom more than 0.3 K above the interface temperature.

---

[5] This definition is somewhat arbitrary but follows current cooling tower design practice. Some texts define the number of transfer units as $N_{tu} = g_m S/\dot{m}_G$.

Example 3.10 Use of Merkel's Method

A wet cooling tower is required to cool water from 40°C to 26°C when the inlet air is at 10°C, 1 atm, and saturated (a miserable rainy day!). Calculate the required number of transfer units for balanced flow, that is, $\dot{m}_G/\dot{m}_L = 1$.

Solution

Given: Wet countercurrent cooling tower with balanced flow.
Required: Number of transfer units for specified performance.
Assumptions: Merkel's method is appropriate.
Equation (3.96) is to be integrated numerically, with h_G obtained from Eq. (3.94).

$$N_{tu} = \int_{h_{L,\text{out}}}^{h_{L,\text{in}}} \frac{dh_L}{h_s - h_G}; \quad h_G = h_{G,\text{in}} + \frac{\dot{m}_L}{\dot{m}_G}(h_L - h_{L,\text{out}})$$

Using Table A.31, $h_{G,\text{in}} = h_{\text{sat}}(10°C) = 29.15$ kJ/kg, $h_{L,\text{out}} = h_L(26°C) = 109.07$ kJ/kg. Hence,

$$h_G = 29.15 + (h_L - 109.07)$$

Choosing 2°C intervals for convenient numerical integration, the following table is constructed, with h_L and $h_s = h_s(T_L)$ also obtained from Table A.31.

T_L °C	h_L kJ/kg	h_G kJ/kg	h_s kJ/kg	$h_s - h_G$ kJ/kg	$\frac{1}{h_s - h_G}$
26	109.07	29.15	79.12	49.97	0.02001
28	117.43	37.51	87.86	50.35	0.01986
30	125.79	45.87	97.36	51.49	0.01942
32	134.15	54.23	107.67	53.44	0.01871
34	142.50	62.58	118.89	56.31	0.01776
36	150.86	70.94	131.10	60.16	0.01662
38	159.22	79.30	144.39	65.09	0.01536
40	167.58	87.66	158.86	71.20	0.01404

Using the trapezoidal rule,

$$\int_{h_{L,\text{out}}}^{h_{L,\text{in}}} \frac{dh_L}{h_s - h_G} = \frac{8.36}{2}[0.02001 + 2(0.01986 + 0.01942 + 0.01871$$

$$+ \ 0.01776 + 0.01662 + 0.01536) + 0.01404]$$

$$= 1.043$$

From Eq. (3.96), $N_{tu} = 1.043$. Also, using Table A.31, $T_{G,\text{out}} = 27.9°C$ for saturated outlet air.

Comments

1. Under usual operating conditions, the air stream becomes saturated a short distance up the packing. However, it continues to take up more water vapor as it flows upward because its temperature (and hence $P_{1,\text{sat}}$) increases.

2. Greater accuracy can be obtained by using smaller integration steps (CTOWER uses 14 steps). However, evaluation of the number of transfer units is not the major source of error in cooling tower design. There is usually a much greater uncertainty associated with the packing mass transfer correlations and the effects of nonuniform water and air flows.   ▲

3.5.3 Cross-Flow Cooling Towers

Figure 3.23 shows examples of natural- and mechanical-draft cross-flow cooling towers. Figure 3.24 shows a schematic of a cross-flow tower packing and the coordinate system. To model this cross-flow exchanger, we will assume that both the liquid and gas streams are unidirectional, and that there is no mixing in either stream. The packing has dimensions $X \times W \times H$, and a $[m^2/m^3]$ is the contact area per unit volume, that is, the area of liquid-gas interface per unit volume. Figure 3.24 also shows an elemental control volume $\Delta x \Delta y \Delta z$. Superficial mass velocities for the gas and liquid streams are defined as

$$G = \frac{\dot{m}_G}{WH}; \quad L = \frac{\dot{m}_L}{WX} \tag{3.98a}$$

Mass conservation for the gas stream requires that

$$(G|_{x+\Delta x} - G|_x)\Delta y \Delta z = j_{1,s} a \Delta x \Delta y \Delta z$$

Dividing by $\Delta x \Delta y \Delta z$ and letting Δx, Δy, $\Delta z \to 0$, we obtain

$$\frac{\partial G}{\partial x} = a j_{1,s} \tag{3.98b}$$

Similarly, mass conservation for the liquid stream requires that

$$(L|_{y+\Delta y} - L|_y)\Delta x \Delta z = -j_{1,s} a \Delta x \Delta y \Delta z$$

Dividing by $\Delta x \Delta y \Delta z$, and letting Δx, Δy, $\Delta z \to 0$ we obtain

$$\frac{\partial L}{\partial y} = -a j_{1,s} \tag{3.99}$$

The species conservation equation for the gas stream, and the energy conservation equations for the gas and liquid streams are derived in a similar manner:

$$\frac{\partial}{\partial x}(Gm_1) = a j_{1,s} \tag{3.100}$$

$$\frac{\partial}{\partial x}(Gh_G) = a[g_h c_{pG}(T_s - T_G) + j_{1,s}h_{1,s}] \tag{3.101}$$

$$\frac{\partial}{\partial y}(Lh_L) = -a[g_h c_{pG}(T_s - T_G) + j_{1,s}h_{1,s}] \tag{3.102}$$

(a)

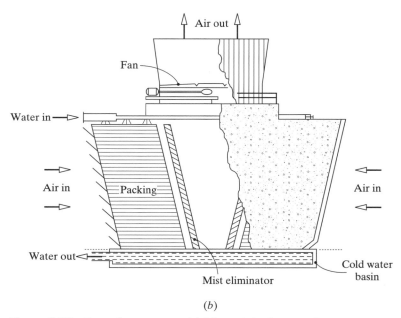

(b)

Figure 3.23 Cross-flow towers. (a) Natural-draft tower for a power plant. (b) A mechanical-draft tower for an oil refinery.

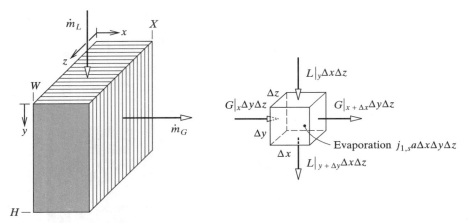

Figure 3.24 Schematic of a cross-flow cooling tower packing showing the coordinate system and mass balances on an elemental volume $\Delta x \Delta y \Delta z$.

The energy equations can be manipulated as was done for the counterflow case in Section 3.5.2, to obtain

$$\frac{\partial h_G}{\partial x} = \frac{g_m a}{G}(h_s - h_G) \tag{3.103}$$

$$\frac{\partial h_L}{\partial y} = -\frac{g_m a}{L}(h_s - h_G) \tag{3.104}$$

Also, as before, we assume

$$h_s = h_s(h_L) \tag{3.105}$$

The required boundary conditions are the inlet conditions of the streams,

$$x = 0: h_G = h_{G,\text{in}} \tag{3.106a}$$

$$y = 0: h_L = h_{L,\text{in}} \tag{3.106b}$$

Equations (3.103) and (3.104) are a pair of first-order partial differential equations in the two dependent variables h_G and h_L, and can be solved numerically. (Exercise 3–54 requires the formulation of a suitable finite-difference solution procedure.) To determine the tower performance, the average enthalpy of the exit liquid must be determined:

$$\bar{h}_{L,\text{out}} = \frac{1}{X} \int_0^X h_{L,\text{out}}\, dx \tag{3.107}$$

Substituting in an exchanger energy balance on the liquid stream then gives the heat transfer as

$$\dot{Q} = \dot{m}_L(h_{L,\text{in}} - \bar{h}_{L,\text{out}}) \tag{3.108}$$

which is the desired result.

3.5.4 Thermal-Hydraulic Design of Cooling Towers

The thermal-hydraulic design of a mechanical-draft cooling tower is relatively straight-forward. The flow rate ratio $\dot{m}_L/\dot{m}_G$ can be specified and varied parametrically to obtain an optimal design, for which the size and cost of the packing is balanced against fan power requirements and operating costs. Data are required for mass transfer conductances and friction for candidate packings. Tables 3.5*a* and 3.5*b* give correlations for a selection of packings. In Table 3.5*b*, the mass transfer conductance is correlated as $g_m a/L$, where a is the transfer area per unit volume and $L = \dot{m}_L/A_{\mathrm{fr}}$ is the superficial mass velocity of the water flow (also called the *water loading* on the packing). Similarly, we define $G = \dot{m}_G/A_{\mathrm{fr}}$. Typical water loadings are 1.8–2.7 kg/m^2 s, and superficial air velocities fall in the range 1.5–4 m/s. No attempt is made to correlate g_m and a separately, as was done for the mass-exchanger packings considered in Section 3.4.2. The number of transfer units of a packing of height H is then

$$N_{\mathrm{tu}} = \frac{g_m S}{\dot{m}_L} = \frac{g_m a H}{L} \qquad (3.109)$$

The correlations are in terms of dimensionless mass velocities L^+ and G^+, and a *hot water correction* T_{HW}^+. The hot water correction accounts for a number of factors, such as errors associated with Merkel's method, deviations from low mass transfer rate theory at higher values of T_s, and fluid property dependence on temperature. Frictional resistance to air flow through the packings is correlated as a *loss coefficient* $N = \Delta P/(\rho V^2/2)$ per unit height or depth of packing, as a function of L^+ and G^+. The velocity V is the superficial gas velocity. No hot water correction is required.

Table 3.5*a* Packings for counterflow and cross-flow cooling towers: Designations and descriptions.

Counterflow Packings
1. Flat asbestos sheets, pitch 4.45 cm
2. Flat asbestos sheets, pitch 3.81 cm
3. Flat asbestos sheets, pitch 3.18 cm
4. Flat asbestos sheets, pitch 2.54 cm
5. 60° angle corrugated plastic, Munters M12060, pitch 1.17 in
6. 60° angle corrugated plastic, Munters M19060, pitch 1.8 in
7. Vertical corrugated plastic, American Tower Plastics Coolfilm, pitch 1.63 in
8. Horizontal plastic screen, American Tower Plastics Cooldrop, pitch 8 in, 2 in grid
9. Horizontal plastic grid, Ecodyne shape 10, pitch 12 in
10. Angled corrugated plastic, Marley MC67, pitch 1.88 in
11. Dimpled sheets, Toschi Asbestos-Free Cement, pitch 0.72 in
12. Vertical plastic honeycomb, Brentwood Industries Accu-Pack, pitch 1.75 in
Cross-Flow Packings
1. Doron V-bar, 4 in × 8 in spacing
2. Doron V-bar, 8 in × 8 in spacing
3. Ecodyne T-bar, 4 in × 8 in spacing
4. Ecodyne T-bar, 8 in × 8 in spacing
5. Wood lath, parallel to air flow, 4 in × 4 in spacing
6. Wood lath, perpendicular to air flow, 4 in × 4 in spacing
7. Marley α-bar, perpendicular to air flow, 16 in × 4 in spacing
8. Marley ladder, perpendicular to air flow, 8 in × 2 in spacing

Table 3.5b Mass transfer and pressure drop correlations for cooling towers. Data from Lowe and Christie [22] for counterflow packings 1 through 4[a]; all other data from EPRI GS–6370 [23].

Correlations (SI units)

Mass transfer: $\dfrac{g_m a}{L\,[\text{kg/m}^2\,\text{s}]} = C_1 (L^+)^{n_1} (G^+)^{n_2} (T_{HW}^+)^{n_3}$

where $L^+ = \dfrac{L}{L_0}$, $G^+ = \dfrac{G}{G_0}$, $T_{HW}^+ = \dfrac{1.8 T_{L,\text{in}}\,[°\text{C}] + 32}{110}$

Pressure drop: $\dfrac{N}{H \text{ or } X} = C_2 (L^+)^{n_4} (G^+)^{n_5}$

Packing Number	C_1, m^{-1}	n_1	n_2	n_3	C_2, m^{-1}	n_4	n_5
Counterflow Packings: $L_0 = G_0 = 3.391$ kg/m^2 s							
1	0.289	−0.70	0.70	0.00	2.72	0.35	−0.35
2	0.361	−0.72	0.72	0.00	3.13	0.42	−0.42
3	0.394	−0.76	0.76	0.00	3.38	0.36	−0.36
4	0.459	−0.73	0.73	0.00	3.87	0.52	−0.36
5	2.723	−0.61	0.50	−0.34	19.22	0.34	0.19
6	1.575	−0.50	0.58	−0.40	9.55	0.31	0.05
7	1.378	−0.49	0.56	−0.35	10.10	0.23	−0.04
8	0.558	−0.38	0.48	−0.54	4.33	0.85	−0.60
9	0.525	−0.26	0.58	−0.45	2.36	1.10	−0.64
10	1.312	−0.60	0.62	−0.60	8.33	0.27	−0.14
11	0.755	−0.51	0.93	−0.52	1.51	0.99	0.04
12	1.476	−0.56	0.60	−0.38	6.27	0.31	0.10
Cross-Flow Packings $L_0 = 8.135$ kg/m^2 s, $G_0 = 2.715$ kg/m^2 s							
1	0.161	−0.58	0.52	−0.44	1.44	0.66	−0.73
2	0.171	−0.34	0.32	−0.43	1.97	0.72	−0.82
3	0.184	−0.51	0.28	−0.31	1.38	1.30	0.22
4	0.167	−0.48	0.20	−0.29	1.25	0.89	0.07
5	0.171	−0.58	0.28	−0.29	3.18	0.76	−0.80
6	0.217	−0.51	0.47	−0.34	4.49	0.71	−0.59
7	0.213	−0.41	0.50	−0.42	3.44	0.71	−0.85
8	0.233	−0.45	0.45	−0.48	4.82	0.59	0.16

[a] For packings 1–4, CTOWER calculates pressure drop by interpolation in the original data (which are very sparse). C_2, n_4, and n_5 were obtained from an approximate curve fit of the data, and are not used in CTOWER.

In a natural-draft tower, the thermal and hydraulic performance of the tower are coupled, and the flow rate ratio $\dot{m}_L/\dot{m}_G$ ($= L/G$) cannot be specified a priori. The buoyancy force producing the air flow depends on the state of the air leaving the packing, which in turn depends on L/G and the inlet air and water states. An iterative solution is required to find the operating point of the tower. The buoyancy force available to overcome the shell and packing pressure drops is

$$\Delta P^B = g(\rho_a - \rho_{G,\text{out}})H \tag{3.110}$$

where ρ_a is the ambient air density, and H is usually taken as the distance from the bottom of the packing to the top of the shell. The various pressure drops are conveniently expressed as

$$\Delta P_i = N_i \frac{\rho_{Gi} V_i^2}{2} \tag{3.111}$$

where N_i is the loss coefficient and V_i is the air velocity at the corresponding location. The pressure drops are associated with the shell, the packing, the mist eliminators, supports and pipes, and the water spray below the packing. Some sample correlations are given in Table 3.6.

Water loadings in natural-draft towers typically range from 0.8 to 2.4 kg/m^2 s, and superficial air velocities range from 1 to 2 m/s. The ratio of base diameter to height may be 0.75 to 0.85, and the ratio of throat to base diameter 0.55 to 0.65. The height of the air inlet is usually 0.10 to 0.12 times the base diameter to facilitate air flow into the tower. When a counterflow packing is used, the air flow distribution is not very uniform; for this reason, cross-flow packings are becoming more widely used. However, the assumption of uniform air and water flows in our model of counterflow packing is adequate for most design purposes.

Table 3.6 Pressure drop correlations for cooling tower shells, sprays, supports, and mist eliminators. N is the loss coefficient defined by Eq. (3.111), with velocity based on cross-sectional area for air flow underneath the packing in items 1–4.

1. Shell (natural-draft counterflow) [22]:

$$N = 0.167 \left(\frac{D_B}{b}\right)^2$$

where D_B is the diameter of the shell base and b is the height of the air inlet
2. Spray (natural-draft counterflow) [24]:

$$N = 0.526(Z_p \text{ [m]} + 1.22)(\dot{m}_L/\dot{m}_G)^{1.32}$$

3. Mist eliminators:

$$N = 2 - 4(N = 3 \text{ is the nominal value in CTOWER})$$

4. Support columns, pipes, etc. (natural-draft counterflow) [24]:

$$N = 2 - 6(N = 4 \text{ is the nominal value in CTOWER})$$

5. Fan exit losses for mechanical-draft towers (velocity based on fan exit area)

$$N = 1.0, \text{ forced draft}$$

$$\simeq 0.5, \text{ induced draft, depending on diffuser design}$$

6. Miscellaneous losses for mechanical-draft towers (velocity based on packing cross-sectional area):

$$N \simeq 3$$

($N = 3.5$ is the nominal value in CTOWER to cover items 5 and 6 for mechanical-draft counterflow and cross-flow towers)

The Computer Program CTOWER

CTOWER calculates the thermal-hydraulic performance of cooling towers. The options are:

1. Counterflow, mechanical draft or natural draft.
2. Cross-flow, mechanical draft.

The number of transfer units is calculated using Merkel's assumptions and numerical integration. The property data are taken from the ASHRAE Handbook of Fundamentals [25]. Packing performance data are taken from EPRI GS–6370 [23] and Lowe and Christie [22], as given in Table 3.5. Other pressure drops are calculated from the correlations in Table 3.6. For the counterflow towers, the "design" problem is solved; that is, the tower duty is specified, and the required height of packing is calculated. For cross-flow towers, the "rating" problem is solved; that is, for given inlet stream conditions and packing dimensions, the outlet stream conditions are calculated.

CTOWER can be used to perform parametric studies and so help the student understand the response of cooling towers to environmental, duty, and design changes. However, CTOWER should not be used without some thought given to important characteristics of cooling tower behavior. For this purpose, it is useful to consider a graphical representation of Merkel's theory for a counterflow tower. Figure 3.25 shows a chart with moist air enthalpy plotted versus water enthalpy (or, equivalently, water temperature) at 1 atm pressure. The *saturation curve* $h_s(h_u)$, or, equivalently, $h_s(T_s)$, is the enthalpy of saturated air tabulated in the last column of Table A.31. The *operating lines* $h_G(h_L)$ are given by Eq. (3.94) and relate the air enthalpy to the water enthalpy at each location in the packing. The slope of an operating line is L/G. Since the assumption $T_s = T_L$ is made in Merkel's method, vertical lines on the chart connect h_s and h_G at each location in the packing. The driving force for the enthalpy transfer, $(h_s - h_G)$, is the vertical distance between the saturation curve and the operating line. The integral in Eq. (3.96) averages the reciprocal of this distance. Using this chart, a number of observations about cooling tower behavior can be made.

1. Figure 3.25 shows the effect of L/G for fixed water inlet and outlet temperatures, and fixed inlet air temperature and humidity. If we imagine L to be fixed as well, we see that as G decreases, the driving forces decrease, and so a larger NTU is required.
2. The minimum NTU required corresponds to $L/G = 0$, that is, an infinite air flow rate, for which the operating line is horizontal.
3. Due to the curvature of the operating line, it is possible for the operating line to be tangent to the saturation curve. The indicated NTU is then infinite, which tells us that the air flow rate must be increased in order to achieve the desired water cooling range.
4. For a mechanical-draft tower, the optimal value of L/G lies between the two limits described in items 2 and 3 above. If L/G is large, the required height

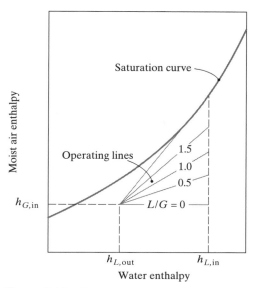

Figure 3.25 Counterflow cooling tower operating lines for various water-to-air flow-rate ratios shown on an enthalpy chart.

of packing is large, and the capital cost will be excessive. If L/G is small, the required height of packing is small, and the capital cost will be small, but the fan power will be excessive (since fan power is proportional to air volume flow rate times pressure drop). These features are illustrated in Example 3.12.

5. Figure 3.26a shows an interesting situation, in which water inlet and outlet temperatures, the air inlet condition, and both flow rates are specified. The exchanger energy balance, Eq. (3.74), could show that the outlet air state is thermodynamically possible. However, the operating line must cross the saturation curve to join the inlet and outlet states, so that the mass transfer process required is impossible. Because of the curvature of the saturation line, thermodynamically admissible end states may not be physically attainable.

6. Figure 3.26b shows how a nonnegligible water-side heat transfer resistance is represented on the chart. Since the energy transferred to the air stream equals the rate at which heat is transferred from the bulk water to the interface,

$$g_m(h_s - h_G) \simeq h_{cL}(T_L - T_s) = g_{hL}(h_L - h_u)$$

if the specific heat of the water c_{pL} is taken to be constant. Thus, the slope of a *tie line* joining corresponding locations on the operating line and saturation curve is

$$\frac{h_s - h_G}{h_u - h_L} = -\frac{g_{hL}}{g_m} \tag{3.112}$$

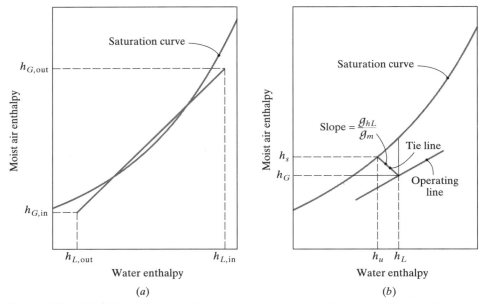

Figure 3.26 (a) Thermodynamically admissable end states for counterflow cooling tower operation that are not physically attainable. (b) The tie line for counterflow cooling tower operation when the water-side heat transfer resistance is not negligible.

For g_{hL} very large, the tie line is vertical, and $h_u = h_L$; that is, $T_s = T_L$, as assumed in Merkel's method. If the process is liquid-side controlled—that is, g_m is very large—then the tie lines are horizontal. It is easy to see how a graphical method could be devised to obtain the NTUs without assuming $T_s = T_L$. It would involve dividing the h_L range into increments and drawing tie lines of the correct slope on the chart to obtain $(h_s - h_G)$ for use in Eq. (3.96). Exercise 3–48 requires the development of a numerical method for this purpose.

Range and Approach

Cooling tower designers and utility engineers have traditionally used two temperature differences to characterize cooling tower operation. The *range* is the difference between the water inlet and outlet temperatures (also called simply the hot and cold water temperatures). The *approach* is the difference between the outlet water temperature and the wet-bulb temperature of the entering (ambient) air. The approach characterizes cooling tower performance; for a given inlet condition, a larger packing will produce a smaller approach to the ambient wet-bulb temperature, and hence a lower water outlet temperature. The approach concept is useful because the ambient dry-bulb temperature has little effect on performance at usual operating conditions (for a specified wet-bulb temperature). Exercise 3–49 is a check of this feature of cooling tower performance.

Cooling Demand Curves

Electric utility engineers have found it convenient to use charts of *cooling demand curves* to evaluate packing specifications. Figure 3.27 is an example of such a chart, on which the required NTU, for a given inlet air wet-bulb temperature and range, is plotted versus L/G with the approach as a parameter. Such a plot is possible since the inlet air dry-bulb temperature has only a small effect under usual operating conditions. Now, if it is possible to correlate the mass transfer conductance as

$$\frac{g_m a}{L} = C\left(\frac{L}{G}\right)^{-n} \tag{3.113}$$

the NTU of a packing of height H is

$$\frac{g_m S}{\dot{m}_L} = \frac{g_m a H}{L} = C\left(\frac{L}{G}\right)^{-n} H \tag{3.114}$$

Then Eq. (3.114) can also be plotted on the chart to give the *packing capability line*. For a required approach, the *operating point* of the tower is the intersection of the cooling demand curve and packing capability line. Charts of cooling demand curves have been prepared by the Cooling Tower Institute [26] and by Kelly [27].

Correlations for g_m in the form of Eq. (3.113) do not necessarily fit experimental data well. A dependence $g_m a \propto L^{1-n} G^n$ is implied and, in the past, experimental data were often forced to fit such a relation. The correlations for the flat asbestos sheet counterflow packings, items 1–4 in Table 3.5, are examples. An examination of the correlations for the remaining counterflow packings in Table 3.5 show some large deviations: for example, packing 11 gives $n_1 + 1 = -0.51 + 1 = 0.49$, which is much different than $n_2 = 0.93$. (However, $n_2 = 0.93$ is abnormally high and possibly the

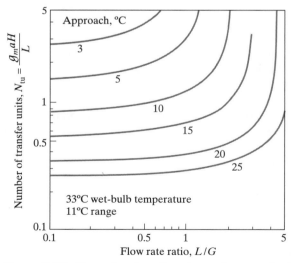

Figure 3.27 Example of cooling demand curves for a specified wet-bulb temperature and range: NTU versus flow rate ratio for a fixed approach.

experimental data are unreliable. If the $g_m a$ correlation does not have the form of Eq. (3.113), the NTU cannot be plotted as a line on a cooling demand chart.

With the almost universal use of computers and the availability of suitable computer programs, one can expect to see less use of cooling demand charts in the future. Other than CTOWER, computer programs in current use include ESC [28], TEFRI [29], FACTS [30], VERA2D [31], and STAR [32]. These computer programs are based on mathematical models of varying degrees of complexity. Both ESC and TEFRI have one-dimensional models similar to those analyzed in Sections 3.5.2 and 3.5.3, and used by CTOWER. (Even for cross-flow the model can be viewed as one-dimensional since no transverse mixing of flow is allowed.) VERA2D models the water flow as one-dimensional, but allows for two-dimensional air flow: an appropriate form of the momentum conservation equation is solved to determine the flow pattern in the cooling tower. STAR solves complete two-dimensional conservation equations but is a research tool rather than a design tool. Minor differences in the heat and mass transfer models used by the programs are of little consequence. The major sources of error in the predictions made by these programs are related to nonuniform air and water flow, and the correlations of packing mass transfer and pressure drop experimental data. The experimental data are obtained in small-scale test rigs, in which it is impossible to simulate many features of full-size towers—for example, nonuniform flow due to entrance configuration, nonuniform wetting of the packing, and, in the case of counterflow towers, the effect of spray above the packing and rain below the packing. Furthermore, since testing of packings in small-scale test rigs is itself not easy, considerable scatter is seen in such test data. Correlations of the data typically have root mean square errors of 10–20%.

Notwithstanding the limitations of the computer programs described above, it is certain that cooling demand curve charts will soon become obsolete and be replaced by interactive computer software, of which CTOWER is a modest example.

Legionnaires' Disease

Legionnaires' disease is a form of pneumonia caused by a strain of *legionnella* bacteria (sero group I). Smokers and sick people are particularly vulnerable to the disease. Major outbreaks have occurred at conventions and in hospitals, for which the source of the bacteria has been traced to cooling towers of air-conditioning systems. The bacteria require nutrients such as algae or dead bacteria in sludge, and thrive if iron oxides are present. However, properly designed, installed, and maintained cooling towers have never been implicated in an outbreak of the disease. Key requirements to be met include the following:

1. Mist (drift) eliminators should be effective.
2. The tower should be located so as to minimize the possibility of mist entering a ventilation system.
3. Corrosion in the tower and water lines should be minimized by use of glass fiber, stainless steel, and coated steel.
4. The design should facilitate inspection and cleaning, to allow early detection and remedy of sludge buildup.

5. Water treatment and filtration procedures should meet recommended standards.

Example 3.11   A Natural-Draft Cooling Tower for a Power Plant

A hyperbolic-shell, natural-draft counterflow cooling tower has the following specifications:

Tower height $= 125$ m

Packing diameter $= 85$ m

Water loading on the packing $= 1.75$ kg/m^2 s

Height of packing above basin $= 11$ m

Height of air inlet $= 10$ m

Water inlet temperature $= 45°$C

Cooling range $= 17°$C for ambient air at 5°C, 35% RH, 1000 mbar

Determine the packing height as a function of packing pitch for flat asbestos sheet packing.

Solution

Given: A natural-draft counterflow cooling tower.
Required: Packing height as a function of packing pitch for flat asbestos sheet packing.
Assumptions:

1. Merkel's method for calculating NTU is adequate.
2. The conductance and pressure drop correlations in CTOWER are adequate.

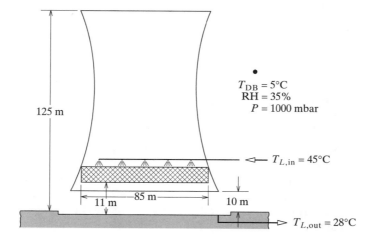

$T_{DB} = 5°C$
$RH = 35\%$
$P = 1000$ mbar

$T_{L,in} = 45°C$

125 m

11 m   85 m   10 m

$T_{L,out} = 28°C$

CTOWER will be used to perform the computations. A water loading of $1.75 \, \text{kg/m}^2 \, \text{s}$ on an 85 m–diameter packing gives a water flow rate of $\dot{m}_L = (1.75)(\pi/4)(85)^2 = 9930 \, \text{kg/s}$. The required inputs are:

> Packings 1, 2, 3, and 4
>
> Height of cooling tower $= 125 \, \text{m}$
>
> Base diameter $= 85 \, \text{m}$
>
> Height of air inlet $= 10 \, \text{m}$
>
> Height of spray effect $= 11 \, \text{m}$
>
> Flow rate of water $= 9930 \, \text{kg/s}$
>
> Inlet temperature of water $= 45°\text{C}$
>
> Outlet temperature of water $= 28°\text{C}$
>
> Barometric pressure $= 1000 \, \text{mbar}$
>
> Dry-bulb temperature $= 5°\text{C}$
>
> Relative humidity $= 35\%$

The required results are summarized below.

Packing Number	Pitch, p cm	N_{tu}	Height, H m	H/p
1	4.45	0.683	1.633	36.7
2	3.87	0.682	1.310	33.9
3	3.18	0.683	1.180	37.1
4	2.54	0.682	1.024	40.3

Comments

1. The packing height H decreases as the pitch decreases because the contact area increases linearly with decreasing pitch.
2. The cost of the packing is proportional to the area of the plates but will also increase somewhat with decreasing pitch (due to more fastenings, etc.). The area of the plates is proportional to H/p, which is seen to be relatively constant. The irregular behavior is probably due to a lack of precision in the mass transfer conductance correlations.
3. It appears that secondary factors will dictate the choice of packing pitch. ▲

Example 3.12   A Mechanical-Draft Cooling Tower for an Oil Refinery

A 5 m–square mechanical-draft counterflow cooling tower is required to cool 45 kg/s of water from $40°\text{C}$ to $20°\text{C}$ when the ambient air is at $10°\text{C}$, 1 atm, and 35% relative humidity. The packing under consideration is vertical corrugated plastic at 1.63 in pitch, manufactured by American Tower. As a function of air flow rate, determine the required packing height and fan power. Use a fan mechanical efficiency of 75%.

Solution

Given: A mechanical-draft counterflow cooling tower.
Required: Packing height and packing fan power as a function of air flow rate for a corrugated plastic packing.
Assumptions:
   1. Merkel's method for calculating the NTU is adequate.
   2. The packing correlations in CTOWER based on EPRI GS–6370 are adequate.

CTOWER will be used to perform the computations. The packing is listed as item 7 on the packing menu. The air flow rate is varied from 30 kg/s to 90 kg/s and the results tabulated. The fan power is equal to the air volume flow rate times pressure drop divided by fan efficiency. Taking $\rho_{\text{air}} \simeq 1.25$ kg/m^3 gives power $\simeq 1.067 \dot{m}_G \Delta P$.

$\dot{m}_G$		$T_{G,\text{out}}$	H	ΔP	Power
kg/s	N_{tu}	°C	m	Pa	W
30	4.35	37.7	4.06	24.7	791
40	2.27	32.6	1.80	21.2	905
50	1.84	28.9	1.29	25.1	1340
60	1.65	26.1	1.05	30.5	1950
70	1.54	23.8	0.90	36.8	2750
80	1.47	22.0	0.79	44.0	3760
90	1.42	20.5	0.72	51.8	4970

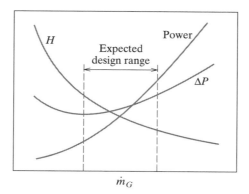

Comments

   1. The required packing height steadily decreases as the air flow rate increases.
   2. The pressure drop first decreases and then increases with air flow rate, owing to the competing effects of air velocity and packing height.
   3. Fan power increases with air flow rate, with the rate of increase becoming rapidly larger at high air flow rates.

4. The design range will reflect a trade-off between packing cost (proportional to height) and operating cost (of which an important component is the fan power).

5. A constant air density of 1.25 kg/m^3 was used to calculate the fan power. In fact, $\rho_{out} \simeq \rho_{sat}(T_{G,out})$ varies with air flow rate, but a constant value is adequate for the purposes of this example. ▲

Example 3.13   A Cross-Flow Cooling Tower for an Air-Conditioning System

A large building is to have a new air-conditioning system installed. Since the existing cooling tower is in good working order, it is planned to retain the tower. Performance data for the tower are to be made available to the contractor. The packing is Doron V-bar with a 4 in by 8 in spacing, and is 5 m high, 20 m wide, and 2 m deep. The nominal water and air flow rates are both 100 kg/s. Performance data are required for a water inlet temperature of 38°C and a nominal air inlet condition of 1 atm, 28°C, and 60% RH. Determine the effect of air temperature and humidity on the outlet water temperature.

Solution

Given: A cross-flow mechanical-draft cooling tower.
Required: Effect of inlet air temperature and humidity on outlet water temperature.
Assumptions:

1. Both streams are unmixed.
2. Merkel's approximation is valid.
3. The conductance correlations in CTOWER (based on EPRI GS–6370) are adequate.

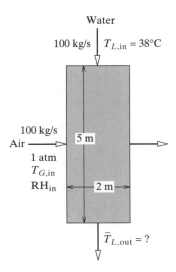

CTOWER can be used for the calculations. The required input is:

> Packing 1
>
> Height of packing $= 5$ m
>
> Width of packing $= 20$ m
>
> Depth of packing $= 2$ m
>
> Flow rate of air $= 100$ kg/s
>
> Flow rate of water $= 100$ kg/s
>
> Inlet temperature of water $= 38.0°$C
>
> Barometric pressure $= 1013$ mbar
>
> Dry-bulb temperature $= 28°$C (nominal value)
>
> Relative humidity $= 60\%$ (nominal value)

The number of transfer units is $N_{tu} = g_m S / \dot{m}_L = 0.396$, and the average outlet water temperatures are tabulated below for a range of inlet air conditions.

	$T_{G,\text{in}}$		
RH_{in}	$25°$C	$28°$C	$31°$C
30%	27.2	27.8	28.5
60%	28.8	29.7	30.8
90%	30.4	31.6	33.0

Comments

1. The cooling tower is but one unit in the air-conditioning system. These data allow the efficiency of the system as a whole to be determined.
2. Study the CTOWER output to see the behavior of the outlet air temperature.
3. The pressure drop for this packing is very small; the fan exit and miscellaneous losses are substantially greater.   ▲

3.6 CLOSURE

A great variety of mass exchangers are used in engineering practice. For purposes of analysis, we chose to differentiate single-stream and two-stream steady-flow exchangers. In addition, we differentiated exchangers that transfer mass only and exchangers that transfer both heat and mass. The formula for the effectiveness of a single-stream mass exchanger was found to be identical to that for a single-stream heat exchanger, irrespective of the nature of the physical transfer process—catalytic conversion, scrubbing, precipitating, filtering, or humidifying. In order to analyze two-stream mass exchangers, it was necessary to define an overall mole transfer conductance that properly accounted for a concentration discontinuity at the liquid-gas interface. Special attention was given the role played by the Henry number (or a more general solubility relation) in determining whether the mass transfer processes were gas-side or liquid-side controlled. A special class of two-stream mass exchangers was analyzed, namely, gas scrubbers with gas absorption described by a linear solubility law. An analytical result for the effectiveness of a countercurrent unit was obtained and was of identical form to the corresponding result for a two-stream heat exchanger.

The only example of a two-stream simultaneous heat and mass exchanger considered was the cooling tower, in both counterflow and cross-flow configurations. The analysis was simplified using Merkel's approximation, but, nevertheless, numerical methods were required to determine the final solution. For the counterflow configuration, the "design" problem was solved; that is, the required water inlet temperature and cooling range was specified, and the number of transfer units required to do the task was calculated. For the cross-flow configuration it was more convenient to solve the "rating" problem; that is, the packing and its dimensions were specified, and outlet water and gas stream temperatures were calculated.

Three computer programs accompany Chapter 3: FILTER, SCRUB, and CTOWER. These programs should prove useful for making parametric studies of filter, gas scrubber, and cooling tower performance. These programs are primarily intended as educational tools but may also find use for preliminary design.

The exchangers considered in Chapter 3 could all be analyzed using low mass transfer rate theory, with the analysis of the mass and heat transfer processing following the procedures described in Sections 1.4 and 1.5. An important example of a simultaneous heat and mass exchanger that cannot be analyzed using low mass transfer rate theory is the power plant condenser. As shown in Example 2.9, a small amount of noncondensable gas in steam markedly reduces the condensation rate. In Example 2.9, condensation on a single tube was considered, whereas a typical condenser has many thousands of tubes, with complex steam flow patterns through the tube bank (see Fig. 1.20 of *Heat Transfer*). Reliable analysis of such condensers requires numerical solution of multidimensional governing equations, and has yet to be successfully accomplished.

REFERENCES

1. Davies, C. N., "Definitive equations for the fluid resistance of spheres," *Proc. Roy. Soc. London,* ser. A, 57, 259–269 (1945).

2. White, H. J., *Industrial Electrostatic Precipitation*, Addison-Wesley, Reading, Mass. (1963).

3. Smith, W. B., and MacDonald, J. R., "Development of a theory for charging of particles by unipolar ions," *J. Aerosol Sci.*, 7, 151–166 (1976).

4. Tsai, R., and Mills, A. F., "Modeling of electrostatic precipitators," *PHOENICS Journal*, 785–810 (1990).

5. Kihm, K., Mitchner, M., and Self, S. A., "Comparison of wire-plate and plate-plate electrostatic precipitators in turbulent flow," *J. Electrostatics*, 19, 21–32 (1987).

6. Fernandez de la Mora, J., "Inertia and interception in the deposition of particles from boundary layers," *Aerosol Sci. Technol.*, 5, 261–266 (1986).

7. Pich, J., "Gas filtration theory," Chap. 1 in *Filtration: Principles and Practice*, eds. M. J. Matteson and C. Orr., Marcel Dekker, New York (1987).

8. Lee, K. W., and Liu, B. Y. H., "Theoretical study of aerosol filtration by fibrous filters," *Aerosol Sci. Technol.*, 1, 147–161 (1982).

9. Onda, K., Takeuchi, H., and Okumoto, Y., "Mass transfer coefficients between gas and liquid phases in packed columns," *J. Chem. Eng. Jpn.*, 1, 56–62 (1968).

10. Perry, R. H., and Chilton, C. H., *Chemical Engineers Handbook*, 5th ed., pp. 18–35–18–39, McGraw-Hill, New York (1973).

11. Foust, A. S., Wenzel, L. A., Clump, C. W., Maus, L., and Anderson, L. B., *Principles of Unit Operations*, 2d ed., John Wiley & Sons, New York (1980).

12. Treybal, R. E., *Mass-Transfer Operations*, 3d ed., McGraw-Hill, New York (1980).

13. Emmert, R. E., and Pigford, R. L., "Interface resistance study of gas absorption in falling liquid films," *Chem. Eng. Prog.*, 50, 87–93 (1954).

14. Won, Y. S., and Mills, A. F., "Correlation of the effects of viscosity and surface tension on gas absorption rates into freely falling turbulent liquid films," *Int. J. Heat Mass Transfer*, 25, 223–229 (1982).

15. W. Brotz, "Über die vorausberechnung der absorptions geschwindigkeit von gasen in strämen der Füssigkeitsschichten," *Chem. Eng. Tech.*, 26, 470–478 (1954).

16. Eckert, J. S., "How tower packings behave," *Chem. Eng.*, 82, 70–76, April 14 (1975).

17. Bharathan, D., and Wallis, G. B., "Air-water countercurrent annular flow," *Int. J. Multiphase Flow*, 9, 349–366 (1983).

18. Tien, C. L., Chung, K. S., and Liu, C. P., "Flooding in two-phase countercurrent flows: I. Analytical modeling," *PhysicoChemical Hydrodynamics*, 1, 195–207 (1980).

19. Chung, K. S., Liu, C. P., and Tien, C. L., "Flooding in two-phase countercurrent flows: II. Experimental investigations," *PhysicoChemical Hydrodynamics*, 1, 209–220 (1980).

20. Merkel, F., "Verdunstungskühlung," *Forschungsarb. Ing. Wes.*, no. 275 (1925).

21. Spalding, D. B., *Convective Mass Transfer*, McGraw-Hill, New York (1963).

22. Lowe, H. J., and Christie, D. G., "Heat transfer and pressure drop data on cooling tower packings, and model studies of the resistance of natural draft towers to airflow," Paper 113, *International Developments in Heat Transfer*, Proceedings of the International Heat Transfer Conference, Boulder, Colo., August 1961, ASME, New York.

23. Johnson, B. M., ed., *Cooling Tower Performance Prediction and Improvement*, vols. 1 and 2, EPRI GS–6370, Electric Power Research Institute, Palo Alto, Calif. (1990).

24. Singham, J. R., "Natural draft towers," in Hewitt, G. E., coord. ed., *Hemisphere Handbook of Heat Exchanger Design*, Sec. 3.12.3, Hemisphere, New York (1990).

25. *ASHRAE Handbook: 1989 Fundamentals, ASHRAE*, American Society of Heating, Refrigerating, and Air Conditioning Engineers, Atlanta (1989).

26. *Cooling Tower Performance Curves*, Cooling Tower Institute, Houston (1967).

27. Kelly, N. W., *Kelly's Handbook of Cross-flow Cooling Tower Performance*, Neil W. Kelly and Associates, Kansas City, Missouri (1976).

28. Baker, D., *Cooling Tower Performance*, Chemical Publishing Co., New York (1984).

29. Bourillot, C., *TEFRI: Numerical Model for Calculating the Performance of an Evaporative Cooling Tower*, EPRI CS–3212-SR, Electric Power Research Institute, Palo Alto, Calif. (1983).

30. Benton, D. J., *A Numerical Simulation of Heat Transfer in Evaporative Cooling Towers*, WR28–1–900–110, Tennessee Valley Authority (1983).

31. Majumdar, A. K., Singhal, A. K., and Spalding, D. B., *VERA2D–84: A Computer Program for 2-D Analysis of Flow, Heat and Mass Transfer in Evaporative Cooling Towers*, EPRI CS–4073, Electric Power Research Institute, Palo Alto, Calif. (1985).

32. Caytan, Y., "Validation of the two-dimensional numerical model STAR developed for cooling tower design," Proceedings of the 3rd Cooling Tower Workshop, International Association for Hydraulic Research, Budapest, Hungary (1982).

33. Levich, V. G., *Physicochemical Hydrodynamics*, Prentice–Hall, Englewood Cliffs, N.J. (1962).

34. Haberman, W. L., and Morton, R. K., *Experimental Investigation of the Drag and Shape of Air Bubbles Rising in Various Liquids*, U.S. Department of the Navy, Report No. 802, Washington, D.C. (1953).

35. Oldenkamp, R. D., and Katz, B., "Integration of molten carbonate process for control of sulfur oxide emissions into a power plant," ASME Paper 69-WA/APC–6, Winter Annual Meeting, Nov. 16–29 (1969).

EXERCISES

3-1 Propane (C_3H_8) can be used as a fuel in internal combustion engines. Consider an automobile that gets 8 km/liter of liquid propane when cruising at 90 km/h.

  i. Write down the main combustion reaction.

  ii. Calculate the rates at which fuel and oxygen are consumed and at which water vapor and carbon dioxide are produced, and the nitrogen throughput.

  iii. Calculate the rate at which nitric oxide is produced, assuming an emissions concentration of 600 ppm for an engine operating at the stoichiometric air/fuel ratio. Take the specific gravity of propane as 0.5.

3-2 In Section 3.2.4 we obtained an overall balance for an urban air volume under conditions in which pure air flowed into the system at a rate $\dot{m}$, absorbed and mixed with a pollutant species 1 that was generated within the system at a rate $\dot{r}_1$, and then flowed out of the considered volume with a pollutant concentration given by

$$(\dot{m}m_1)_{\text{out}} = (\dot{m}m_1)_{\text{in}} + \dot{r}_1$$

or, since $m_{1,\text{in}}$ was presumed to be zero and $\dot{m}$ was essentially constant, $m_{1,\text{out}} = \dot{r}_1/\dot{m}$. Consider now a balance on an urban air volume over a community that is downwind of a major polluter (this frequently is the case when a sea breeze carries Los Angeles smog inland over Pasadena, California). Suppose a layer of air 600 m thick advances on the community at 3 m/s. When the concentration of CO in the incoming air is 7.5 ppm, calculate the maximum rate (per square kilometer) at which CO may be generated within the community if the CO concentration within its air volume is not to exceed 10.0 ppm. (Take the effective depth of the community as 5000 m.)

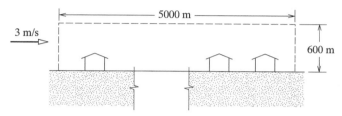

3-3 A demonstration hydrogen-fueled automobile consumes 2.9 kg of fuel per 100 km while traveling at 80 km/h. If the fuel injection system is set to maintain a stoichiometric air/fuel ratio,

   i. write down the combustion reaction.

   ii. calculate the rates at which fuel and oxygen are consumed, the rate of water vapor production, and the nitrogen throughput.

   iii. calculate the allowable mass fraction of NO_x in the exhaust if the automobile is to meet the 1993–94 standards of the U.S. Environmental Protection Agency for California.

3-4 Methanol is a promising alternative fuel for automobiles, since fewer smog-causing pollutants are produced. It is currently more expensive than gasoline, but Catalytica Inc. of Mountain View, California, has discovered a promising process to convert inexpensive natural gas to methanol using a mercury-based catalyst (*New York Times*, January 17, 1993). If the process can be successfully commercialized, increased use of methanol as a fuel is likely. Consider an automobile traveling at 80 km/h consuming 21 liters of methanol per 100 km. If the fuel-injection system is controlled to maintain stoichiometric combustion,

   i. write down the combustion reaction.

   ii. calculate the rates at which fuel and oxygen are consumed, the rates of carbon dioxide and water vapor production, and the nitrogen throughput.

   iii. calculate the allowable mass fraction of NO_x in the exhaust if the automobile is to meet the 1993–94 standards of the U.S. Environmental Protection Agency for California.

The specific gravity of methanol can be taken as 0.796.

3-5 A 1000-MW coal-fired power plant burns a coal that has an elemental composition by mass of 3.8% hydrogen, 81.4% carbon, 1.6% nitrogen, and 3.5% oxygen and that also contains 8% ash and 1.7% sulfur. The overall efficiency of the plant is 38%, and the heat of combustion of the coal is 32.3×10^6 J/kg. Ten percent excess air is used to reduce CO emissions. For the effluent from the boiler unit, estimate

   i. the flow rates of the major components.

   ii. the concentration of sulfur oxides, in parts per million by volume, if design guidelines for power plant boilers suggest oxides production of (38 × % sulfur content) lb/ton coal burned.

   iii. the flow rate of particulates if design guidelines for pulverized coal boiler units suggest particulates production of (16 × % ash content) lb/ton coal burned.

3-6 In an open-cycle ocean thermal-energy conversion pilot plant, warm surface seawater at 300 K enters an upcomer 25 m below sea level and is delivered to the evaporator at a little less than the barometric height ($\simeq 10$ m). As the water flows up the upcomer, air comes out of solution: bubbles originally in the water grow, and new bubbles nucleate on suitable nucleation sites. At a height of $\simeq 9$ m, a primary holding tank is situated and is maintained at a pressure of 10,000 Pa by a vacuum exhaust system. Estimate the composition and volume (m^3/m^3 water) of the vapor-gas mixture vented if the maximum possible amount of air is removed. The warm seawater can be taken to be saturated with air at 1 bar, and dissolved air in seawater can be modeled as a single species with an effective Henry constant of 85,000 bar at 300 K.

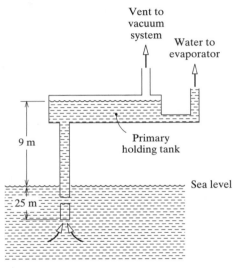

3-7 A clean room for an electronics manufacturing process is of length 20 m, width 10 m, and height 3 m. The air-conditioning system supplies 100 m^3/min of fresh, clean air to the room. At time $t = 0$, a faulty unit commences releasing a contaminant at a rate of 600 μg/min. Assuming the air in the room to be well mixed, determine the concentration of contaminants in the air as a function of time, and its final steady value.

3-8 Determine the surface areas required for the catalytic reactor described in Example 3.1 if

    i. the passage width is 0.7 mm.

    ii. the required effectiveness is 0.999.

    iii. the vehicle gets only 5 km/liter.

Compare your results with the result obtained in Example 3.1, and sketch qualitatively the effects of the three parameters cited above.

3-9 Exhaust gases from an automobile flow through the first stage of a catalytic converter at a rate of 0.04 kg/s. The inlet gases contain 1000 ppm NO and an estimated 5–10% CO content. Catalytic reduction of the NO takes place in 0.4 cm–diameter spherical porous catalyst pellets with an effective rate constant based on s-surface area of $k''_{\text{eff}} = 0.2$ m/s (see Exercise 1–55) packed with a volume void fraction of 0.3. Space considerations suggest a 15 cm outside diameter for the converter. Assuming an operating temperature and pressure of 850 K, 1 atm, and a mean molecular weight for the exhaust gases of 29, determine

    i. the NTU required to reduce 99% of the NO.

    ii. the overall mass transfer coefficient.

iii. the required s-surface transfer area.

iv. the required length of converter.

v. the inlet and outlet NO mass fractions.

vi. Why are the catalyst pellets made porous? What information would you require in order to calculate the effective rate constant?

3-10 A catalytic converter for an automobile contains a matrix formed of porous ceramic coated with a metallic catalyst. The flow passages are approximately 1.5 mm square, and the walls are 0.2 mm thick. At 80 km/h the gasoline consumption is 14.3 liters/100 km, and the exhaust gases are at 800 K and 1.2 bar pressure and contain 0.13% NO by weight. Suggest suitable dimensions for a converter that will reduce the NO mass fraction to 0.008%. The effective rate constant k''_{eff} for the reaction at 800 K is 0.15 m/s, which accounts for the additional catalyst area in the porous ceramic. Take an air/fuel ratio of 14:1.

3-11 Determine the length of a 10 cm–diameter catalytic converter necessary to obtain a 20-fold reduction in CO emission from an automobile engine as it develops 34.0 kW at a specific fuel consumption of 0.2 kg/kW h and an air/fuel ratio of 14:1. The unit is to operate at 850 K, the concentration of CO in the exhaust is 1% by volume, and the exhaust back pressure is 120 kPa. Air is injected into the converter at 10% of the exhaust flow rate. The catalyst is in the form of 0.3 cm–diameter spherical pellets of copper oxide on alumina with 100 m^2/cm^3 catalytic surface area. The pellets are packed to give a volume void fraction of 0.25. Assume a pellet effectiveness of 2.0%.

3-12 A catalytic converter for an automobile has passages of equilateral triangular cross section with 1 mm sides. The porous passage walls are 0.2 mm thick and have a void fraction of 0.75, a tortuosity factor of 5.0, a mean pore radius of 1 μm, and a surface area per unit volume of 6×10^5 cm^2/cm^3. The converter is 10 cm in diameter and 10 cm long. The converter is tested with an exhaust flow rate of 0.060 kg/s at 1 atm and 900 K. The platinum-based catalyst reduces nitric oxide in a reaction that may be approximately modeled as first-order with an activation energy of 17 kcal/mol and a preexponential factor of 70 m/s. Determine the decontamination factor for the converter $[\mathrm{DF} = 1/(1 - \varepsilon)]$.

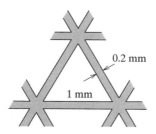

3-13 A catalytic converter for a luxury automobile has a matrix formed of porous ceramic coated with a metallic catalyst. The flow passages are 1.7 mm square, and the walls are 0.25 mm thick. At 90 km/h the gasoline consumption is 20.9 liters/100 km, and the exhaust gas is at 850 K, 1.15 bar, and contains 3.5% CO by weight. Suggest suitable dimensions for a converter that will reduce the CO concentration to 0.1%. The effective rate constant for the reaction at 850 K is $k''_{\mathrm{eff}} = 0.106$ m/s, which accounts for the additional catalyst area in the porous walls. The air/fuel ratio is 13.2:1, with air injected into the reactor at 10% of the exhaust flow rate. Also calculate the pressure drop. Take a fuel density of 705 kg/m^3.

3-14 A catalytic converter for removing CO from automobile engine exhaust is to be fabricated from oxidized copper-clad metal foil wound in spiral form. The design condition is 80 km/h, 20 liters/100 km, and 10% excess air. The diameter of the unit must not exceed 15 cm, and the flow in the converter should be laminar. The converter is to be 98% effective when operating at 1150 K and 1 atm. Determine the required length of the converter, and the pressure drop, as a function of foil pitch. Take the rate constant for CO oxidation as $500 \exp(-9400/T)$ m/s for T in kelvins, and an air/fuel ratio of 15:1.

3-15 A 99.5% effective converter is required for an automobile engine when it develops 36.4 kW at a specific fuel consumption of 0.18 kg/kW h and an air/fuel ratio of 13.4:1. The converter is to operate at 820 K, the concentration of CO in the exhaust is 3% by volume, and the exhaust back pressure is 115 kPa. Air is injected into the converter at 10% of the exhaust flow rate. The catalyst is in the form of 0.25 cm–diameter spherical pellets of copper oxide on alumina with 100 m^2/cm^3 catalytic surface area. The pellets are packaged in a 12 cm–diameter can to give a volume void fraction of 0.27. Determine the length of the packing. Assume a pellet effectiveness of 2.5%.

3-16 In an experimental rig, gas is absorbed from a CO_2 stream into a turbulent water film falling inside a vertical tube. The tube has a 2.05 cm–inner diameter and a height of 1.98 m. For a particular test, the system was maintained at 25°C, and the water flow rate was 5.3×10^{-5} m^3/s. The bulk inlet and outlet CO_2 concentrations in the water were measured as 0.0153 $kmol/m^3$ and 0.0242 $kmol/m^3$, respectively. The CO_2 gas was at 1 atm pressure and can be assumed to be saturated with water vapor. Calculate the average mole transfer conductance $\overline{G}_m$ for mass transfer into the liquid film. Equation (3.46) can be used to calculate the average film thickness.

3-17 In a pilot plant for open-cycle ocean thermal-energy conversion (Claude cycle), warm water enters an upcomer 30 m below sea level and flows at 3 m/s to the evaporator located at the barometric height of approximately 10 m above sea level. The inlet seawater can be assumed to be saturated with air at 300 K and 1 atm pressure and contains air bubbles of radius $R_0 = 20 - 150$ μm. As the bubbles travel through the upcomer, air is absorbed into the water below sea level and subsequently desorbed above sea level. In addition, the bubbles grow due to the decrease in hydrostatic pressure. By performing an analysis on a molar basis, obtain an expression for bubble radius $R(z)$ as a function of height above the upcomer inlet. Hence prepare graphs of $R(z)$ for $20 < R_0 < 150$ μm. Since seawater is relatively unclean, small bubbles tend to behave as solid spheres; thus the Levich expression for the liquid-side mole transfer conductance is appropriate [33],

$$\text{Sh} = (2\text{Pe}/\pi)^{1/2}; \quad \text{Pe} \gg 1$$

where the Sherwood and Peclet numbers have diameter as characteristic length. For very small bubbles Stokes' law, Eq. (4.73) of *Heat Transfer*, can be used to give the velocity of bubbles relative to the seawater. However, for larger bubbles ($R = 2–5$ mm) Stokes' law does not apply, and a constant value of 0.23 m/s is appropriate [34].

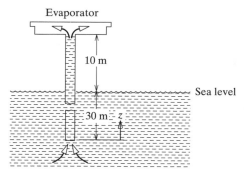

Evaporator

10 m

Sea level

30 m z

3-18 Water at 330 K flows down the inside of a 4 cm–I.D., 2 m–high tube at 223 kg/h. Air at 1 atm pressure flows up the tube at 0.1 m/s and contains 5% carbon monoxide by volume at the inlet. If the inlet water contains no CO, determine the bulk concentration of CO in the outlet water. See Section 3.4.2 for appropriate liquid-side mole transfer conductance correlations.

3-19 A laboratory desorption column consists of a bundle of 15, 2 cm–O.D., thin-wall stainless steel tubes, 0.7 m high, in a glass shell, with the liquid able to flow down both the insides and outsides of the tubes. Water at 300 K containing 0.0238 kmol/m^3 dissolved CO_2 is fed to the tower at a rate of 400 kg/h. If a vacuum pump maintains the pressure inside the tower at 50 mm Hg absolute, calculate the column effectiveness and the concentration of CO_2 in the outlet water. See Section 3.4.2 for appropriate liquid-side mole transfer conductance correlations.

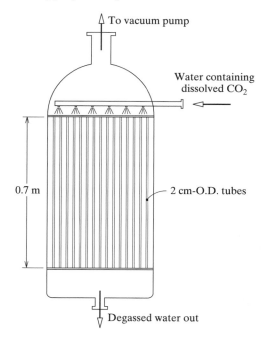

To vacuum pump

Water containing
dissolved CO_2

0.7 m

2 cm-O.D. tubes

Degassed water out

3-20 An electrostatic precipitator has a plate spacing of 0.25 m and is 25 m long. Gas at 1 atm and 300 K flows at a bulk velocity of 2 m/s through the precipitator. Calculate the effectiveness of the unit for removing particles of size

    i. 0.3 μm.

    ii. 1 μm.

    iii. 3 μm.

Use the migration velocity from Fig. 3.6, with $j = 54 \times 10^{-9}$ A/cm^2.

3-21 An electrostatic precipitator was designed to be 98.5% effective for an effective migration velocity of 6 cm/s and a gas velocity of 1.5 m/s. Determine the ineffectiveness of the unit when the following changes are made to the operating conditions:

    i. The migration velocity is decreased to 4 cm/s.

    ii. The migration velocity is increased to 8 cm/s.

    iii. The gas velocity is decreased to 1 m/s.

    iv. The gas velocity is increased to 2 m/s.

3-22 An electrostatic precipitator is to treat a gas stream flowing at 6000 m^3/min. If it is to be 98% effective in removing 0.5 μm size particles, calculate the required plate area, and the number of plates, if plates are available in 3 m lengths and are 10 m high. The operating current density is 60×10^{-9} A/cm^2 at 30 kV. The gas velocity should not exceed 2.0 m/s, and a plate spacing of 0.25 m is appropriate.

3-23 A filter has a cross-sectional area of 1 m^2, has a void fraction of 0.8, and is 0.5 m thick. Air at 290 K and 1 atm containing particles flows through the filter at 20 m^3/min. Prepare a graph of decontamination factor versus fiber diameter ($30 < d_f < 100$ μm) for particle diameters of 0.2, 0.5, and 1.5 μm. Assume a particle density of 4800 kg/m^3 and a fiber density of 1800 kg/m^3.

3-24 A filter containing 25 μm fibers, packed at a density of 380 kg/m^3, is required to remove particulate matter from an air flow at 320 K, 1 atm. The flow rate is 22 m^3/min, and the filter must be accommodated in a 1.3 m–diameter duct. Prepare a graph of decontamination factor versus filter thickness (2 cm $< L <$ 30 cm) for particle diameters of 0.1, 0.5, 1, and 2 μm. Assume a fiber density of 2000 kg/m^3 and a particle density of 5100 kg/m^3.

3-25 A fiberglass filter is required to process 40 m^3/min of air at 310 K and 1.03 bar, and yield a decontamination factor of 15 for 0.5 μm particles of 5500 kg/m^3 density. The packing available has 40 μm–diameter fibers packed at a density of 450 kg/m^3. The air flows in a 1.2 m–square duct, and the filter should fill the duct to minimize pressure drop. Determine the required thickness of filter.

3-26 Rederive Eq. (3.29) by eliminating x_u in Eq. (3.28). Also show that Eq. (3.29) is valid even if the Henry number is not constant—that is, if the relation is nonlinear in form, say $y_s = f(x_u)x_u$.

3-27 Paralleling the derivations in Section 3.4.3, develop the ε–N_{tu} relations for a cocurrent mass exchanger.

3-28 Ammonia gas is to be scrubbed from an air stream by a countercurrent flow of water in a packed column. The column is 1 m in diameter and 3 m high, and is packed with 0.038 m–nominal size ceramic Raschig rings. The gas and liquid flow rates are 2500 and 5000 kg/h, respectively, and the column operates at 310 K and 1 atm. The entering air stream contains 5% ammonia by volume, whereas the entering water is pure.

i. Calculate the column effectiveness and the outlet ammonia concentration.

ii. How much is the outlet ammonia concentration reduced if the column height is doubled?

iii. How much is the outlet ammonia concentration reduced if the outlet gas stream is passed through a second 3 m–high column, supplied with pure water?

3-29 Experiments with a new type of packing in a packed column were performed with dilute NH_3 in air absorbed into water at 300 K and 1 atm. At the test flow rates, $\dot{M}_G/\dot{M}_L = 2$ and the countercurrent effectiveness is measured to be 0.81. Estimate the number of transfer units of the column for the following two situations.

i. Absorption of acetone from a dilute mixture with air into water at 300 K. Use an effective Henry number of 1.2.

ii. Absorption of carbon monoxide from a dilute mixture with air into water at 330 K.

3-30 A falling-film tower has a shell diameter of 2.5 m; contains 900, 4.75 cm–I.D. tubes; and is to be used for scrubbing SO_2 from an air stream. Each tube is to process 50 kg/h of gas in countercurrent flow with 600 kg/h of water. The tower operates at 1 atm pressure, and cooling water circulated in the shell maintains the film temperature at approximately 290 K. If the inlet water is pure and the inlet air contains 3% SO_2 by volume, determine the tower height required to reduce the outlet SO_2 mole fraction to 0.006. Since Henry's law is invalid over this concentration range, use an appropriate numerical integration procedure.

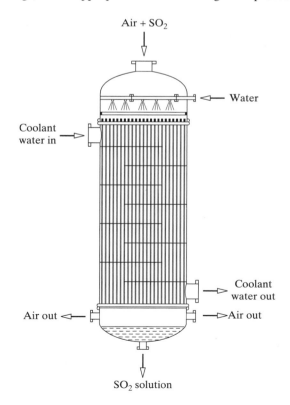

3-31 An existing countercurrent column for recovery of SO_2 from industrial exhaust gas is a packed tower 20 m high. The gas contains 6% SO_2 by volume, and is scrubbed at atmospheric pressure by water at 295 K flowing over the packing at a rate 50% greater than that which would be required by an infinite countercurrent unit to recover 85% of the SO_2. Experimental data indicate that the existing unit recovers 85% of the SO_2 in the entering gas stream. A study is being made of ways to reduce the air pollution still further. One proposal is to build a second, duplicate tower alongside the first and have gas from the top of the first tower enter the top of the second tower and flow cocurrent with the falling water in the new unit. It is proposed to continue the same gas and water flow rates, but the water will enter the top of the new tower, be collected at its base, and be pumped to the top of the existing one. The entering gas will continue to flow into the base of the existing unit and will exit from the base of the new one. Find the outlet SO_2 mole fraction and the percent recovery that would be obtained if the proposal were adopted.

3-32 Reevaluate the performance of the ammonia scrubber in Example 3.7 for cocurrent flow.

3-33 An absorption column is to be designed to recover 99% of the ammonia in an air–ammonia stream by absorption into water. The entering stream contains 3.0% NH_3 by volume. The column is maintained at 290 K by cooling coils, and the pressure is 1 atm.

    i. What is the minimum water flow rate in kmol/kmol of entering gas stream?

    ii. For a water flow rate 50% greater than the minimum, how many transfer units are required for a countercurrent unit?

3-34 Air containing 10% NH_3 by volume is to be scrubbed by water in a countercurrent column 2 m high and with a 1.2 m^2 cross-sectional area. The flow rates of the gas and liquid are 1.0 kg/s and 3.0 kg/s, respectively, and the column operates at 310 K, 1 atm. Calculate the column effectiveness, outlet NH_3 concentration, and pressure drop, for 0.025 m–nominal size

    i. ceramic Raschig rings.

    ii. steel Raschig rings.

    iii. ceramic Berl saddles.

    iv. polyvinylchloride Interlox saddles.

Repeat for a gas flow rate of 2.0 kg/s.

3-35 A mixture of air and 5% inert species by volume is to be scrubbed by water in a 1 m–diameter, 3 m–high countercurrent column packed with 0.025 m–nominal size ceramic Raschig rings. The column operates at 295 K, 1 atm, and the flow rates of gas and liquid are 0.8 kg/s and 2.1 kg/s, respectively. Determine the Henry number, C_{min}, relative resistances to mass transfer in each phase ($1/\mathcal{G}_G$ and $He/\mathcal{G}_L$), the overall mole transfer conductance $\mathcal{G}_G^{oa}$, the number of transfer units N_{tu}, and the column effectiveness ε, when the inert species is

    i. carbon monoxide.

    ii. carbon dioxide.

    iii. hydrogen sulfide.

    iv. sulfur dioxide.

    v. ammonia.

Tabulate your results and comment on significant trends.

3-36 Water at 290 K containing 1.11×10^{-4} kmol/m^3 O_2 is to be oxygenated by a pure oxygen stream at 1.02 atm. An available packed column has a 0.5 m^2 cross-sectional area, is 2.4 m high, and is packed with 0.013 m–nominal size ceramic Raschig rings. Prepare performance maps in which the column effectiveness and the gas-side pressure drop are tabulated as a function of water and gas flow rates. Hence recommend operating parameters that will likely be economical and feasible. (The density and viscosity of the oxygen stream can be approximated by air values in order to use SCRUB.)

3-37 Oldenkamp [35] has described a "dry" scrubbing process to remove SO_2 from flue gas, in which the SO_2 is absorbed into a molten mixture of carbonates of lithium (32%), sodium (33%), and potassium (35%) at 700 K. The absorption step is followed by a complex sequence of operations to recover the carbonates in essentially pure form: here we consider only the design of the SO_2 absorber. The table below gives equilibrium data for the air-SO_2-M_2CO_3 system, where M_2CO_3 designates the mixture of carbonates.

x_u	$He \times 10^4$	$y_s \times 10^4$
0	3.12	0
0.1	3.47	0.347
0.2	3.90	0.784
0.3	4.46	1.34
0.4	5.20	2.08
0.5	6.24	3.12
0.6	7.80	4.68

Notice that the Henry number is much less than unity, suggesting a gas side-controlled mass transfer process; also, He is not constant.

Design a SO_2 absorber to reduce the SO_2 in the flue gas of the oil-fired power plant described in Section 3.2.3, from a mole fraction of 5.2×10^{-4} to 1.5×10^{-4}. The packing should be in the form of closely packed concentric cylinders with the M_2CO_3 liquid films flowing downward, and countercurrent flow of flue gas up through the annular spaces. In order to fit into the stack of the power plant, the exchanger cannot exceed 15 m in diameter. Laminar flow of the flue gas would be desirable to have a low pressure drop. It will be sufficient to use a constant value of the Henry number. The absorber is to operate at 700 K, and the flue gas is at 1 atm pressure.

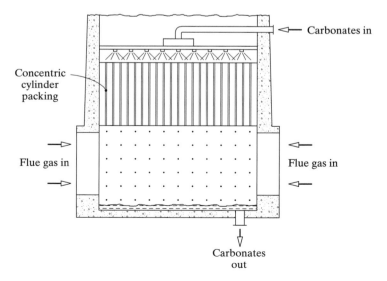

3-38 Redesign the absorber of Exercise 3–37 for cocurrent operation.

3-39 A low-pressure-drop humidifier has a packing of vertical plastic plates with falling water films. The plate center-to-center spacing is 8 mm, and the combined width of one plate plus two liquid films can be taken as 2 mm. A 70 m^3/min flow of air at 1 atm pressure, 300 K, and 30% relative humidity is to be processed. Suggest dimensions for the unit if it is to be 85% effective, have a height equal to twice its width, and have laminar air flow.

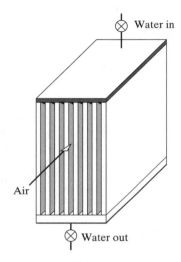

3-40 In an experimental rig, a water film flows down the inside of a 4 cm–I.D. vertical tube, 3 m high. Air flows up the tube at 3 m/s and enters at 318 K, 10% RH, and 1 atm. The water inlet temperature is adjusted to equal the wet-bulb temperature of the inlet air, which is 294 K. The outside of the tube is well insulated. If the mass fraction of water vapor in the outlet air is measured to be 0.0142, determine

i. the effectiveness of the exchanger.

ii. the number of transfer units.

iii. the constant C in the correlation $\overline{Sh}_D = C\,Re_D^{0.8}Sc^{0.4}$ for mass transfer between the water surface and the air stream.

iv. an estimate of the makeup water required for steady operation.

3-41 A cross-flow humidifier has a packing of 1 m–high vertical plates with falling water films. A 100 m^3/min flow of air at 310 K, 1 atm, and 15% RH is to be processed. If an 85% effective unit is required, determine possible dimensions and pressure drop as a function of plate pitch (5 mm $< p <$ 10 mm). The combined width of one plate and two films can be taken as 2 mm.

3-42 An existing evaporative cooler for a solar air-conditioning process has a cross section of 1.0 m × 0.50 m (chosen to match an associated heat exchanger) and has 18 in–thick Munters Humi–Kool packing made out of CELdek material (a cellulose paper impregnated with insoluble anti-rot salts, rigidifying saturants, and wetting agents). The packing has a cross-fluted configuration, which induces good mixing between the water and air. The manufacturer's data for temperature effectiveness, $\varepsilon = (T_{in} - T_{out})/(T_{in} - T_s)$, and air flow pressure drop are given in the figure. Since pressure drop is a key factor in the system design, it has been suggested that a laminar-flow humidifier, with cross-flow of air through a parallel-plate packing that supports vertical falling water films, may prove to be a superior design. Investigate this possibility for a volumetric air flow rate of 56 m³/min at 1 atm, and inlet air at 37°C and 40% RH.

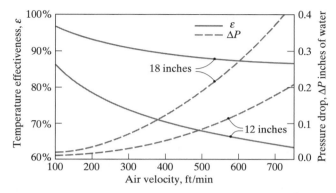

3-43 Show that the liquid-side heat transfer coefficient for a laminar falling film on a cooling tower packing is approximately $k/(\delta/2)$ where δ is the film thickness. Hence, determine the difference betwen T_L and T_s for the following conditions: $\mathrm{Re}_L = 300$, $T_L = 310$ K, $g_{m1} = 0.01$ kg/m² s, $T_G = 295$ K, $P = 1$ atm. [*Hint:* Use the solution of an appropriate transient heat conduction problem.]

3-44 Air, at an average temperature of 310 K and 1 atm, flows countercurrent to water at an average temperature of 320 K in a cooling tower packed with parallel plates 2 mm thick, at a pitch of 15 mm. If the superficial air velocity is 5 m/s and the water flow rate is 20 kg/s per square meter of tower cross-sectional area, determine the pressure drop of the air for a 2 m–high packing. Also check to see that the flooding limit is not exceeded. Use the correlations in Section 3.4.4.

3-45 Perform a hand calculation using Merkel's method to determine the number of transfer units for Example 3.12 when $\dot{m}_G/\dot{m}_L = 1$.

3-46 An operating hyperbolic-shell, natural-draft cooling tower is 110 m high, with an air inlet 11 m high. The packing has diameter 72 m and is 15 m above the catch basin. The packing is 2 m–high flat asbestos sheets at 4.45 cm (1 3/4 inches) pitch. The water has an inlet temperature of 44°C and is loaded onto the packing at 2.1 kg/s per square meter of cross-sectional area. Determine the outlet water temperature for an ambient air condition of $P = 990$ mbar, $T = 16$°C, and RH = 40%.

3-47 A 5 m–square mechanical-draft counterflow cooling tower is required to cool 45 kg/s of water from 40°C to 20°C, when the ambient air is at 1 atm, 10°C, and 35% relative humidity. As a function of air flow rate, determine the required packing height and fan power. The packing is in the form of asbestos sheets at a pitch of 3.81 cm (1.5 in). Use a fan efficiency of 80%.

3-48 In Section 3.5.2 a method was developed for determining the number of transfer units of a counterflow cooling tower. A key assumption was that the liquid-side heat transfer resistance was negligible. Modify the method to account for a finite liquid-side resistance.

3-49 A 5 m–square mechanical-draft counterflow cooling tower is required to cool 45 kg/s of water from 40°C to 20°C with an air flow of 50 kg/s. The packing is vertical corrugated plastic at 1.63 in pitch, manufactured by American Tower. Prepare a map of required packing height versus inlet dry- and wet-bulb temperatures in the range 0–20°C at 1000 mbar pressure. Comment on the significance of the results.

3-50 Use CTOWER to generate a cooling demand curve for the following conditions: $P = 1000$ mbar, $T_{WB,in} = 10°C$, range $= 15°C$, approach $= 5°C$, and an air superficial mass velocity of 1 kg/m^2 s. Show that $T_{DB,in}$ has little effect on the curve. Also show the packing capability line for 6.0 m–high asbestos sheets of pitch 3.18 cm (item 3 of Table 3.5), and hence determine the corresponding operating point.

3-51 A cross-flow cooling tower for an engineering school air-conditioning system is to be retrofitted with a new packing. The space available for the packing in one module of a four-module system is 4 m high, 5 m wide, and 2 m deep. The nominal water and air flow rates are 25 kg/s and 30 kg/s, respectively. For a water inlet temperature of 40°C and an air inlet condition of 1000 mbar, $T_{DB} = 25°C$, $T_{WB} = 23°C$, investigate the performance of the eight cross-flow packings listed in Table 3.5. Tabulate the cooling range achieved and the air flow pressure drop. Comment on the results.

3-52 A new type of cooling tower packing has been tested in a small-scale counterflow test rig containing a 28 cm × 28 cm square cross-section, 0.9 m–high packing element. The following data were measured: $\dot{m}_{G,in} = 0.178$ kg/s; $\dot{m}_{L,in} = 0.550$ kg/s; $T_{DB,in} = 24.3°C$, $T_{WB,in} = 14.5°C$; $T_{WB,out} = 25.2°C$; $T_{L,in} = 28.90°C$, $T_{L,out} = 26.03°C$. Also, $P = 1$ atm.

   i. Perform an energy balance to assess the accuracy of the data.

   ii. Determine the NTU obtained in the test.

   iii. The manufacturer has proposed the following correlation for the mass transfer conductance times specific area product:

$$g_m a = 1.46 L^{0.39} G^{0.5} \text{ kg/m}^3 \text{ s}$$

Compare the measured $g_m a$ with the manufacturer's value.

3-53 A closed-form analytical solution can be obtained for the performance of a counterflow cooling tower if the saturation curve is linearized over the interval of concern. Then we can write $h_s = \xi + \eta h_u$, where ξ and η can be determined from the chord joining $h_{s,in}$ and $h_{s,out}$, or as a tangent to the h_s curve. Derive an expression for the number of transfer units that is similar in form to Eq. (3.63) for a countercurrent mass exchanger.

3-54 Develop a finite-difference numerical scheme to solve Eqs. (3.103) and (3.104) for a cross-flow cooling tower. Supply all necessary auxiliary data so that a computer programmer could write a program similar to the cross-flow option of CTOWER, based on the information you provide.

3-55 Data for the performance of cooling tower packings are often not of high precision. The test rigs are, of necessity, rather complex, and some of the measurements are difficult to make. Tabulated below are sample counterflow test data for a commercial packing 0.915 m high, with a frontal area of 0.093 m^2. Also given are estimated errors in each measurement.

	$\dot{m}_{L,in}$ kg/s	$\dot{m}_{G,in}$ kg/s	$T_{L,in}$ °C	$T_{L,out}$ °C	$T_{DB,in}$ °C	$T_{WB,in}$ °C	P mbar
1	0.232	0.161	31.1	23.8	22.9	16.2	990
2	0.232	0.088	31.1	26.8	23.2	16.3	990
3	0.579	0.136	31.0	28.1	23.6	16.6	990
Error	±2%	±3%	±0.2 K	±0.5 K	±0.2 K	±0.2 K	±5 mbar

The error in $T_{L,\text{out}}$ is large because air- and water-flow maldistribution causes the outlet water temperature to be nonuniform, and an appropriate average value must be obtained. Using CTOWER, propagate the measurement errors to obtain the expected errors in $g_m a/L$. Discuss the trends that can be observed.

3-56 Using the result of Exercise 3–48, rework Example 3.10 accounting for the liquid-side heat transfer resistance. Consider a parallel-plate packing at a pitch of 3.18 cm, with $L = G = 2.5 \, \text{kg/m}^2 \, \text{s}$. Use Eqs. (3.40) and (3.44) for the gas-side and liquid-side resistances, respectively. Evaluate all properties at 300 K and 1 atm, and take the width of the plates plus two films as 6 mm.

3-57 An air-cooled condenser consists of a tube bank located in the air duct with air flow transverse to the tubes, in which a hydrocarbon vapor condenses at pressures corresponding to saturation temperatures in the range 25°C–40°C. In order to increase the performance of the condenser, it is proposed to have a water feed system that maintains a thin film of water on the tubes, and so augment the sensible heat loss with an evaporative heat loss. The unevaporated water is recirculated. If it is reasonable to assume that the water-film, tube-wall, and condensing-side thermal resistances are all negligible, the surface of the water film is then at a constant temperature $T_s = T_{\text{sat}}(P)$ of the condensing hydrocarbon vapor. Use Merkel's assumptions to analyze this single-stream simultaneous heat and mass exchanger, and thereby obtain an enthalpy-based ε-N_{tu} relation. Hence, determine the heat transfer rate, condensation rate, and outlet air composition and temperature.

3-58 An *evaporative condenser* (see Exercise 3–57) has a tube bank with 17 rows of 1 m–long, 15 mm–O.D. tubes in a duct of 1 m × 0.5 m cross section. The tubes are in a staggered arrangement, with both transverse and longitudinal pitches of 25 mm. A water feed system maintains a thin film of water on the surface of the tubes. Air at 10°C, 50% RH, and 1 atm flows along the duct at 1 kg/s. A hydrocarbon vapor condenses inside the tubes at 40°C. Determine the heat transfer rate, vapor condensation rate, outlet air condition, and make-up water supply rate. The enthalpy of vaporization of the hydrocarbon vapor at 40°C is 3.10×10^5 J/kg.

3-59 An air cooler for an air conditioning system has a tube bank with 15 rows of 1 m–long, 15 mm–O.D. tubes in a duct of 1 m × 0.5 m cross section. The tubes are in a staggered arrangement, with both transverse and longitudinal pitches of 25 mm. Air at 40°C, 10% RH, and 1 atm flows into the cooler at 1 kg/s. Refrigerant evaporates in the tubes at 15°C.

  i. Determine the air outlet temperature. Assume that the tube-wall and evaporating-side thermal resistances are negligible.

  ii. A proposal has been made to improve the air cooling by installing a water feed system that maintains a thin water film on the surface of the tubes. As a consulting engineer you have been asked to evaluate this proposal. Will there be greater cooling of the air, or does the proposal show a lack of understanding of the basic principles of evaporative cooling? In making your evaluation, notice that the analysis of the evaporative condenser required in Exercise 3–57 applies to this "evaporative" evaporator as well: also use the results of the analysis of the adiabatic humidifier in Section 3.5.1.

3-60 Moist air at 25°C, 50% RH, and 1 atm is to be cooled in a *cooler-condenser* by flowing it upwards through Munters M12060 packing, onto which is fed water at 10°C. The air flow rate is 0.325 kg/s, $\dot{m}_L/\dot{m}_G = 0.5$, and the packing cross-sectional area is 0.1 m^2. The outlet water is required for another process for which its temperature must be 15.0°C.

  i. Sketch a chart of moist air enthalpy versus water enthalpy showing the saturation curve and operating line.

ii. Calculate the number of transfer units required, and the outlet air temperature.

iii. Calculate the corresponding packing height.

3-61 A counterflow cooler-condenser has 0.4 m^2 cross-sectional area of Munters M12060 packing, onto which is fed water at 10°C. The air flow is 1.3 kg/s and $\dot{m}_L/\dot{m}_G = 0.9$. The air enters at 25°C, 50% RH, and 1 atm, and it is required to cool the air to within 0.75°C of the inlet water temperature. Determine the required packing height.

3-62 A mechanical-draft counterflow cooling tower for a hospital is required to cool 30 kg/s of water from 40°C to 20°C when the ambient air is at 10°C, 35% RH, and 1 atm. Space considerations suggest a packing 4 m square, and a supply of American Tower 1.63 in pitch vertical corrugated plastic packing is available at a good price. In order to restrict noise levels, a maximum inlet air velocity of 1 m/s is allowed.

i. Determine the packing height if a single stage tower is used.

ii. How much is the required height of packing reduced if two 4 m–square towers are used, each ingesting ambient air but with the water cooled in two stages? Choose the intermediate water temperature to give equal packing heights.

iii. Compare the fan power for each case for a fan mechanical efficiency of 75%.

3-63 Nitrogen is to be used as a carrier gas for a solvent in a semiconductor fabrication process. Pure gas at 25°C is bubbled through a pool of solvent maintained at 18°C and exits the pool 99% saturated.

i. Estimate the exit temperature of the gas.

ii. At what rate must the pool be cooled (heated) to maintain a steady temperature of 18°C when the N$_2$ flow rate is 9.358×10^{-3} kg/s. For the solvent take $h_{fg} = 0.53 \times 10^6$ J/kg, Sc $= 1.5$, $M = 50$, $P_{\text{sat}}(18°C) = 3300$ Pa.

3-64 Rework Exercise 3–60 for a water-to-air flow rate of 0.9.

3-65 Large air-cooled heat exchangers are used extensively in the process industries. During periods of high ambient temperature, their performance may become marginal. One method to improve performance during each period is to increase the fan speed; however, a variable-speed fan motor is required, and there is an increased cost in fan power. An alternative method is to adiabatically precool the air by injecting a spray of water droplets into the inlet air. Rotary atomizers have have been developed for this purpose and can produce 60 μm–diameter droplets with an acceptable power consumption. It is important to ensure that the droplets completely evaporate before the air contacts the finned heat transfer surfaces to avoid corrosion problems. For inlet air conditions of $P = 882$ mbar, $T_{DB} = 36°C$, $T_{WB} = 18°C$, and an air approach velocity of 3 m/s, determine the length of precooler section required to have a 90% effective evaporator with complete evaporation of 60 μm droplets.

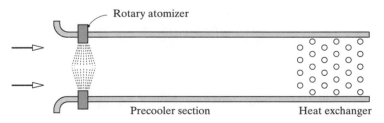

Property Data

LIST OF TABLES

Table A.1a Solid metals: Melting point and thermal properties at 300 K

Metal (% composition)	T_{MP} K	ρ kg/m^3	c J/kg K	k W/m K	α^a m^2/s $\times 10^6$
Aluminum					
Pure	933	2702	903	237	97.1
Duralumin	775	2770	875	174	71.8
(4.4 Cu, 1.0 Mg, 0.75 Mn, 0.4 Si)					
Alloy 195, cast		2790	883	168	68.1
(4.5 Cu)					
Beryllium	1550	1850	1825	200	59.2
Bismuth	545	9780	122	7.9	6.59
Cadmium	594	8650	231	97	48.4
Copper					
Pure	1358	8933	385	401	117
Electrolytic tough pitch		8950	385	386	112
(Cu + Ag 99.90 minimum)					
Commercial bronze	1293	8800	420	52	14.1
(10A1)					
Brass	1188	8530	380	111	34.2
(30 Zn)					
German silver		8618	410	116	32.8
(15 Ni, 22 Zn)					
Constantan		8920	420	22.7	6.06
(40 Ni)					
Constantan		8860		23	
(45 Ni)					
Gold	1336	19300	129	317	127
Iron					
Pure	1810	7870	447	80.2	22.8
Armco		7870	447	72.7	20.7
(99.75 pure)					
Cast		7272	420	51	16.7
(4 C)					
Carbon steels					
AISI 1010		7830	434	64	18.8
(0.1 C, 0.4 Mn)					
AISI 1042, annealed		7840	460	50	13.9
(0.42 C, 0.64 Mn, 0.063 Ni, 0.13 Cu)					
AISI 4130, hardened and tempered		7840	460	43	11.9
(0.3 C, 0.5 Mn, 0.3 Si, 0.95 Cr, 0.5 Mo)					
Stainless steels					
AISI 302 (18–8)		8055	480	15	3.88
(0.15 C, 2 Mn, 1 Si, 16–18 Cr, 6–8 Ni)					
AISI 304	1670	7900	477	15	3.98
(0.08 C, 2 Mn, 1 Si, 18–20 Cr, 8–10 Ni)					

(Continued)

Table A.1a (Concluded)

Metal (% composition)	T_{MP} K	ρ kg/m^3	c J/kg K	k W/m K	α^a m^2/s $\times 10^6$
Stainless steels (continued)					
AISI 316		8238	468	13	3.37
(0.08 C, 2 Mn, 1 Si, 16–18 Cr, 10–14 Ni, 2–3 Mo)					
AISI 410		7770	460	25	7.00
(0.15 C, 1 Mn, 1 Si, 11.5–13 Cr)					
Lead	601	11340	129	35.3	24.1
Magnesium					
Pure	929	1740	1024	156	87.6
Alloy A8					
(8 Al, 0.5 Zn)					
Molybdenum	2894	10240	251	138	53.6
Nickel					
Pure	1728	8900	444	91	23.0
Inconel-X–750	1665	8510	439	11.7	3.13
(15.5 Cr, 1 Nb, 2.5 Ti, 0.7 Al, 7 Fe)					
Nichrome	1672	8314	460	13	3.40
(20 Cr)					
Nimonic 75		8370	461	11.7	3.03
(20 Cr, 0.4 Ti)					
Hasteloy B		9240	381	12.2	3.47
(38 Mo, 5 Fe)					
Cupro-Nickel		8800	421	19.5	5.26
(50 Cu)					
Chromel-P		8730		17	
(10 Cr)					
Alumel		8600		48	
(2 Mn, 2 Al)					
Palladium	1827	12020	244	71.8	24.5
Platinum					
Pure	2045	21450	133	71.6	25.1
60 Pt–40 Rh	1800	16630	162	47	174
(40 Rh)					
Silicon	1685	2330	712	148	89.2
Silver	1235	10500	235	429	174
Tantalum	3269	16600	140	57.5	24.7
Tin	505	7310	227	66.6	40.1
Titanium					
Pure	1993	4500	522	21.9	9.32
Ti–6Al–4V		4420	610	5.8	2.15
Ti–2Al–2 Mn		4510	466	8.4	4.0
Tungsten	3660	19300	132	174	68.3
Zinc	693	7140	389	116	41.8
Zirconium					
Pure	2125	6570	278	22.7	12.4
Zircaloy–4		6560	285	14.2	7.60
(1.2–1.75 Sn, 0.18–0.24 Fe, 0.07–0.13 Cr)					

^{a} This table and subsequent ones are to read as $\alpha \times 10^6 = 97.1$, that is, $\alpha = 97.1 \times 10^{-6}$ m^2/s.

351

Table A.1b Solid metals: Temperature dependence of thermal conductivity k [W/m K] (see Table A.1a for metal compositions)

Metal	Temperature, K								
	200	300	400	500	600	800	1000	1200	1500
Aluminum									
Pure	237	237	240	236	231	218			
Duralumin	138	174	187	188					
Alloy 195, cast		168	174	180	185				
Copper									
Pure	413	401	393	386	379	366	352	339	
Commercial bronze	42	52	52	55					
Brass	74	111	134	143	146	150			
German silver		116	135	145	147				
Gold	323	317	311	304	298	284	270	255	
Iron									
Armco	81	73	66	59	53	42	32	29	31
Cast		51	44	39	36	27	23		
Carbon steels									
AISI 1010		64	59	54	49	39	31		
AISI 1042		52	50	48	45	37	29	26	30
AISI 4130		43	42	41	40	37	31	27	31
Stainless steels									
AISI 302		15	17	19	20	23	25		
AISI 304	13	15	17	18	20	23	25		
AISI 316		13	15	17	18	21	24		
AISI 410	25	25	26	27	27	29			
Lead	37	35	34	33	31				
Magnesium									
Pure	199	156	153	151	149	146			
Alloy A 8			84						
Nickel									
Pure	105	91	80	72	66	68	72	76	83
Inconel-X–750	10.3	11.7	13.5	15.1	17.0	20.5	24.0	27.6	30.0
Nichrome		13	14	16	17	21			
Platinum	73	72	72	72	73	76	79	83	90
Silver	420	429	425	419	412	396	379	361	
Tantalum	58	58	58	59	59	59	60	61	62
Tin	73	67	62	60					
Titanium									
Pure	25	22	20	20	19	19	21	22	25
Ti–6Al–4V		5.8							
Tungsten	185	174	159	146	137	125	118	112	106
Zirconium									
Pure	25	23	22	21	21	21	23	26	29
Zircaloy–4	13.3	14.2	15.2	16.2	17.2	19.2	21.2	23.2	

Table A.1c Solid metals: Temperature dependence of specific heat capacity c [J/kg K] (see Table A.1a for metal compositions)

Metal	Temperature, K								
	200	300	400	500	600	800	1000	1200	1500
Aluminum									
Pure	798	903	949	996	1033	1146			
Duralumin		875							
Alloy 195, cast		883							
Copper									
Pure	356	385	397	412	417	433	451	480	
Commercial bronze	785	420	460	500					
Brass	360	380	395	410	425				
German silver		410							
Gold	124	129	131	133	135	140	145	155	
Iron									
Armco	384	447	490	530	574	680	975	609	634
Cast		420							
Carbon steels									
AISI 1010		434	487	520	559	685	1168		
AISI 1042			500	530	570	700	1430		
AISI 4130			500	530	570	690	840		
Stainless steels									
AISI 302		480	512	531	559	585	606		
AISI 304	402	477	515	539	557	582	611	640	682
AISI 316		468	504	528	550	576	602		
AISI 410		460							
Lead	125	129	132	136	142				
Magnesium									
Pure	934	1024	1074	1170	1170	1267			
Alloy A8			1000						
Nickel									
Pure	383	444	485	500	512	530	562	594	616
Inconel-X–750	372	439	473	490	510	546	626		
Nichrome			480	500	525	545			
Platinum	125	133	136	139	141	146	152	157	165
Silver	225	232	239	244	250	262	277	292	
Tantalum	133	140	144	145	146	149	152	155	160
Tin	215	227	243						
Titanium									
Pure	405	522	551	572	591	633	675	680	686
Ti–6Al–4V		610							
Tungsten	122	132	137	140	142	145	148	152	157
Zirconium									
Pure	264	278	300	312	322	342	362	344	344
Zircaloy–4			300	314	327	348	369		

Table A.2   Solid dielectrics: Thermal properties

Dielectric	T K	ρ kg/m^3	c J/kg K	k W/m K	α m^2/s $\times 10^6$
Aluminum oxide, Al$_2$O$_3$					
Sapphire	300	3970	765	46	15.2
Alumina	300	3970	765	36	11.9
	400		940	27	7.2
	600		1110	16	3.6
	1000		1225	7.6	1.6
	1500			5.4	
Carbon					
Diamond (type IIb)	300	3300	510	1300	772
ATJ-S graphite	300	1810	1300	98	42
	1000		1926	55	16
	2000		2139	38	9.8
	3000		2180	33	8.4
Pyrolytic graphite	300	2210	709	1950	1240
k parallel to layers	600		1406	892	287
	1000		1793	534	135
	2000		2043	262	58
k perpendicular to layers	300	2210	709	5.7	4.1
	600		1406	2.68	2.1
	1000		1793	1.60	1.67
	2000		2043	0.81	1.51
Graphite fiber epoxy	200	1400	640	8.7	9.7
(25% volume) composite	300		935	11.1	8.5
k parallel to fibers	400		1220	13.0	7.6
k perpendicular to fibers	200	1400	640	0.68	0.76
	300		935	0.87	0.66
	400		1220	1.1	0.64
Carbon-carbon weave	300	1860	810	110	73
	1000		1800	56	17
	2000		2140	36.5	9.2
	3000		2220	34.5	8.4
	4000		2260	34	8.1
	5000		2270	34	8.1
Ice	273	910	1930	2.22	1.26
Plastics					
Cellulose acetate	300	1300	1510	0.24	0.12
Neoprene rubber	300	1250	1930	0.19	0.079
Phenolic, filled	300	1760	1260	0.50	0.23
Polyamide (nylon)	300	1140	1670	0.24	0.13
Polyethylene (high density)	300	960	2090	0.33	0.16
Polypropylene	300	1170	1930	0.17	0.075
Polyvinylchloride	300	1714	1050	0.092	0.051
Teflon	300	2200	1050	0.35	0.15
	400			0.45	0.19

(Continued)

Table A.2   (Concluded)

Dielectric	T K	ρ kg/m^3	c J/kg K	k W/m K	α m^2/s $\times 10^6$
Silicon dioxide, SiO$_2$	200	2650		16.4	
Crystalline (quartz)	300		745	10.4	5.3
k parallel to c-axis	400		885	7.6	3.2
	600		1075	5.0	1.7
k perpendicular to c-axis	200	2650		9.5	
	300		745	6.2	3.1
	400		885	4.7	2.0
	600		1075	3.4	1.2
Polycrystalline (fused silica glass)	300	2220	745	1.38	0.83
	400		905	1.51	0.75
	600		1040	1.75	0.75
	800		1105	2.17	0.88
	1000		1155	2.87	1.11
	1200		1195	4.0	1.51
Titanium dioxide, TiO$_2$ (rutile)	300	4157	710	8.4	2.8
	600		880	5.0	1.4
	1200		945	3.3	0.84
Uranium oxide, UO$_2$	300	10890	240	7.9	3.0
	500		265	6.0	2.1
	1000		305	3.9	1.2
	1500		325	2.6	0.79
	2000		355	2.3	0.59
	2500		405	2.5	0.57

Table A.3 Insulators and building materials: Thermal properties

	T K	ρ kg/m^3	c J/kg K	k W/m K	α m^2/s $\times 10^6$
Asbestos paper, laminated and corrugated	300	190		0.078	
4 ply	320			0.085	
	340			0.091	
	360			0.097	
	380			0.101	
8 ply	300	300		0.068	
	320			0.073	
	340			0.077	
	360			0.080	
	380			0.083	
Brick					
B&W K–28 insulating	600			0.03	
	1300			0.04	
Chrome	400	3010	835	2.3	0.92
	800			2.5	
	1200			2.0	
Fireclay	400	2645	960	0.9	0.35
	800			1.4	
	1200			1.7	
	1600			1.8	
Common	300	1920	835	0.72	0.45
Face	300	2083		1.3	
Concrete					
Stone 1–2–4 mix	300	2100	880	1.4	0.75
Cork	300	160	1680	0.043	0.16
Cotton	300	80	1300	0.06	0.58
Glass					
Fused silica	300	2220	745	1.38	0.83
Borosilicate (Pyrex)	300	2640	800	1.09	0.51
Soda-lime (25 Na$_2$O, 10 CaO)	300	2400	840	0.88	0.44
Cellular glass	240			0.048	
	260			0.051	
	280			0.054	
	300	145		0.058	
	320			0.063	
	340			0.067	
Fiberglass, paper-faced batt	300	16	835	0.046	3.4
	300	40		0.035	
	260			0.029	
	280			0.033	
	300	28		0.038	
	320			0.043	
	340			0.048	
	360			0.054	
	380			0.060	
	400			0.066	

(Continued)

Table A.3   (Concluded)

	T K	ρ kg/m^3	c J/kg K	k W/m K	α m^2/s $\times 10^6$
Loose fill					
Cellulose, wood or paper pulp	290			0.038	
	300	45		0.039	
	310			0.042	
Vermiculite, expanded	240			0.058	
	260			0.061	
	280			0.064	
	300	122		0.069	
	320			0.074	
	240			0.052	
	260			0.056	
	280			0.059	
	300	80		0.063	
	320			0.068	
Magnesia	300	270		0.062	
(85 %)	350			0.068	
	400			0.073	
	450			0.078	
	500			0.082	
Paper	300	930	2500	0.13	0.056
Polystyrene, rigid	240			0.023	
	260			0.024	
	280			0.026	
	300	30–60	1210	0.028	0.4–0.8
	320			0.030	
Polyurethane, rigid foam	300	70		0.026	
Rubber					
Hard	270	1200	2010	0.15	0.062
Neoprene	300	1250	1930	0.19	0.079
Rigid foamed	260			0.028	
	280			0.030	
	300	70		0.032	
	320			0.034	
Snow	273	110		0.049	
		500		0.190	
Soil					
Dry	300	1500	1900	1.0	0.35
Wet	300	1900	2200	2.0	0.5
Woods					
Oak, parallel to grain	300	820	2400	0.35	0.18
perpendicular to grain	300	820	2400	0.21	0.11
White Pine, parallel to grain	300	500	2800	0.24	0.17
perpendicular to grain	300	500	2800	0.10	0.071
Wool, sheep	300	145		0.05	

Table A.4   Thermal conductivity of selected materials at cryogenic temperatures

Material	Temperature, K				
	5	10	30	100	200
Metals					
Aluminum (2024–T4)	3.5	7.7	21	50	72
Brass				71	94
Copper (OFHC)			950	430	400
Carbon steel (C1020)	4.3	12	34	64	70
Stainless steel (303)	0.29	0.71	3.5	9.0	12
(304)	0.16	0.82	3.3	9.5	13
Titanium	0.15	0.80	2.0	4.5	6.5
Nonmetals					
Diamond (type 2A)	45	310	330	10000	
Glass, Pyrex				0.56	0.89
Glass, Phoenix		0.11	0.17	0.55	
Nylon–66	0.018	0.033	0.21		
Polyethylene	0.040	0.15	0.75		
Silicon	350	2000	4700	500	
Silicone rubber				0.18	0.21
Teflon			0.18	0.62	1.0
Vacuum grease					
(Dow-Corning silicone)	0.021				
Varnish (G.E. #7031, thermosetting)	0.065	0.075	0.15	0.25	0.36

Table A.5a   Total hemispherical emittance at $T_s \simeq 300$ K, and solar absorptance[a]

Material and Surface Condition	Total Hemispherical Emittance	Solar Absorptance
Aluminum		
Foil, as received	0.05	
Foil, bright dipped	0.03	0.10
Vacuum-deposited on duPont Mylar	0.03	0.10
Alloy 6061, as received	0.04	0.37
Alloy 7075–T6, sandblasted with		
60 mesh silicon carbide grit	0.30	0.55
Weathered alloy 75S–T6	0.20	0.54
Aluminized silicone resin paint,		
Dow-Corning XP–310	0.20	0.27
Hard-anodized	0.80	0.23
Soft-anodized	0.76	0.55
Roofing	0.24	
Asbestos		
Board	0.93	
Cloth	0.87	
Slate	0.94	
Asphalt	0.88	
Brass		
Oxidized	0.60	
Polished	0.04	
Brick	0.90	0.63
Carbon		
Graphite, crushed on sodium silicate	0.88	0.96
Lampblack	0.92	
Chromium		
Bright plate	0.16	
Heated 50 hr at 870 K	0.18	0.78
Coal	0.78	
Concrete, rough	0.91	0.60
Copper		
Electroplated	0.03	0.47
Black oxidized in Ebanol C	0.16	0.91
Oxidized plate	0.76	
Earthenware		
Glazed	0.90	
Matte	0.93	
Frost, rime	0.99	
Glass		
Polished	0.87–0.92	
Pyrex	0.80	
Smooth	0.91	
Second-surface mirror	0.81	0.13
Gold		
On stainless steel	0.09	
On 3M tape Y8194	0.025	

(Continued)

Table A.5a (Continued)

Material and Surface Condition	Total Hemispherical Emittance	Solar Absorptance
Granite	0.44	
Gravel	0.30	
Ice		
Crystal	0.96	
Smooth	0.97	
Inconel X		
Bright	0.21	0.90
Oxidized 4 hr at 1270 K	0.72	
Oxidized 10 hr at 980 K	0.79	
Iron		
ARMCO, bright	0.12	
ARMCO, oxidized	0.30	
Cast, oxidized	0.57	
Rusted	0.83	
Wrought, polished	0.29	
Wrought, dull	0.91	
Limestone	0.92	
Magnesium oxide	0.72	
Marble		
Polished	0.89	
Smooth	0.56	
White	0.92	
Mortar, lime	0.90	
Mylar, duPont film, aluminized on second surface		
6 μm thick	0.37	
25 μm thick	0.63	
75 μm thick	0.81	
Nickel		
Electroplated	0.03	0.22
Tabor solar absorber, electro-oxidized on copper		
110–30	0.05	0.85
125–30	0.11	0.85
Oak, planed	0.88	
Paints		
Black		
Parson's optical	0.92	0.97
Silicone high heat	0.90	0.94
Epoxy	0.87	0.95
Gloss	0.90	
Enamel, heated 1000 hr at 650 K	0.80	
Silver Chromatone	0.24	0.20
White		
Acrylic resin	0.90	0.26
Gloss	0.85	
Epoxy	0.85	0.25

(Continued)

Table A.5a (Concluded)

Material and Surface Condition	Total Hemispherical Emittance	Solar Absorptance
Paper		
Roofing	0.88	
White	0.86	
Plaster, rough	0.89	
Platinum-coated stainless steel	0.12	
Refractory		
Black	0.94	
White	0.90	
Rubber	0.88	
Sand	0.75	
Sandstone, red	0.59	
Silica		
Sintered, powdered, fused	0.82	0.08
Second-surface mirror		
Aluminized	0.81	0.14
Silvered	0.81	0.07
Silver		
Polished	0.02	
Plated on nickel on stainless steel	0.08	
Heated 300 hr at 650 K	0.15	
Slate	0.85	
Snow, fresh	0.82	0.13
Soil	0.94	
Spruce, sanded	0.80	
Stainless steel		
AISI 312, heated 300 hr at 530 K	0.26	
AISI 301, with armco black oxide	0.75	0.89
AISI 410, heated to 980 K	0.15	0.76
AISI 303, sandblasted heavily with 80 mesh		
aluminum oxide grit	0.42	0.85
Teflon	0.85	0.12
Titanium		
75 A	0.12	
75 A, oxidized 300 hr at 730 K in air	0.21	0.80
C–110 M, oxidized 100 hr at 700 K in air	0.06	0.52
C–110 M, oxidized 300 hr at 730 K in air	0.20	0.77
Tungsten, polished	0.03	
Water	0.90	0.98
White potassium zirconium silicate spacecraft coating	0.87	0.13
Zinc, blackened by Tabor solar collector		
electrochemical treatment, 120–20	0.14	0.89

[a] Since the solar spectrum outside the earth's atmosphere is different from that at ground level, appropriate values of solar absorptance are a little different. The values in this table are for extraterrestrial conditions, except those for brick, concrete, snow, and water.

Table A.5b   Temperature variation of total hemispherical emittance for selected surfaces

Material and Surface Condition	Temperature, K						
	200	400	600	800	1000	1200	1400
Aluminum, polished foil	0.03	0.05	0.06	0.07			
Aluminum alloy 245T, polished	0.03	0.04	0.05	0.06	0.08		
Aluminum oxide, Al_2O_3		0.78	0.69	0.61	0.54	0.49	0.42
Cadmium sulphide, CdS	0.56	0.27	0.16				
Carbon		0.83	0.82	0.81	0.81	0.80	0.80
Copper, polished	0.02	0.03	0.03	0.04	0.04	0.05	0.05
Copper, oxidized at 1000 K			0.50	0.58	0.80		
Gold, polished foil	0.02	0.03	0.05	0.06	0.07	0.08	
Iron, polished	0.05	0.09	0.14	0.20	0.25	0.30	0.35
Iron oxides					0.82	0.82	0.80
Magnesium oxide, MgO	0.72	0.73	0.62	0.52	0.47	0.41	0.36
Molybdenum, polished	0.05	0.07	0.08	0.10	0.12	0.14	0.17
Monel metal, polished	0.14	0.15	0.17	0.19	0.22	0.26	0.34
Sodium chloride, NaCl	0.44	0.24					
Nickel, polished	0.08	0.10	0.11	0.12	0.15	0.17	0.21
Platinum, polished	0.05	0.07	0.09	0.11	0.13	0.15	0.17
Silver, polished	0.02	0.02	0.03	0.03	0.03	0.04	0.04
Silicon carbide, SiC	0.84	0.84	0.83	0.83	0.83	0.83	0.83
Silicon oxide, SiO_2, fused	0.72	0.79					
Tantalum, polished	0.01	0.03	0.05	0.08	0.11	0.13	0.16
Titanium alloy A110-AT, polished	0.14	0.17	0.21	0.23	0.26	0.28	0.29
Tungsten, polished		0.03	0.05	0.07	0.09	0.12	0.16
Zirconium oxide, ZrO_2	0.84	0.73	0.63	0.51	0.44	0.42	0.40
Zinc sulphide	0.56	0.30					

Table A.6a   Spectral and total absorptances of metals for normal incidence

Spectral absorptance, normal incidence, $1 < \lambda < 25\mu m$

$$\alpha(\lambda, T_s) \simeq A \left[\frac{(1 + \lambda^2/\lambda_{12}^2)^{1/2} - 1}{\lambda^2/2\lambda_{12}^2} \right]^{1/2} + \frac{B}{C + \lambda^2}$$

Total absorptance, normal incidence, $330\ K < T_e < 2200\ K$

$$\alpha(T_s, T_e) \simeq A \left[\frac{(1 + 0.078C_2^2/\lambda_{12}^2 T_e^2)^{1/2} - 1}{0.039C_2^2/\lambda_{12}^2 T_e^2} \right]^{1/2} + \frac{18.6(BT_e^2/C_2^2)}{1 + 18.6(CT_e^2/C_2^2)}$$

where $C_2 = \hbar c_o/k = 149{,}387\ \mu m\ K$

Metal	Parameter ($T_s \simeq 300\ K$)			
	A	B	C	λ_{12}
Aluminum foil	0.0165	0.23	8.9	14
Cadmium, 99.99%, rolled plate	0.054	2.15	3.2	9
Chromium, polished electroplate	0.076	1.58	3.9	3
Columbium, 99.99%, rolled plate	0.15	0.29	$\simeq 0$	1
Copper, 99.99%, polished	0.018	0.077	3.2	45
Gold, 99.99%, polished	0.020	0.056	1.4	45
Indium, 99.99%, scraped	0.060	0.24	1.3	6
Inconel X, rolled plate	0.44	0.036	$\simeq 0$	1
Lead, 99.99%, scraped	0.16	0.39	1.1	4
Manganese, 99.99%, polished	0.19	4.8	11	8
Molybdenum, 99.99%	0.033	0.36	$\simeq 0$	7
Nickel, 99.99%, polished	0.029	0.83	2.4	5
Platinum, 99.99%, cold rolled	0.038	0.42	$\simeq 0$	4
Rhodium, polished electroplate	0.06	1.27	10	6
Silver, polished electroplate	0.011	0.16	11	70
Stainless steel, 303, lapped	$\simeq 0.71$	$\simeq 0$	$\simeq 0$	$\simeq 0.125$
Tin, 99.99%, rolled plate	0.052	0.56	0.8	7
Titanium, polished electroplate	0.13	2.9	8.3	12
Titanium, 99%, lapped	0.09	6.5	15	12
Tungsten, 99.99%, lapped	0.05	0.49	0.3	3
Vanadium, 99.99%, rolled plate	0.17	0.66	0.93	1
Zinc, 99.9%	0.036	0.26	$\simeq 0$	8
Zirconium, 99.99%, rolled plate	0.64	4	35	1

From *Advances in Thermophysical Properties at Extreme Temperatures and Pressures,* Am. Soc. Mech. Engrs., New York: 1965, pp. 189–199.

Table A.6b   Spectral absorptances at room temperature and an angle of incidence of 25° from the normal [for nonconductors $\alpha(25°) \simeq \alpha$ (hemispherical)]

	Wavelength, λ [μm]										
	0.3	0.35	0.4	0.45	0.5	0.6	0.7	0.8	1.0	1.5	2.0
Bright metals:											
Aluminum	0.05	0.05	0.07	0.07	0.08	0.11	0.12	0.14	0.08	0.04	0.035
Chromium	0.52	0.48	0.43	0.40	0.39	0.37	0.37	0.37	0.40	0.34	0.26
Copper					0.53	0.23	0.14	0.10	0.06	0.032	0.029
Gold	0.80	0.78	0.75	0.74	0.60	0.17	0.11	0.08	0.043	0.034	0.027
Stainless steel		0.61	0.57	0.54	0.53	0.49	0.46	0.44	0.34	0.28	0.25
Titanium	0.71	0.65	0.59	0.56	0.53	0.48	0.47	0.43	0.44	0.40	0.34
Paints and coatings:											
3M black velvet	0.97	→	→	→	→	→	→	→	→	→	0.96
Hard-anodized aluminum	0.95	0.94	0.93	→	→	→	→	→	0.92	0.90	0.85
Anodized titanium					0.53	0.48	0.48	0.48	0.50	0.50	0.52
White epoxy paint			0.60	0.12	0.10	0.15	0.21	0.09	0.08	0.30	0.43
Flame sprayed alumina	0.60	0.48	0.34	0.29	0.27	0.24	0.23	0.23	0.25	0.32	0.51
Aluminum paint			0.25	0.25	0.25	0.26	0.29	0.31	0.28	0.25	0.23

	Wavelength, λ [μm]										
	3	4	5	6	8	10	12	15	20	30	40
Bright metals:											
Aluminum	0.029	0.026	0.023	0.021	0.019	0.018	0.017	0.015	0.014	0.013	0.012
Chromium	0.19	0.145	0.110	0.088	0.078	0.065	0.059	0.05	0.047	0.036	0.030
Copper	0.029	0.022	0.021	0.020	0.020	0.018	0.018	0.018	0.018		
Gold	0.025	0.023	0.023	0.022	0.022	0.020	0.020	0.020	0.020		
Stainless steel	0.20	0.17	0.15	0.14	0.12	0.11	0.10	0.09	0.077	0.062	0.053
Titanium	0.29	0.24	0.22	0.20	0.17	0.15	0.14	0.13	0.11	0.09	0.08
Paints and coatings:											
3M black velvet	0.96	0.96	0.95	0.96	0.91	0.95	0.95	0.94	0.94	0.97	0.97
Hard-anodized aluminum	0.92	0.74	0.70	0.83	0.96	0.98	0.84	0.83	0.80	0.80	
Anodized titanium	0.89	0.76	0.76	0.82	0.83	0.90	0.91	0.88	0.85		
White epoxy paint	0.93	0.90	0.90	0.91	0.93	0.91	0.93	0.90	0.84	0.81	0.80
Flame sprayed alumina	0.73	0.53	0.62	0.88	0.98	0.98	0.74	0.79	0.75		
Aluminum paint	0.23	0.23	0.22	0.22	0.26	0.22	0.21	0.20	0.19		

Table A.7   Gases^a: Thermal properties

Gas	T K	k W/m K	$\rho^{\,b}$ kg/m^3	c_p J/kg K	$\mu \times 10^{6\,c}$ kg/m s	$\nu \times 10^{6\,c}$ m^2/s	Pr
Air	150	0.0158	2.355	1017	10.64	4.52	0.69
(82 K BP)	200	0.0197	1.767	1009	13.59	7.69	0.69
	250	0.0235	1.413	1009	16.14	11.42	0.69
	260	0.0242	1.360	1009	16.63	12.23	0.69
	270	0.0249	1.311	1009	17.12	13.06	0.69
	280	0.0255	1.265	1008	17.60	13.91	0.69
	290	0.0261	1.220	1007	18.02	14.77	0.69
	300	0.0267	1.177	1005	18.43	15.66	0.69
	310	0.0274	1.141	1005	18.87	16.54	0.69
	320	0.0281	1.106	1006	19.29	17.44	0.69
	330	0.0287	1.073	1006	19.71	18.37	0.69
	340	0.0294	1.042	1007	20.13	19.32	0.69
	350	0.0300	1.012	1007	20.54	20.30	0.69
	360	0.0306	0.983	1007	20.94	21.30	0.69
	370	0.0313	0.956	1008	21.34	22.32	0.69
	380	0.0319	0.931	1008	21.75	23.36	0.69
	390	0.0325	0.906	1009	22.12	24.42	0.69
	400	0.0331	0.883	1009	22.52	25.50	0.69
	500	0.0389	0.706	1017	26.33	37.30	0.69
	600	0.0447	0.589	1038	29.74	50.50	0.69
	700	0.0503	0.507	1065	33.03	65.15	0.70
	800	0.0559	0.442	1089	35.89	81.20	0.70
	900	0.0616	0.392	1111	38.65	98.60	0.70
	1000	0.0672	0.354	1130	41.52	117.3	0.70
	1500	0.0926	0.235	1202	53.82	229.0	0.70
	2000	0.1149	0.176	1244	64.77	368.0	0.70
Ammonia	250	0.0198	0.842	2200	8.20	9.70	0.91
(239.7 K BP)	300	0.0246	0.703	2200	10.1	14.30	0.90
	400	0.0364	0.520	2270	13.8	26.60	0.86
	500	0.0511	0.413	2420	17.6	42.50	0.83
Argon	150	0.0096	3.28	527	12.5	3.80	0.68
(77.4 K BP)	200	0.0125	2.45	525	16.3	6.65	0.68
	250	0.0151	1.95	523	19.7	10.11	0.68
	300	0.0176	1.622	521	22.9	14.1	0.68
	400	0.0223	1.217	520	28.6	23.5	0.67
	500	0.0265	0.973	520	33.7	34.6	0.66
	600	0.0302	0.811	520	38.4	47.3	0.66
	800	0.0369	0.608	520	46.6	76.6	0.66
	1000	0.0427	0.487	520	54.2	111.2	0.66
	1500	0.0551	0.324	520	70.6	218.0	0.67

(Continued)

Table A.7 (Continued)

Gas	T K	k W/m K	ρ [b] kg/m^3	c_p J/kg K	$\mu \times 10^{6}$ [c] kg/m s	$\nu \times 10^{6}$ [c] m^2/s	Pr
Carbon dioxide	250	0.01435	2.15	782	12.8	5.97	0.70
(195 K subl.)	300	0.01810	1.788	844	15.2	8.50	0.71
	400	0.0259	1.341	937	19.6	14.6	0.71
	500	0.0333	1.073	1011	23.5	21.9	0.71
	600	0.0407	0.894	1074	27.1	30.3	0.71
	800	0.0544	0.671	1168	33.4	49.8	0.72
	1000	0.0665	0.537	1232	38.8	72.3	0.72
	1500	0.0945	0.358	1329	51.5	143.8	0.72
	2000	0.1176	0.268	1371	61.9	231.0	0.72
Refrigerant–22	240	0.00744	4.13	629	10.32	2.351	0.87
(232.2 K BP)	250	0.00808	4.22	655	10.78	2.558	0.87
	260	0.00871	4.05	687	11.25	2.775	0.89
	270	0.00932	3.90	726	11.72	3.003	0.91
	280	0.00993	3.76	773	12.21	3.245	0.95
	290	0.0105	3.63	829	12.74	3.505	1.00
	300	0.0111	3.51	897	13.31	3.790	1.07
	310	0.0117	3.40	978	13.96	4.106	1.16
	320	0.0123	3.29	1074	14.70	4.464	1.28
Refrigerant–134a	240	0.00886	5.18	762	9.49	1.832	0.82
(246.9 K BP)	250	0.00979	4.97	795	9.92	1.995	0.81
	260	0.0107	4.78	831	10.36	2.166	0.81
	270	0.0115	4.61	871	10.81	2.347	0.82
	280	0.0124	4.44	915	11.27	2.538	0.83
	290	0.0133	4.29	966	11.77	2.745	0.86
	300	0.0142	4.15	1025	12.31	2.970	0.89
	310	0.0153	4.01	1096	12.91	3.219	0.93
	320	0.0164	3.89	1186	13.60	3.500	0.98
	330	0.0177	3.77	1308	14.42	3.827	1.06
	340	0.0192	3.66	1492	15.47	4.230	1.20
	350	0.0209	3.55	1816	16.91	4.760	1.47
Helium	50	0.046	0.974	5200	6.46	6.63	0.73
(4.3 K BP)	100	0.072	0.487	5200	9.94	20.4	0.72
	150	0.096	0.325	5200	13.0	40.0	0.70
	200	0.116	0.244	5200	15.6	64.0	0.70
	250	0.133	0.195	5200	17.9	92.0	0.70
	300	0.149	0.1624	5200	20.1	124.0	0.70
	400	0.178	0.1218	5200	24.4	200.0	0.71
	500	0.205	0.0974	5200	28.2	290.0	0.72
	600	0.229	0.0812	5200	31.7	390.0	0.72
	800	0.273	0.0609	5200	37.8	620.0	0.72
	1000	0.313	0.0487	5200	43.3	890.0	0.72

(Continued)

Table A.7   (Continued)

Gas	T K	k W/m K	$\rho^{\,b}$ kg/m^3	c_p J/kg K	$\mu \times 10^{6\,c}$ kg/m s	$\nu \times 10^{6\,c}$ m^2/s	Pr
Hydrogen	20	0.0158	1.219	10400	1.08	0.893	0.72
(20.3 K BP)	40	0.0302	0.6094	10300	2.06	3.38	0.70
	60	0.0451	0.4062	10660	2.87	7.06	0.68
	80	0.0621	0.3047	11790	3.57	11.7	0.68
	100	0.0805	0.2437	13320	4.21	17.3	0.70
	150	0.125	0.1625	16170	5.60	34.4	0.73
	200	0.158	0.1219	15910	6.81	55.8	0.68
	250	0.181	0.0975	15250	7.91	81.1	0.67
	300	0.198	0.0812	14780	8.93	109.9	0.67
	400	0.227	0.0609	14400	10.8	177.6	0.69
	500	0.259	0.0487	14350	12.6	258.1	0.70
	600	0.299	0.0406	14400	14.3	350.9	0.69
	800	0.385	0.0305	14530	17.4	572.5	0.66
	1000	0.423	0.0244	14760	20.5	841.2	0.72
	1500	0.587	0.0164	16000	25.6	1560	0.70
	2000	0.751	0.0123	17050	30.9	2510	0.70
Mercury	650	0.0100	3.761	104	64.08	17.04	0.67
(630 K BP)	700	0.0108	3.493	104	69.25	19.83	0.67
	800	0.0124	3.056	104	79.45	26.00	0.67
	900	0.0139	2.716	104	89.30	32.87	0.67
	1000	0.0154	2.445	104	98.67	40.36	0.67
	1200	0.0181	2.037	104	115.9	56.93	0.67
	1400	0.0206	1.746	104	132.1	75.68	0.67
	1600	0.0231	1.528	104	148.3	97.11	0.67
	1800	0.0258	1.358	104	165.1	121.5	0.67
	2000	0.0282	1.222	104	180.9	148.0	0.67
Nitrogen	150	0.0157	2.276	1050	10.3	4.53	0.69
(77.4 K BP)	200	0.0197	1.707	1045	13.1	7.65	0.69
	250	0.0234	1.366	1044	15.5	11.3	0.69
	300	0.0267	1.138	1043	17.7	15.5	0.69
	400	0.0326	0.854	1047	21.5	25.2	0.69
	500	0.0383	0.683	1057	25.1	36.7	0.69
	600	0.044	0.569	1075	28.3	49.7	0.69
	800	0.055	0.427	1123	34.2	80.0	0.70
	1000	0.066	0.341	1167	39.4	115.6	0.70
	1500	0.091	0.228	1244	51.5	226.0	0.70
	2000	0.114	0.171	1287	61.9	362.0	0.70

(Continued)

Table A.7   (Continued)

Gas	T K	k W/m K	ρ [b] kg/m^3	c_p J/kg K	$\mu \times 10^6$ [c] kg/m s	$\nu \times 10^6$ [c] m^2/s	Pr
Oxygen	150	0.0148	2.60	890	11.4	4.39	0.69
(90.2 K BP)	200	0.0192	1.949	900	14.7	7.55	0.69
	250	0.0234	1.559	910	17.8	11.4	0.69
	300	0.0274	1.299	920	20.6	15.8	0.69
	400	0.0348	0.975	945	25.4	26.1	0.69
	500	0.042	0.780	970	29.9	38.3	0.69
	600	0.049	0.650	1000	33.9	52.5	0.69
	800	0.062	0.487	1050	41.1	84.5	0.70
	1000	0.074	0.390	1085	47.6	122.0	0.70
	1500	0.101	0.260	1140	62.1	239	0.70
	2000	0.126	0.195	1180	74.9	384	0.70
Saturated steam	273.15	0.0182	0.0048	1850	7.94	1655	0.81
(not at 1 atm)	280	0.0186	0.0076	1850	8.29	1091	0.83
	290	0.0192	0.0142	1860	8.69	612	0.84
	300	0.0198	0.0255	1870	9.09	356.5	0.86
	310	0.0204	0.0436	1890	9.49	217.7	0.88
	320	0.0210	0.0715	1890	9.89	138.3	0.89
	330	0.0217	0.1135	1910	10.3	90.7	0.91
	340	0.0223	0.1741	1930	10.7	61.4	0.92
	350	0.0230	0.2600	1950	11.1	42.6	0.94
	360	0.0237	0.3783	1980	11.5	30.4	0.96
	370	0.0246	0.5375	2020	11.9	22.1	0.98
	373.15	0.0248	0.5977	2020	12.0	20.1	0.98
	380	0.0254	0.7479	2057	12.3	16.4	1.00
Superheated	400	0.0277	0.555	1900	14.0	25.2	0.96
steam	500	0.0365	0.441	1947	17.7	40.1	0.94
(373.2 K BP)	600	0.046	0.366	2003	21.4	58.5	0.93
	800	0.066	0.275	2130	28.1	102.3	0.91
	1000	0.088	0.220	2267	34.3	155.8	0.88
	1500	0.148	0.146	2594	49.1	336.0	0.86
	2000	0.206	0.109	2832	62.7	575.0	0.86

[a] At 1 atm pressure unless otherwise noted.
[b] Calculated using the ideal gas law.
[c] This table and subsequent ones are to be read as $\nu \times 10^6 = 4.52$, that is, $\nu = 4.52 \times 10^{-6}$ m^2/s.

Table A.8   Dielectric liquids: Thermal properties

Saturated Liquid (Melting point) (Boiling point) (Latent heat at BP)	T K	k W/m K	ρ kg/m^3	c_p J/kg K	$\mu \times 10^4$ kg/m s	$\nu \times 10^6$ m^2/s	Pr
Ammonia	220	0.547	705	4480	3.35	0.475	2.75
(195 K MP)	230	0.547	696	4480	2.82	0.405	2.31
(240 K BP)	240	0.547	683	4480	2.42	0.355	1.99
(1.37×10^6 J/kg)	250	0.547	670	4500	2.14	0.320	1.76
	260	0.544	657	4550	1.93	0.293	1.61
	270	0.540	642	4620	1.74	0.271	1.49
	280	0.533	631	4710	1.60	0.253	1.41
	290	0.522	616	4800	1.44	0.234	1.33
	300	0.510	602	4900	1.31	0.217	1.26
	310	0.496	587	4990	1.19	0.202	1.19
	320	0.481	572	5080	1.08	0.188	1.14
Carbon dioxide	220	0.080	1170	1850	1.39	0.119	3.22
(195 K subl.)	230	0.096	1130	1900	1.33	0.118	2.64
(0.57×10^6 J/kg)	240	0.1095	1090	1950	1.28	0.117	2.27
	250	0.1145	1045	2000	1.21	0.1155	2.11
	260	0.113	1000	2100	1.14	0.1135	2.11
	270	0.1075	945	2400	1.04	0.1105	2.33
	280	0.100	885	2850	0.925	0.1045	2.64
	290	0.090	805	4500	0.657	0.094	3.78
	300	0.076	670	11000	0.549	0.082	7.95
Engine oil, unused	280	0.147	895	1810	21900	2450	27000
(SAE 50)	290	0.146	889	1850	10900	1230	13900
	300	0.1445	883	1900	5030	570	6600
	310	0.1435	877	1950	2500	285	3400
	320	0.1425	871	1990	1370	157	1910
	330	0.1415	865	2030	796	92	1140
	340	0.1405	859	2070	515	60	760
	350	0.139	854	2120	350	41	530
	360	0.138	848	2160	255	30.1	400
	370	0.137	842	2200	189	22.5	300
	380	0.136	837	2250	147	17.6	245
	390	0.135	832	2290	112	13.5	191
	400	0.134	826	2330	88.4	10.7	154
	410	0.133	820	2380	71.3	8.7	128
	420	0.132	815	2420	57.9	7.1	106

(Continued)

Table A.8 (Continued)

Saturated Liquid (Melting point) (Boiling point) (Latent heat at BP)	T K	k W/m K	ρ kg/m^3	c_p J/kg K	$\mu \times 10^4$ kg/m s	$\nu \times 10^6$ m^2/s	Pr
Refrigerant–22 (CHClF$_2$)	220	0.120	1444	1090	3.54	0.245	3.22
(115.6 K MP)	230	0.115	1416	1098	3.26	0.230	3.11
(232.2 K BP)	240	0.111	1386	1108	2.97	0.214	2.97
(0.234 × 10^6 J/kg)	250	0.106	1356	1120	2.69	0.198	2.84
	260	0.102	1324	1136	2.42	0.183	2.70
	270	0.0974	1292	1158	2.17	0.168	2.58
	280	0.0932	1258	1185	1.95	0.155	2.48
	290	0.0891	1222	1220	1.75	0.143	2.40
	300	0.0850	1183	1263	1.57	0.133	2.33
	310	0.0810	1143	1315	1.41	0.123	2.29
	320	0.0770	1098	1378	1.27	0.116	2.27
Refrigerant–134a	210	0.123	1478	1219	7.45	0.504	7.42
(CF$_3$CH$_2$F)	220	0.118	1451	1226	6.21	0.428	6.45
(172 K MP)	230	0.113	1423	1239	5.21	0.366	5.69
(246.9 K BP)	240	0.109	1394	1255	4.62	0.331	5.33
(0.217 × 10^6 J/kg)	250	0.104	1364	1275	3.82	0.280	4.67
	260	0.0997	1339	1299	3.35	0.251	4.37
	270	0.0951	1302	1326	2.94	0.226	4.10
	280	0.0904	1270	1357	2.58	0.203	3.87
	290	0.0857	1235	1392	2.31	0.187	3.75
	300	0.0810	1199	1434	2.09	0.174	3.70
	310	0.0763	1160	1484	1.86	0.160	3.62
	320	0.0715	1118	1546	1.64	0.147	3.55
Nitrogen	70	0.151	841	2025	2.17	0.258	2.91
(63.3 K MP)	77.4	0.137	809	2060	1.62	0.200	2.43
(77.4 K BP)	80	0.132	796	2070	1.48	0.186	2.32
(0.200 × 10^6 J/kg)	90	0.114	746	2130	1.10	0.147	2.05
	100	0.097	689	2310	0.87	0.126	2.07
	110	0.080	620	2710	0.71	0.115	2.42
	120	0.063	525	4350	0.48	0.091	3.30
Oxygen	60	0.19	1280	1660	5.89	0.46	5.1
(55 K MP)	70	0.17	1220	1666	3.78	0.31	3.7
(90 K BP)	80	0.16	1190	1679	2.50	0.21	2.6
(0.213 × 10^6 J/kg)	90	0.15	1140	1694	1.60	0.14	1.8
	100	0.14	1110	1717	1.22	0.11	1.50

(Continued)

Table A.8 (Concluded)

Saturated Liquid (Melting point) (Boiling point) (Latent heat at BP)	T K	k W/m K	ρ kg/m³	c_p J/kg K	$\mu \times 10^4$ kg/m s	$\nu \times 10^6$ m²/s	Pr
Therminol 60[a]	230	0.132	1040	1380	6210	597	6490
(205 K MP)	250	0.131	1030	1460	686	66.6	765
(561 K 10% BP)	300	0.129	995	1640	63.8	6.41	81.1
	350	0.125	960	1820	21.5	2.24	31.3
	400	0.120	924	1990	10.8	1.17	17.9
	450	0.115	888	2160	6.62	0.745	12.4
	500	0.108	849	2320	4.59	0.541	9.86
	550	0.100	808	2470	3.47	0.429	8.57
Water	275	0.556	1000	4217	17.00	1.70	12.9
(273 K MP)	280	0.568	1000	4203	14.50	1.45	10.7
(373 K BP)	285	0.580	1000	4192	12.50	1.25	9.0
(2.26×10^6 J/kg)	290	0.591	999	4186	11.00	1.10	7.8
	295	0.602	998	4181	9.68	0.97	6.7
	300	0.611	996	4178	8.67	0.87	5.9
	310	0.628	993	4174	6.95	0.70	4.6
	320	0.641	989	4174	5.84	0.59	3.8
	330	0.652	985	4178	4.92	0.50	3.2
	340	0.661	980	4184	4.31	0.44	2.7
	350	0.669	973	4190	3.79	0.39	2.4
	360	0.676	967	4200	3.29	0.34	2.0
	370	0.680	960	4209	2.95	0.31	1.81
	373.15	0.681	958	4212	2.85	0.30	1.76
	380	0.683	953	4220	2.67	0.28	1.65
	390	0.684	945	4234	2.44	0.26	1.51
	400	0.685	937	4250	2.25	0.24	1.40
	420	0.684	919	4290	1.93	0.21	1.21
	440	0.679	899	4340	1.71	0.19	1.09
	460	0.670	879	4400	1.49	0.17	0.98
	480	0.657	857	4490	1.37	0.16	0.94
	500	0.638	837	4600	1.26	0.15	0.91
	520	0.607	820	4770	1.15	0.14	0.90
	540	0.577	806	5010	1.05	0.13	0.91
	560	0.547	796	5310	0.955	0.12	0.93
	580	0.516	787	5590	0.866	0.11	0.94

[a] Registered trademark of Monsanto Chemical Company, St. Louis; also sold under the brand name "Santotherm."

Table A.9 Liquid metals: Thermal properties

Liquid metal (Melting point) (Boiling point) (Latent heat at BP)	T K	k W/m K	ρ kg/m^3	c_p J/kg K	$\mu \times 10^4$ kg/m s	$\nu \times 10^6$ m^2/s	Pr
Lead	650	16.7	10530	158	23.9	0.227	0.023
(600 K MP)	700	17.5	10470	156	21.1	0.202	0.019
(2020 K BP)	800	19.0	10350	155	17.2	0.166	0.014
(0.850 × 10^6 J/kg)	900	20.4	10230	155	14.9	0.146	0.011
Lithium	500	43.7	514	4340	5.31	1.033	0.053
(453 K MP)	600	46.1	503	4230	4.26	0.847	0.039
(1613 K BP)	700	48.4	493	4190	3.58	0.726	0.031
(19.5 × 10^6 J/kg)	800	50.7	483	4170	3.10	0.642	0.025
	900	55.9	473	4160	2.47	0.522	0.018
Mercury	300	8.4	13530	140	14.9	0.110	0.025
(234 K MP)	400	9.8	13280	140	11.3	0.085	0.016
(630 K BP)	500	11.0	13040	140	9.78	0.075	0.012
(0.292 × 10^6 J/kg)	600	12.1	12780	140	8.31	0.065	0.010
Potassium	400	45.5	814	800	4.9	0.60	0.0086
(337 K MP)	500	43.6	790	790	2.8	0.35	0.0050
(1049 K BP)	600	41.6	765	780	2.1	0.28	0.0040
(2.02 × 10^6 J/kg)	700	39.5	741	770	1.9	0.25	0.0036
	800	36.8	717	750	1.6	0.23	0.0034
	900	34.4	692	740	1.5	0.21	0.0031
Sodium	500	79.2	900	1335	4.2	0.47	0.0071
(371 K MP)	600	74.7	868	1310	3.1	0.36	0.0055
(1156 K BP)	700	70.1	840	1280	2.5	0.30	0.0046
(3.86 × 10^6 J/kg)	800	65.7	813	1260	2.2	0.27	0.0042
	900	62.1	792	1255	2.0	0.25	0.0040
	1000	59.3	772	1255	1.8	0.23	0.0038
	1100	56.7	753	1255	1.6	0.21	0.0035

Table A.10a Volume expansion coefficients for liquids

Liquid	T K	$\beta \times 10^3$ 1/K	Liquid	T K	$\beta \times 10^3$ 1/K
Ammonia	293	2.45	Hydrogen	20.3	15.1
Engine oil	273	0.70	Mercury	273	0.18
(SAE 50)	430	0.70		550	0.18
Ethylene glycol	273	0.65	Nitrogen	70	4.9
$C_2H_4(OH)_2$	373	0.65		77.4	5.7
Refrigerant–22	250	2.27		80	5.9
	260	2.41		90	7.2
	270	2.58		100	9.0
	280	2.78		110	12
	290	3.03		120	24
	300	3.35	Oxygen	89	2.0
	310	3.75	Sodium	366	0.27
	320	4.30	Therminol 60	230	0.79
	330	5.09		250	0.75
	340	6.34		300	0.70
	350	8.64		350	0.70
				400	0.76
Refrigerant–134a	230	2.00		450	0.84
	240	2.09		500	0.96
	250	2.20		550	1.1
	260	2.32			
	270	2.47			
	280	2.65			
	290	2.86			
	300	3.13			
	310	3.48			
	320	3.95			
	330	4.61			
	340	5.60			
	350	7.32			
Glycerin	280	0.47			
$C_3H_5(OH)_3$	300	0.48			
	320	0.50			

Table A.10 b   Density and volume expansion coefficient of water

T K	ρ kg/m^3	$\beta \times 10^6$ 1/K
273.15	999.8679	−68.05
274.00	999.9190	−51.30
275.00	999.9628	−32.74
276.00	999.9896	−15.30
277.00	999.9999	1.16
278.00	999.9941	16.78
279.00	999.9727	31.69
280.00	999.9362	46.04
285.00	999.5417	114.1
290.00	998.8281	174.0
295.00	997.8332	227.5
300.00	996.5833	276.1
310.00	993.4103	361.9
320.00	989.12	436.7
330.00	984.25	504.0
340.00	979.43	566.0
350.00	973.71	624.4
360.00	967.12	697.9
370.00	960.61	728.7
373.15	957.85	750.1
380.00	953.29	788
390.00	945.17	841
400.00	937.21	896
450.00	890.47	1129
500.00	831.26	1432

Table A.11 Surface tension

Fluid	T K	$\sigma \times 10^3$ N/m	Fluid	T K	$\sigma \times 10^3$ N/m
Water	275	75.3	Oxygen (continued)	90	13.5
	280	74.8		100	11.1
	290	73.7	Potassium	400	110
	300	71.7		500	105
	310	70.0		600	97
	320	68.3		700	90
	330	66.6		800	83
	340	64.9		900	76
	350	63.2	Sodium	500	175
	360	61.4		700	160
	370	59.5		900	140
	373.15	58.9		1100	120
	380	57.6	Nitrogen	68	11.00
	390	55.6		70	10.53
	400	53.6		72	10.07
	420	49.4		74	9.62
	440	45.1		76	9.16
	460	40.7		77.4	8.85
	480	36.2		78	8.72
	500	31.6		80	8.27
	550	19.7		82	7.84
	600	8.4		84	7.42
	647.30	0.0		86	6.99
Ammonia	220	39		88	6.57
	240	34		90	6.16
	360	30	Refrigerant–22	220	20.2
	280	25		230	18.6
	300	20		240	16.9
	320	16		250	15.3
Carbon dioxide	248	9.1		260	13.7
	293	1.2		270	12.2
Hydrogen	15	2.8		280	10.7
	20	2.0		290	9.2
	25	1.1		300	7.8
Lead	600	470		310	6.4
	700	452		320	5.1
	800	437	Refrigerant–134a	210	21.5
	900	421		220	19.9
Lithium	500	390		230	18.3
	700	360		240	16.7
	900	335		250	15.1
Mercury	300	470		260	13.6
	400	450		270	12.1
	500	430		280	10.7
	600	400		290	9.28
	700	380		300	7.92
Oxygen	60	20.7		310	6.60
	70	18.3		320	5.33
	80	16.0		330	4.12

Table A.12 a   Thermodynamic properties of saturated steam

T K	$P \times 10^{-5}$ Pa	v m^3/kg	ρ kg/m^3	$h_{fg} \times 10^{-6}$ J/kg
273.15	0.00610	206.4	0.00484	2.501
274.00	0.00649	194.6	0.00514	2.499
275.00	0.00698	181.8	0.00550	2.496
276.00	0.00750	169.8	0.00589	2.494
277.00	0.00805	158.8	0.00630	2.492
278.00	0.00863	148.6	0.00673	2.490
279.00	0.00925	139.1	0.00719	2.488
280.00	0.00991	130.4	0.00767	2.486
281.00	0.01061	122.2	0.00818	2.484
282.00	0.01136	114.6	0.00873	2.482
283.00	0.01215	107.6	0.00929	2.479
284.00	0.01299	101.0	0.00990	2.476
285.00	0.01388	94.75	0.01055	2.473
286.00	0.01482	89.06	0.01123	2.471
287.00	0.01582	83.73	0.01194	2.468
288.00	0.01688	78.75	0.01270	2.466
289.00	0.01800	74.09	0.01350	2.463
290.00	0.01918	69.74	0.01434	2.461
291.00	0.02043	65.68	0.01523	2.459
292.00	0.02176	61.89	0.01616	2.456
293.00	0.02315	58.35	0.01714	2.454
294.00	0.02463	55.05	0.01817	2.451
295.00	0.02619	51.96	0.01925	2.449
296.00	0.02783	49.07	0.02038	2.447
297.00	0.02957	46.37	0.02157	2.444
298.00	0.03139	43.82	0.02282	2.442
299.00	0.03331	41.42	0.02414	2.439
300.00	0.03533	39.15	0.02554	2.437
301.00	0.03746	37.05	0.02700	2.434
302.00	0.03971	35.07	0.02851	2.432
303.00	0.04206	33.21	0.03011	2.430
304.00	0.04454	31.46	0.03179	2.427
305.00	0.04714	29.81	0.03355	2.425
306.00	0.04987	28.26	0.03539	2.423
307.00	0.05274	26.81	0.03730	2.421
308.00	0.05576	25.44	0.03931	2.418
309.00	0.05892	24.16	0.04139	2.416
310.00	0.06224	22.95	0.04357	2.414
311.00	0.06572	21.81	0.04585	2.412
312.00	0.06936	20.73	0.04824	2.409

(Continued)

Table A.12 a   Thermodynamic properties of saturated steam

T K	$P \times 10^{-5}$ Pa	v m³/kg	ρ kg/m³	$h_{fg} \times 10^{-6}$ J/kg
313.00	0.07318	19.72	0.05071	2.407
314.00	0.07717	18.75	0.05333	2.404
315.00	0.08135	17.83	0.05609	2.401
316.00	0.08573	16.97	0.05893	2.399
317.00	0.09031	16.16	0.06188	2.396
318.00	0.09511	15.39	0.06498	2.394
319.00	0.10012	14.66	0.06821	2.391
320.00	0.10535	13.98	0.07153	2.389
321.00	0.11082	13.33	0.07502	2.387
322.00	0.11652	12.72	0.07862	2.384
323.00	0.12247	12.14	0.08237	2.382
324.00	0.12868	11.59	0.08628	2.379
325.00	0.13514	11.06	0.09042	2.377
326.00	0.14191	10.56	0.09470	2.375
327.00	0.14896	10.09	0.09911	2.372
328.00	0.15630	9.644	0.1037	2.370
329.00	0.16395	9.219	0.1085	2.367
330.00	0.17192	8.817	0.1134	2.365
331.00	0.18021	8.434	0.1186	2.363
332.00	0.18885	8.072	0.1239	2.360
333.00	0.19783	7.727	0.1294	2.358
334.00	0.20718	7.400	0.1351	2.355
335.00	0.2169	7.090	0.1410	2.353
336.00	0.2270	6.794	0.1472	2.351
337.00	0.2375	6.512	0.1536	2.348
338.00	0.2484	6.244	0.1602	2.346
339.00	0.2597	5.987	0.1670	2.343
340.00	0.2715	5.741	0.1742	2.341
341.00	0.2837	5.509	0.1815	2.339
342.00	0.2964	5.288	0.1891	2.336
343.00	0.3096	5.077	0.1970	2.334
344.00	0.3233	4.876	0.2051	2.332
345.00	0.3375	4.684	0.2135	2.329
346.00	0.3521	4.500	0.2222	2.326
347.00	0.3673	4.325	0.2312	2.324
348.00	0.3831	4.158	0.2405	2.321
349.00	0.3994	3.999	0.2501	2.319
350.00	0.4164	3.847	0.2599	2.316
351.00	0.4339	3.701	0.2702	2.313
352.00	0.4520	3.562	0.2807	2.311

(Continued)

Table A.12 a Thermodynamic properties of saturated steam

T K	$P \times 10^{-5}$ Pa	v m³/kg	ρ kg/m³	$h_{fg} \times 10^{-6}$ J/kg
353.00	0.4708	3.429	0.2916	2.308
354.00	0.4902	3.301	0.3029	2.306
355.00	0.5103	3.179	0.3146	2.303
356.00	0.5310	3.062	0.3266	2.301
357.00	0.5525	2.951	0.3389	2.299
358.00	0.5747	2.844	0.3516	2.296
359.00	0.5976	2.742	0.3647	2.294
360.00	0.6213	2.644	0.3782	2.291
361.00	0.6457	2.550	0.3922	2.288
362.00	0.6710	2.460	0.4065	2.285
363.00	0.6970	2.373	0.4214	2.283
364.00	0.7240	2.291	0.4365	2.280
365.00	0.7518	2.212	0.4521	2.277
366.00	0.7804	2.136	0.4682	2.274
367.00	0.8100	2.063	0.4847	2.272
368.00	0.8405	1.993	0.5018	2.269
369.00	0.8719	1.925	0.5195	2.267
370.00	0.9044	1.861	0.5373	2.265
371.00	0.9377	1.798	0.5562	2.263
372.00	0.9722	1.738	0.5754	2.260
373.00	1.0076	1.681	0.5949	2.257
373.15	1.0133	1.673	0.5977	2.257
380.00	1.2875	1.337	0.7479	2.238
390.00	1.7952	0.9800	1.020	2.211
400.00	2.4563	0.7308	1.368	2.183
410.00	3.303	0.5535	1.807	2.154
420.00	4.371	0.4254	2.351	2.124
430.00	5.701	0.3311	3.020	2.093
440.00	7.335	0.2609	3.833	2.059
450.00	9.322	0.2082	4.803	2.025
460.00	11.708	0.1671	5.984	1.990
470.00	14.551	0.1353	7.391	1.953
480.00	17.908	0.1109	9.017	1.914
490.00	21.839	0.09172	10.90	1.872
500.00	26.401	0.07573	13.20	1.827
510.00	31.676	0.06374	15.69	1.779
520.00	37.726	0.05427	18.43	1.729
530.00	44.618	0.04639	21.56	1.676
540.00	52.420	0.03919	25.52	1.621
550.00	61.200	0.03175	31.50	1.563

Table A.12 b Thermodynamic properties of saturated ammonia

T K	P kPa	v m^3/kg	$h_{fg} \times 10^{-6}$ J/kg
224	42.98	2.5055	1.414
226	48.27	2.2479	1.409
228	54.09	2.0212	1.404
230	60.48	1.8213	1.398
232	67.46	1.6445	1.392
234	75.10	1.4878	1.387
236	83.42	1.3487	1.381
238	92.50	1.2248	1.375
240	102.29	1.1143	1.369
242	112.96	1.0155	1.363
244	124.52	0.9271	1.358
246	137.05	0.8478	1.351
248	150.58	0.7765	1.345
250	165.07	0.7123	1.339
252	180.68	0.6544	1.333
254	197.50	0.6021	1.327
256	215.47	0.5548	1.320
258	234.80	0.5118	1.314
260	255.41	0.4728	1.307
262	277.46	0.4374	1.301
264	300.98	0.4051	1.294
266	326.04	0.3757	1.287
268	352.80	0.3487	1.280
270	381.12	0.3241	1.273
272	411.23	0.3016	1.267
274	443.24	0.2809	1.259
276	477.14	0.2619	1.252
278	512.97	0.2445	1.245
280	550.86	0.2285	1.238
282	590.87	0.2136	1.230
284	633.16	0.2000	1.223
286	677.70	0.1873	1.215
288	724.66	0.1756	1.207
290	774.15	0.1648	1.199
292	826.05	0.1548	1.191
294	880.67	0.1455	1.183
296	938.00	0.1369	1.175
298	998.03	0.1289	1.167
300	1061.35	0.1214	1.159
302	1127.39	0.1144	1.150

(Continued)

Table A.12b   (Concluded)

T K	P kPa	v m³/kg	$h_{fg} \times 10^{-6}$ J/kg
304	1195.88	0.1079	1.141
306	1268.41	0.1019	1.133
308	1344.02	0.0962	1.124
310	1423.12	0.0909	1.115
312	1505.21	0.0860	1.106
314	1590.85	0.0814	1.097
316	1680.86	0.0770	1.087
318	1773.70	0.0729	1.078
320	1871.02	0.0690	1.068

Table A.12c   Thermodynamic properties of saturated nitrogen

T K	P kPa	v m³/kg	$h_{fg} \times 10^{-6}$ J/kg
77.4	101.3	0.2209	0.1995
80	137.8	0.1656	0.1962
85	228.0	0.1028	0.1892
90	358.7	0.06681	0.1813
95	538.1	0.04486	0.1726
100	777.2	0.03123	0.1623
105	1085	0.02227	0.1509
110	1473	0.01613	0.1373
115	1954	0.01156	0.1201
120	2537	0.008101	0.0962
125	3236	0.004889	0.0495
126	3392	0.003216	0

Table A.12 d   Thermodynamic properties of saturated mercury

T K	P MPa	v m^3/kg	$h_{\text{fg}} \times 10^{-6}$ J/kg
390	0.00009	182.0	0.2971
395	0.00011	146.0	0.2970
400	0.00014	117.6	0.2969
405	0.00018	94.76	0.2968
410	0.00022	77.00	0.2967
415	0.00027	63.15	0.2966
420	0.00033	51.84	0.2965
425	0.00041	42.72	0.2964
430	0.00049	35.46	0.2963
435	0.00060	29.51	0.2962
440	0.00073	24.68	0.2961
445	0.00088	20.74	0.2960
450	0.00105	17.53	0.2959
460	0.00150	12.68	0.2957
470	0.00208	9.223	0.2955
480	0.00287	6.863	0.2954
490	0.00390	5.135	0.2952
500	0.00524	3.902	0.2950
520	0.00913	2.331	0.2946
540	0.01527	1.459	0.2942
560	0.02460	0.9370	0.2938
580	0.03815	0.6210	0.2935
600	0.05749	0.4276	0.2931
620	0.08450	0.3010	0.2927
640	0.12117	0.2180	0.2923
660	0.16969	0.1602	0.2920
680	0.23281	0.1201	0.2916
700	0.31357	0.09179	0.2912
720	0.41544	0.07133	0.2908
740	0.54145	0.05629	0.2905
760	0.69616	0.04501	0.2901
780	0.88308	0.03645	0.2897
800	1.10570	0.02989	0.2893
820	1.36881	0.02477	0.2890
840	1.67803	0.02072	0.2886
860	2.03709	0.01750	0.2882
880	2.44991	0.01491	0.2878
900	2.92017	0.01282	0.2874
920	3.45295	0.01109	0.2871
940	4.05586	0.00967	0.2867
960	4.72617	0.00849	0.2863
980	5.47829	0.00749	0.2859
1000	6.30516	0.00665	0.2856

Table A.12e   Thermodynamic properties of saturated refrigerant–22 (chlorodifluoromethane)

T K	P MPa	v m³/kg	$h_{\mathrm{fg}} \times 10^{-6}$ J/kg
175	0.002393	6.9904	0.2686
180	0.003710	4.6346	0.2657
185	0.005596	3.1545	0.2627
190	0.008232	2.1993	0.2597
195	0.011836	1.5673	0.2568
200	0.016663	1.1395	0.2538
205	0.023007	0.8439	0.2508
210	0.031206	0.6355	0.2477
215	0.041634	0.4860	0.2447
220	0.054710	0.3770	0.2416
225	0.070890	0.2962	0.2385
230	0.090670	0.2355	0.2354
235	0.11458	0.1893	0.2321
240	0.14319	0.1537	0.2289
245	0.17709	0.1259	0.2255
250	0.21691	0.1041	0.2221
255	0.26332	0.08665	0.2185
260	0.31699	0.07266	0.2149
265	0.37864	0.06132	0.2111
270	0.44898	0.05206	0.2072
275	0.52878	0.04444	0.2032
280	0.61881	0.03812	0.1990
285	0.71986	0.03284	0.1947
290	0.83274	0.02841	0.1902
295	0.95828	0.02466	0.1854
300	1.0973	0.02148	0.1805
305	1.2508	0.01875	0.1753
310	1.4195	0.01641	0.1698
315	1.6045	0.01438	0.1640
320	1.8067	0.01262	0.1579
325	2.0271	0.01108	0.1514
330	2.2667	0.009731	0.1444
335	2.5268	0.008533	0.1368
340	2.8085	0.007464	0.1285
345	3.1132	0.006503	0.1193
350	3.4425	0.005626	0.1089
355	3.7983	0.004813	0.09678
360	4.1829	0.004034	0.08184
365	4.6000	0.003227	0.06127

Table A.12f   Thermodynamic properties of saturated refrigerant–134a
(tetrafluoroethane)

T K	P MPa	v m^3/kg	$h_{fg} \times 10^{-6}$ J/kg
245	0.0904	0.1804	0.2179
250	0.1134	0.1461	0.2147
255	0.1408	0.1193	0.2115
260	0.1733	0.09827	0.2082
265	0.2113	0.08153	0.2047
270	0.2555	0.06812	0.2011
275	0.3066	0.05728	0.1973
280	0.3653	0.04844	0.1934
285	0.4322	0.04119	0.1894
290	0.5080	0.03519	0.1851
295	0.5936	0.03019	0.1807
300	0.6897	0.02600	0.1761
305	0.7970	0.02246	0.1713
310	0.9165	0.01945	0.1662
315	1.0490	0.01687	0.1609
320	1.1954	0.01466	0.1554
325	1.3567	0.01274	0.1495
330	1.5339	0.01107	0.1433
335	1.7282	0.009611	0.1366
340	1.9410	0.008233	0.1294
345	2.1736	0.007177	0.1217
350	2.4277	0.006151	0.1131
355	2.7056	0.005220	0.1035

Table A.13a   Aqueous ethylene glycol solutions: Thermal properties

T K	Percent Glycol, by Mass (Freezing Point)				
	20 (263.7 K)	30 (257.6 K)	40 (248.7 K)	50 (237.6 K)	60 (213.2 K)
Thermal Conductivity, k [W/m K]					
250	—	—	0.456	0.425	0.400
260	—	0.488	0.456	0.423	0.397
270	0.513	0.488	0.456	0.422	0.394
280	0.519	0.491	0.456	0.419	0.393
290	0.522	0.493	0.456	0.418	0.389
300	0.525	0.494	0.456	0.418	0.388
310	0.531	0.495	0.456	0.417	0.385
320	0.538	0.497	0.456	0.416	0.381
Density, ρ [kg/m^3]					
250	—	—	1069.0	1084.0	1098.5
260	—	1051.6	1066.0	1080.5	1094.0
270	1034.1	1048.1	1061.5	1076.0	1089.0
280	1031.1	1044.1	1057.0	1071.0	1085.0
290	1026.6	1039.1	1052.1	1066.0	1079.0
300	1022.6	1035.6	1047.1	1060.5	1073.0
310	1018.6	1031.1	1043.1	1054.1	1066.5
320	1014.1	1025.6	1037.1	1048.1	1060.0
Specific Heat, c_p [J/kg K]					
250	—	—	—	3480	3260
260	—	—	3710	3490	3300
270	4050	3930	3720	3510	3330
280	4030	3920	3740	3550	3360
290	4020	3930	3750	3570	3400
300	4010	3940	3760	3600	3430
310	4010	3950	3780	3620	3460
320	4020	3960	3800	3650	3500
Dynamic Viscosity, μ [kg/m s $\times 10^3$]					
250	—	—	—	29.0	45.0
260	—	—	11.2	16.2	20.0
270	3.6	5.0	7.0	9.5	14.1
280	2.4	3.4	4.9	6.2	8.8
290	1.8	2.3	3.1	4.1	5.6
300	1.4	1.8	2.3	3.0	4.0
310	1.1	1.4	1.8	2.3	3.0
320	0.9	1.1	1.5	1.8	2.4

Table A.13b Aqueous sodium chloride solutions: Thermal properties

T K	Percent NaCl, by Mass				
	5	10	15	20	25
Freezing Point [K]					
	270.2	266.6	262.2	256.7	264.3
Thermal Conductivity, k [W/m K]					
260	—	—	—	0.434	—
270	—	0.504	0.478	0.449	0.420
280	0.542	0.516	0.490	0.464	0.438
290	0.560	0.534	0.508	0.481	0.454
300	0.575	0.550	0.525	0.498	0.470
Density, ρ [kg/m^3]					
260	—	—	—	1160	—
270	—	1076	1115	1156	1189
280	1035	1073	1111	1153	1184
290	1033	1070	1108	1150	1180
300	1032	1069	1107	1149	1178
Specific Heat, c_p [J/kg K]					
260	—	—	—	3365	—
270	—	3680	3515	3380	3280
280	3920	3705	3535	3395	3290
290	3930	3720	3555	3415	3300
300	3940	3730	3575	3425	3310
Dynamic Viscosity, μ [kg/m s $\times 10^3$]					
260	—	—	—	4.70	—
270	—	2.30	2.60	3.05	3.83
280	1.50	1.65	1.95	2.25	2.60
290	1.17	1.30	1.50	1.75	2.03
300	0.90	1.00	1.18	1.38	1.65

Table A.14a   Dimensions of commercial pipes [mm] (ASA standard)

Nominal Pipe Size ($\simeq$ I.D., in)		Schedule					
		5	10	40	80	160	XX Strong
$\frac{1}{4}$	O.D.	13.716	13.716	13.716	13.716		
	Wall	1.245	1.651	2.235	3.023		
	I.D.	11.227	10.414	9.246	7.671		
$\frac{3}{8}$	O.D.	17.145	17.145	17.145	17.145		
	Wall	1.245	1.651	2.311	3.200		
	I.D.	14.656	13.843	12.522	10.744		
$\frac{1}{2}$	O.D.	21.336	21.336	21.336	21.336	21.336	21.336
	Wall	1.651	2.108	2.769	3.734	4.750	7.468
	I.D.	18.034	17.120	15.799	13.868	11.836	6.401
$\frac{3}{4}$	O.D.	26.670	26.670	26.670	26.670	26.670	26.670
	Wall	1.651	2.108	2.870	3.912	5.534	7.823
	I.D.	23.368	22.454	20.930	18.847	15.596	11.024
1	O.D.	33.401	33.401	33.401	33.401	33.401	33.401
	Wall	1.651	2.769	3.378	4.547	6.350	9.093
	I.D.	30.099	27.864	26.645	24.308	20.701	15.215
$1\frac{1}{2}$	O.D.	48.260	48.260	48.260	48.260	48.260	48.260
	Wall	1.651	2.769	3.683	5.080	7.137	10.160
	I.D.	44.958	42.723	40.894	38.100	33.985	27.940
2	O.D.	60.325	60.325	60.325	60.325	60.325	60.325
	Wall	1.651	2.769	3.912	5.537	8.712	11.074
	I.D.	57.023	54.788	52.502	49.251	42.901	38.176
3	O.D.			88.900	88.900	88.900	88.900
	Wall			5.486	7.62	11.125	15.240
	I.D.			77.927	73.660	66.650	58.420
4	O.D.			114.300	114.300	114.300	114.300
	Wall			6.020	8.560	13.487	17.120
	I.D.			102.260	97.180	87.325	80.061
5	O.D.			141.300	141.300	141.300	141.300
	Wall			6.553	9.525	15.875	19.050
	I.D.			128.194	122.250	109.550	103.200
6	O.D.			168.275	168.275	168.275	168.275
	Wall			7.150	10.973	18.237	21.946
	I.D.			153.975	146.329	131.801	124.384
8	O.D.			219.075	219.075		
	Wall			8.179	12.700		
	I.D.			202.717	193.675		
10	O.D.			273.050	273.050		
	Wall			9.271	15.062		
	I.D.			254.508	242.926		
12	O.D.			323.850	323.850		
	Wall			10.312	17.450		
	I.D.			303.255	288.950		

Table A.14b Dimensions of commercial tubes [mm] (ASTM standard)

Nominal Size ($\simeq$O.D., in)	O.D.	Gage (BWG)	Wall	I.D.
$\frac{3}{16}$	4.775	20	0.889	2.997
$\frac{1}{4}$	6.350	22	0.711	4.928
		20	0.889	4.572
		18	1.245	3.861
$\frac{5}{16}$	7.950	20	0.889	6.172
		18	1.245	5.461
$\frac{3}{8}$	9.525	20	0.889	7.747
		18	1.245	7.036
		16	1.651	6.223
$\frac{1}{2}$	12.700	20	0.889	10.922
		18	1.245	10.211
		16	1.651	9.398
		14	2.108	8.484
$\frac{5}{8}$	15.875	18	1.245	13.386
		16	1.651	12.573
		14	2.108	11.659
$\frac{3}{4}$	19.050	20	0.889	17.272
		18	1.245	16.561
		16	1.651	15.748
		14	2.108	14.834
$\frac{7}{8}$	22.225	16	1.651	18.923
1	25.400	20	0.889	23.622
		18	1.245	22.911
		16	1.651	22.098
		14	2.108	21.184
		12	2.769	19.863
		10	3.404	18.593
$1\frac{1}{4}$	31.750	18	1.245	29.261
		16	1.651	28.448
		14	2.108	27.534
		12	2.769	26.213
		10	3.404	24.943
$1\frac{1}{2}$	38.100	18	1.245	35.611
		16	1.651	34.798
		14	2.108	33.884
		12	2.769	32.563
		10	3.404	31.293
		8	4.191	29.718
2	50.800	18	1.245	48.311
		16	1.651	47.498
		14	2.108	46.584
		12	2.769	45.263
		10	3.404	43.993
		8	4.191	42.418

Table A.14c Dimensions of seamless steel tubes for tubular heat exchangers [mm] (DIN 28 180)

Outside diameter	Wall thickness				
	1.2	1.6	2.0	2.6	3.2
Unalloyed and alloy steel tubes					
16	x	x	x		
20		x	x	x	
25		x	x	x	x
30		x	x	x	x
38			x	x	x
Austenitic stainless steel tubes					
16	x	x	x		
20	x	x	x	x	
25		x	x	x	x
30		x	x	x	x
38		x	x	x	x

Table A.14d Dimensions of wrought copper and copper alloy tubes for condensers and heat exchangers [mm] (DIN 1785–83)

Outside diameter	Wall thickness				
	0.75	1.0	1.25	1.5	2.0
8	x	x	x		
10	x	x	x		
11	x	x	x		
12	x	x	x		
14	x	x	x		
15	x	x	x		
16	x	x	x	x	
18		x	x	x	
19		x	x	x	x
20		x	x	x	x
22		x	x	x	x
23		x	x	x	x
24		x	x	x	x
25		x	x	x	x
28		x	x	x	x
30		x	x	x	x
32		x	x	x	x
35		x	x	x	x
					x

Table A.14e   Dimensions of seamless cold drawn stainless steel tubes [mm] (LN 9398)[a]

Outside diameter	Wall thickness								
	0.5	0.6	0.8	1.0	1.2	1.6	2.0	2.5	3.2
5	x		x						
6	x		x	x					
8	x		x	x	x				
10	x		x		x	x			
12			x		x		x		
14				x					
16			x	x	x		x	x	
18		x				x			
20			x	x	x	x		x	x
22			x						
25				x	x	x		x	
28				x					
32			x		x		x		
36									
40			x			x			
45			x			x	x		
50						x	x		

[a] Sizes in conformance with ISO 2964 R20.

Table A.14f   Dimensions of seamless drawn wrought aluminum alloy tubes [mm] (LN 9223)[a]

Outside diameter	Wall thickness					
	0.8	1.0	1.2	1.6	2.0	2.5
12	x					
14						
16	x					
18						
20	x					
22						
25	x					
28						
32	x	x	x			
36						
40	x					
45		x	x			
50		x	x	x		
56						
63		x	x		x	
70						
80				x		x

[a] Sizes in conformance with ISO 2964, R20.

Table A.15   U.S. standard atmosphere

Altitude m	Temperature K	Pressure Pa	Density Ratio ρ/ρ_0	Gravity m/s^2	Mean Free Path m	Molecular Weight
0	288.150	1.01325+5	1.0000+0	9.8066	6.6328−8	28.964
200	286.850	9.89454+4	9.8094−1	9.8060	6.7617−8	28.964
400	285.550	9.66114+4	9.6216−1	9.8054	6.8936−8	28.964
600	284.250	9.43223+4	9.4366−1	9.8048	7.0288−8	28.964
800	282.951	9.20775+4	9.2543−1	9.8042	7.1672−8	28.964
1000	281.651	8.98762+4	9.0748−1	9.8036	7.3090−8	28.964
2000	275.154	7.95014+4	8.2168−1	9.8005	8.0723−8	28.964
3000	268.659	7.01211+4	7.4225−1	9.7974	8.9361−8	28.964
4000	262.166	6.16604+4	6.6885−1	9.7943	9.9166−8	28.964
5000	255.676	5.40482+4	6.0117−1	9.7912	1.1033−7	28.964
6000	249.187	4.72176+4	5.3887−1	9.7882	1.2309−7	28.964
7000	242.700	4.11052+4	4.8165−1	9.7851	1.3771−7	28.964
8000	236.215	3.56516+4	4.2921−1	9.7820	1.5453−7	28.964
9000	229.733	3.08007+4	3.8128−1	9.7789	1.7396−7	28.964
10,000	223.252	2.64999+4	3.3756−1	9.7759	1.9649−7	28.964
20,000	216.650	5.52930+3	7.2579−2	9.7452	9.1387−7	28.964
30,000	226.509	1.19703+3	1.5029−2	9.7147	4.4134−6	28.964
40,000	250.350	2.87143+2	3.2618−3	9.6844	2.0335−5	28.964
47,400	270.650	1.10220+2	1.1581−3	9.6620	5.7272−5	28.964
50,000	270.650	7.97790+1	8.3827−4	9.6542	7.9125−5	28.964
52,000	270.650	6.22283+1	6.5386−4	9.6481	1.0144−4	28.964
60,000	255.722	2.24606+1	2.4973−4	9.6241	2.6560−4	28.964
70,000	219.700	5.52047+0	7.1457−5	9.5941	9.2821−4	28.964
80,000	180.65	1.0366+0	1.632−5	9.564	4.065−3	28.964
90,000	180.65	1.6438−1	2.588−6	9.535	2.563−2	28.96
100,000	210.02	3.0075−2	4.060−7	9.505	1.629−1	28.88
105,000	233.90	1.4318−2	1.728−7	9.490	3.810−1	28.75
110,000	257.00	7.3544−3	8.024−8	9.476	8.150−1	28.56
115,000	303.78	4.1224−3	3.774−8	9.461	1.719+0	28.32
120,000	349.49	2.5217−3	1.988−8	9.447	3.233+0	28.07
125,000	442.35	1.6863−3	1.041−8	9.432	6.118+0	27.81
130,000	533.80	1.2214−3	6.195−9	9.417	1.019+1	27.58
135,000	624.30	9.3330−4	4.017−9	9.403	1.560+1	27.37
140,000	714.22	7.4104−4	2.770−9	9.388	2.248+1	27.20
145,000	803.74	6.0560−4	2.001−9	9.374	3.095+1	27.05
150,000	892.79	5.0617−4	1.498−9	9.360	4.114+1	26.92
155,000	957.94	4.2992−4	1.181−9	9.345	5.197+1	26.79
160,000	1022.23	3.6943−4	1.159−9	9.331	6.454+1	26.66

Table A.16 Selected physical constants

Universal gas constant
$$\mathcal{R} = 8314.3 \text{ J/kmol K}$$
$$= 1.987 \text{ kcal/kmol K (thermochemical calorie)}$$

Standard gravitational acceleration
$$g = 9.80665 \text{ m/s}^2$$

Atomic mass unit
$$u = 1.66057 \times 10^{-27} \text{ kg}$$

Avogadro's number
$$\mathcal{A} = 6.02252 \times 10^{26} \text{ molecules/kmol}$$

Boltzmann constant
$$k = 1.38054 \times 10^{-23} \text{ J/K molecule}$$

Planck's constant
$$\hbar = 6.6256 \times 10^{-34} \text{ J s}$$

Speed of light *in vacuo*
$$c_0 = 2.997925 \times 10^8 \text{ m/s}$$

Stefan-Boltzmann constant
$$\sigma = 2\pi^5 k^4 / 15 \hbar^3 c_0^2 = 5.670 \times 10^{-8} \text{ W/m}^2\text{K}^4$$

Elementary electric charge
$$e = 1.60 \times 10^{-19} \text{ A s}$$

Faraday's constant
$$F = 9.6487 \times 10^7 \text{ coulumbs/kg equivalent}$$

Table A.17 a   Diffusion coefficients in air at 1 atm $(1.013 \times 10^5 \text{Pa})^a$

T [K]	Binary Diffusion Coefficient [m²/s × 10⁴]							
	O_2	CO_2	CO	C_7H_{16}	H_2	NO	SO_2	He
200	0.095	0.074	0.098	0.036	0.375	0.088	0.058	0.363
300	0.188	0.157	0.202	0.075	0.777	0.180	0.126	0.713
400	0.325	0.263	0.332	0.128	1.25	0.303	0.214	1.14
500	0.475	0.385	0.485	0.194	1.71	0.443	0.326	1.66
600	0.646	0.537	0.659	0.270	2.44	0.603	0.440	2.26
700	0.838	0.684	0.854	0.354	3.17	0.782	0.576	2.91
800	1.05	0.857	1.06	0.442	3.93	0.978	0.724	3.64
900	1.26	1.05	1.28	0.538	4.77	1.18	0.887	4.42
1000	1.52	1.24	1.54	0.641	5.69	1.41	1.06	5.26
1200	2.06	1.69	2.09	0.881	7.77	1.92	1.44	7.12
1400	2.66	2.17	2.70	1.13	9.90	2.45	1.87	9.20
1600	3.32	2.75	3.37	1.41	12.5	3.04	2.34	11.5
1800	4.03	3.28	4.10	1.72	15.2	3.70	2.85	13.9
2000	4.80	3.94	4.87	2.06	18.0	4.48	3.36	16.6

^{a} Owing to the practical importance of water vapor–air mixtures, engineers have used convenient empirical formulas for $\mathcal{D}_{H_2O \text{ air}}$. A formula that has been widely used for many years, and is used in this text is

$$\mathcal{D}_{H_2O \text{ air}} = 1.97 \times 10^{-5} \left(\frac{P_0}{P}\right)\left(\frac{T}{T_0}\right)^{1.685} \text{ m}^2/\text{s}; \qquad 273 \text{ K} < T < 373 \text{ K}$$

where $P_0 = 1$ atm; $T_0 = 256$ K. More recently the following formula has found increasing use [Marrero, T. R., and Mason, E. A., "Gaseous diffusion coefficients," *J. Phys. Chem. Ref. Data,* 1, 3–118 (1972)]:

$$\mathcal{D}_{H_2O \text{ air}} = 1.87 \times 10^{-10} \frac{T^{2.072}}{P}; \qquad 280 \text{ K} < T < 450 \text{ K}$$

$$= 2.75 \times 10^{-9} \frac{T^{1.632}}{P}; \qquad 450 \text{ K} < T < 1070 \text{ K}$$

for P in atmospheres and T in kelvins. Over the temperature range 290–330 K, the discrepancy between the two formulas is less than 2.5%. For small concentrations of water vapor in air, the older formula gives a constant value of $Sc_{H_2O \text{ air}} = 0.61$ over the temperature range 273–373 K. On the other hand, the Marrero and Mason formula gives values of $Sc_{H_2O \text{ air}}$ that vary from 0.63 at 280 K to 0.57 at 373 K.

Table A.17b Schmidt numbers for vapors in dilute mixture in air at normal temperature, enthalpy of vaporization, and boiling point at 1 atm[a]

Vapor	Chemical Formula	Sc[b]	h_{fg} J/kg $\times 10^{-6}$	T_{BP} K
Acetone	CH_3COCH_3	1.42	0.527	329
Ammonia	NH_3	0.61[c]	1.370	240
Benzene	C_6H_6	1.79	0.395	354
Carbon dioxide	CO_2	1.00	0.398	194
Carbon monoxide	CO	0.77	0.217	81
Chlorine	Cl_2	1.42	0.288	238
Ethanol	CH_3CH_2OH	1.32	0.854	352
Helium	He	0.22		4.3
Heptane	C_7H_{16}	2.0	0.340	372
Hydrogen	H_2	0.20	0.454	20.3
Hydrogen sulfide	H_2S	0.94	0.548	213
Methanol	CH_3OH	0.98	1.100	338
Naphthalene[d]	$C_{10}H_8$	2.35		491
Nitric oxide	NO	0.87	0.465	121
Octane	C_8H_{18}	2.66	0.303	399
Oxygen	O_2	0.83	0.214	90.6
Pentane	C_5H_{12}	1.49	0.357	309
Sulfur dioxide	SO_2	1.24	0.398	263
Water vapor	H_2O	0.61	2.257	373

[a] With the Clausius-Clapeyron relation, one may estimate vapor pressure as

$$P_{sat} \simeq \exp\left\{-\frac{Mh_{fg}}{\mathcal{R}}\left(\frac{1}{T} - \frac{1}{T_{BP}}\right)\right\} \text{ atm}, \quad \text{for } T \sim T_{BP}$$

[b] The Schmidt number is defined as $Sc = \mu/\rho\mathcal{D} = \nu/\mathcal{D}$. Since the vapors are in small concentrations, values for μ, ρ, and ν can be taken as pure air values.

[c] Values of Sc from 0.55 to 0.73 are seen in the literature: the cited value of 0.61 may be inaccurate.

[d] From a recent study by Cho, C., Irvine, T. F., Jr., and Karni, J., "Measurement of the diffusion coefficient of naphthalene into air," *Int. J. Heat Mass Transfer,* 35, 957–966 (1992). Also, $h_{sg} = 0.567 \times 10^6$ J/kg at 300 K.

Table A.18   Schmidt numbers for dilute solution in water at 300 K[a]

Solute	Sc	M
Helium	120	4.003
Hydrogen	190	2.016
Nitrogen	280	28.02
Water	340	18.016
Nitric oxide	350	30.01
Carbon monoxide	360	28.01
Oxygen	400	32.00
Ammonia	410	17.03
Carbon dioxide	420	44.01
Hydrogen sulfide	430	34.08
Ethylene	450	28.05
Methane	490	16.04
Nitrous oxide	490	44.02
Sulfur dioxide	520	64.06
Sodium chloride	540	58.45
Sodium hydroxide	490	40.00
Acetic acid	620	60.05
Acetone	630	58.08
Methanol	640	32.04
Ethanol	640	46.07
Chlorine	670	70.90
Benzene	720	78.11
Ethylene glycol	720	62.07
n-Propanol	730	60.09
i-Propanol	730	60.09
Propane	750	44.09
Aniline	800	93.13
Benzoic acid	830	122.12
Glycerol	1040	92.09
Sucrose	1670	342.3

[a] For other temperatures use $Sc/Sc_{300\text{ K}} \simeq (\mu^2/\rho T)/(\mu^2/\rho T)_{300\text{ K}}$, where μ and ρ are for water, and T is absolute temperature. For chemically similar solutes of different molecular weights use $Sc_2/Sc_1 \simeq (M_2/M_1)^{0.4}$. A table of $(\mu^2/\rho T)/(\mu^2/\rho T)_{300\text{ K}}$ for water follows.

T [K]	$(\mu^2/\rho T)/(\mu^2/\rho T)_{300\text{ K}}$
290	1.66
300	1.00
310	0.623
320	0.429
330	0.296
340	0.221
350	0.167
360	0.123
370	0.097

From Spalding, D. B., *Convective Mass Transfer,* McGraw–Hill, New York (1963).

Table A.19   Diffusion coefficients in solids, $\mathcal{D} = \mathcal{D}_0 \exp(-E_a/\mathcal{R}T)$

System	$\mathcal{D}_0$ m²/s	E_a kJ/kmol
Oxygen–Pyrex glass	6.19×10^{-8}	4.69×10^{4}
Oxygen–fused silica glass	2.61×10^{-9}	3.77×10^{4}
Oxygen–titanium	5.0×10^{-3}	2.13×10^{5}
Oxygen–titanium alloy (Ti–6Al–4V)	5.82×10^{-2}	2.59×10^{5}
Oxygen–zirconium	4.68×10^{-5}	7.06×10^{5}
Hydrogen–iron	7.60×10^{-8}	5.60×10^{3}
Hydrogen–α-titanium	1.80×10^{-6}	5.18×10^{4}
Hydrogen–β-titanium	1.95×10^{-7}	2.78×10^{4}
Hydrogen–zirconium	1.09×10^{-7}	4.81×10^{4}
Hydrogen–Zircaloy–4	1.27×10^{-5}	6.05×10^{5}
Deuterium–Pyrex glass	6.19×10^{-8}	4.69×10^{4}
Deuterium–fused silica glass	2.61×10^{-9}	3.77×10^{4}
Helium–Pyrex glass	4.76×10^{-8}	2.72×10^{4}
Helium–fused silica glass	5.29×10^{-8}	2.55×10^{4}
Helium–borosilicate glass	1.94×10^{-9}	2.34×10^{4}
Neon–borosilicate glass	1.02×10^{-10}	3.77×10^{4}
Carbon–FCC iron	2.3×10^{-5}	1.378×10^{5}
Carbon–BCC iron	1.1×10^{-6}	8.75×10^{4}

Various sources.

Table A.20   Selected atomic weights[a]

Aluminum	Al	26.98	Molybdenum	Mo	95.94
Antimony	Sb	121.76	Neodymium	Nd	144.24
Argon	Ar	39.95	Neon	Ne	20.18
Arsenic	As	74.92	Nickel	Ni	58.71
Barium	Ba	137.34	Niobium	Nb	92.91
Beryllium	Be	9.012	Nitrogen	N	14.007
Bismuth	Bi	208.98	Oxygen	O	15.999
Boron	B	10.81	Palladium	Pd	106.4
Bromine	Br	79.90	Phosphorus	P	30.97
Cadmium	Cd	112.40	Platinum	Pt	195.09
Calcium	Ca	40.08	Plutonium	Pu	242
Carbon	C	12.01	Potassium	K	39.10
Cesium	Cs	132.91	Radium	Ra	226
Chlorine	Cl	035.45	Radon	Rn	222
Chromium	Cr	51.996	Rhenium	Re	186.2
Cobalt	Co	58.93	Rhodium	Rh	102.90
Copper	Cu	63.55	Rubidium	Rb	85.47
Fluorine	F	18.998	Selenium	Se	78.96
Gadolinium	Gd	157.25	Silicon	Si	28.09
Gallium	Ga	69.72	Silver	Ag	107.87
Germanium	Ge	72.59	Sodium	Na	22.99
Gold	Au	196.97	Strontium	Sr	87.62
Hafnium	Hf	178.5	Sulfur	S	32.06
Helium	He	4.003	Tantalum	Ta	180.95
Hydrogen	H	1.008	Tellurium	Te	127.60
Indium	In	114.82	Thallium	Tl	204.37
Iodine	I	126.90	Thorium	Th	232.04
Iron	Fe	55.85	Tin	Sn	118.69
Krypton	Kr	83.8	Titanium	Ti	47.90
Lead	Pb	207.19	Tungsten	W	183.85
Lithium	Li	6.939	Uranium	U	238.03
Magnesium	Mg	24.31	Xenon	Xe	131.30
Manganese	Mn	54.94	Zinc	Zn	65.37
Mercury	Hg	200.59	Zirconium	Zr	91.22

[a] Based on the isotope $C^{12} = 12$.
Selected from the *Handbook of Chemistry and Physics,* 50th ed.,
Chemical Rubber Co., Cleveland (1969).

Table A.21 Henry constants C_{He} for dilute aqueous solutions at moderate pressures ($P_{i,s}/x_{i,u}$ in atm, or in bar = 10^5 Pa, within the accuracy of the data)

Solute	290 K	300 K	310 K	320 K	330 K	340 K
H_2S	440	560	700	830	980	1,140
CO_2	1,280	1,710	2,170	2,720	3,220	—
O_2	38,000	45,000	52,000	57,000	61,000	65,000
H_2	67,000	72,000	75,000	76,000	77,000	76,000
CO	51,000	60,000	67,000	74,000	80,000	84,000
Air	62,000	74,000	84,000	92,000	99,000	104,000
N_2	76,000	89,000	101,000	110,000	118,000	124,000

Table A.22a   Equilibrium compositions for the NH_3-water system

$P_{i,s}$ atm	$x_{i,u}$				
	290 K	300 K	310 K	320 K	330 K
0.02	0.030	0.019	0.012	0.008	0.006
0.04	0.056	0.036	0.024	0.016	0.012
0.06	0.078	0.052	0.035	0.024	0.017
0.08	0.096	0.064	0.046	0.032	0.023
0.1	0.11	0.079	0.056	0.040	0.029
0.2	0.18	0.14	0.099	0.057	0.052
0.4	0.26	0.21	0.16	0.12	0.092
0.6	0.31	0.26	0.20	0.16	0.13
0.8	0.35	0.29	0.23	0.19	0.15
1.0	—	0.32	0.27	0.22	0.17

Table A.22b   Equilibrium compositions for the SO_2-water system[a]

$P_{i,s}$ atm	$x_{i,u} \times 10^3$			
	290 K	300 K	310 K	320 K
0.001	0.12	0.084	0.059	0.042
0.003	0.25	0.18	0.13	0.093
0.01	0.62	0.42	0.31	0.22
0.03	1.4	1.1	0.73	0.51
0.1	4.1	2.9	2.0	1.4
0.3	11.0	7.9	5.6	3.9
1.0	33.0	24.0	18.0	12.0

[a] Notice that Henry's law is invalid for the SO_2-water system, even at very dilute concentrations.

Table A.23 Solubility and permeability of gases in solids

Gas	Solid	Temperature K	$S[\text{m}^3(\text{STP})/\text{m}^3\text{atm}]$ or S' [a]	Permeability[b] $\text{m}^3(\text{STP})/\text{m}^2\text{ s (atm/m)}$
H_2	Vulcanized rubber	300	$S = 0.040$	0.34×10^{-10}
	Vulcanized neoprene	290	$S = 0.051$	0.053×10^{-10}
	Silicone rubber	300		4.2×10^{-10}
	Natural rubber	300		0.37×10^{-10}
	Polyethylene	300		0.065×10^{-10}
	Polycarbonate	300		0.091×10^{-10}
	Fused silica	400	$S' \simeq 0.035$	
		800	$S' \simeq 0.030$	
	Nickel	360	$S' = 0.202$	
		440	$S' = 0.192$	
He	Silicone rubber	300		2.3×10^{-10}
	Natural rubber	300		0.24×10^{-10}
	Polycarbonate	300		0.11×10^{-10}
	Nylon 66	300		0.0076×10^{-10}
	Teflon	300		0.047×10^{-10}
	Fused silica	300	$S' \simeq 0.018$	
		800	$S' \simeq 0.026$	
	Pyrex glass	300	$S' \simeq 0.006$	
		800	$S' \simeq 0.024$	
	7740 glass	470	$S = 0.0084$	4.6×10^{-13}
	(94% $SiO_2 + B_2O_3 + P_2O_5$	580	$S = 0.0038$	1.6×10^{-12}
	5% $Na_2O + Li_2 + K_2O$	720	$S = 0.0046$	6.4×10^{-12}
	1% other oxides)			
	7056 glass	390	$S = 0.0039$	1.2×10^{-14}
	(90% $SiO_2 + B_2O_3 + P_2O_5$	680	$S = 0.0059$	1.0×10^{-12}
	8% $Na_2O + Li_2 + K_2O$			
	1% PbO, 5% other oxides)			
O_2	Vulcanized rubber	300	$S = 0.070$	0.15×10^{-10}
	Silicone rubber	300		3.8×10^{-10}
	Natural rubber	300		0.18×10^{-10}
	Polyethylene	300		4.2×10^{-12}
	Polycarbonate	300		0.011×10^{-10}
	Silicone-polycarbonate copolymer (57% silicone)	300		1.2×10^{-10}
	Ethyl cellulose	300		0.09×10^{-10}
N_2	Vulcanized rubber	300	$S = 0.035$	0.054×10^{-10}
	Silicone rubber	300		1.9×10^{-10}
	Natural rubber	300		0.062×10^{-10}
	Silicone-polycarbonate copolymer (57% silicone)	300		0.53×10^{-10}
	Teflon	300		0.019×10^{-10}

(Continued)

Table A.23 (Concluded)

Gas	Solid	Temperature K	$S[m^3(STP)/m^3 atm]$ or S' [a]	Permeability[b] $m^3(STP)/m^2\,s\,(atm/m)$
CO_2	Vulcanized rubber	300	$S = 0.90$	1.0×10^{-10}
	Silicone rubber	300		21×10^{-10}
	Natural rubber	300		1.0×10^{-10}
	Silicone-polycarbonate copolymer (57% silicone)	300		7.4×10^{-10}
	Nylon 66	300		0.0013×10^{-10}
H_2O	Cellophane	310		$0.91 - 1.8 \times 10^{-10}$
Ne	Fused silica	300–1200	$S' \simeq 0.002$	
Ar	Fused silica	900–1200	$S' \simeq 0.01$	

[a] Solubility S = Volume of solute gas ($0°C$, 1 atm) dissolved in unit volume of solid when the gas is at 1 atm partial pressure.
Solubility coefficient $S' = c_{1,u}/c_{1,s}$.
[b] Permeability $\mathcal{P}_{12} = \mathcal{D}_{12}\,S$.
[c] Permeability of a construction material to water vapor is often specified for a given thickness. The *permeance* is defined as permeability/thickness and can be expressed in mass units. Typical values are:

Material	Thickness mm	Permeance $ng/m^2\,s\,Pa$
Gypsum wall board	9.5	2860
Plaster on wood lath	19	630
Plywood	6.4	40–109
Hardboard (standard)	3.2	630
Hardboard (tempered)	3.2	290
Aluminum foil	0.025	0.0
Aluminum foil	0.009	2.9
Polyethylene	0.051	9.1
Polyethylene	0.25	1.7
Exterior acrylic house paint	0.04	31.3
Vapor retarder latex paint	0.07	26

From various sources, including Geankoplis, C. J., *Transport Processes and Unit Operations*, 3rd ed., Prentice-Hall, Englewood Cliffs, NJ (1993); Doremus, R. H., *Glass Science*, Wiley, New York (1973); Altemose, V.O., "Helium diffusion through glass," *J. Appl. Phys.*, 32, 1309–1316 (1961); *ASHRAE Handbook: 1993 Fundamentals*, American Society of Heating, Refrigerating and Air Conditioning Engineers, Atlanta (1993).

Table A.24   Solubility of inorganic compounds in water

| | | | T [K] | | | | | | | | | | | |
Solute	Formula	Solid Phase	273.15	280	290	300	310	320	330	340	350	360	370	373.15
Aluminum sulfate	$Al_2(SO_4)_3$	$18H_2O$	31.2	32.8	35.5	39.1	44.3	50.3	57.0	63.9	70.8	78.3	84.6	89.0
Calcium bicarbonate	$Ca(HCO_3)_2$	—	16.15	16.30	16.53	16.75	16.98	17.20	17.43	17.65	17.88	18.10	18.33	18.40
Calcium chloride	$CaCl_2$	$6H_2O$	59.5	63.3	71.5	93.3	137.2	—	—	—	—	—	—	—
	$CaCl_2$	$2H_2O$	—	—	—	—	—	—	134.6	140.2	145.3	150.9	157.0	159.0
Calcium hydroxide	$Ca(OH)_2$	—	0.185	0.179	0.168	0.157	0.145	0.132	0.120	0.109	0.098	0.088	0.080	0.077
Potassium chloride	KCl	—	27.6	29.9	33.1	36.1	39.1	41.8	44.6	47.4	50.2	53.1	55.8	56.7
Potassium nitrate	KNO_3	—	13.3	18.5	28.2	41.3	58.2	78.7	102.3	129.2	159.2	191.6	232.1	246.0
Potassium sulfate	K_2SO_4	—	7.35	8.63	10.51	12.38	14.20	15.95	17.64	19.25	20.88	22.36	23.69	24.1
Sodium bicarbonate	$NaHCO_3$	—	6.9	7.76	9.14	10.63	12.20	13.90	15.79	—	—	—	—	—
Sodium carbonate	Na_2O_3	$10H_2O$	7	10.8	18.7	33.4	—	—	—	—	—	—	—	—
	Na_2CO_3	$1H_2O$	—	—	—	—	49.1	47.8	46.7	46.2	45.9	45.7	45.55	45.5
Sodium chloride	$NaCl$	—	35.7	35.8	35.9	36.2	36.5	36.9	37.2	37.6	38.2	38.8	39.5	39.8
Sodium nitrate	$NaNO_2$	—	73	78	85	93	101	111	121	132	144	159	175	180
Sodium sulfate	Na_2SO_4	$10H_2O$	5.0	7.7	16.1	34.1	—	—	—	—	—	—	—	—
	Na_2SO_4	$7H_2O$	19.5	26.7	39.6	—	—	—	—	—	—	—	—	—
	Na_2SO_4	—	—	—	—	—	49.6	47.4	45.7	43.7	44.0	43.3	42.7	42.5

[a] Solubility expressed in kilograms of anhydrous substance that is soluble in 100 kg water.
Adapted from Handbook of Chemistry, 10th ed., McGraw-Hill, New York (1961).

Table A.25   Combustion data

Fuel	M	Higher Heat (liquid H_2O) $J/kg \times 10^{-6}$	Lower Heat (vapor H_2O) $J/kg \times 10^{-6}$	Required Air kg/kg_{fuel}			Major Flue Gases kg/kg_{fuel}			
				O_2	N_2	Air	CO_2	H_2O	N_2	SO_2
C	12.01	32.78	32.78	2.66	8.86	11.53	3.66	—	8.86	
H_2	2.016	142.1	120.1	7.94	26.41	34.34	—	8.94	26.41	
CO	28.01	10.11	10.11	0.57	1.90	2.47	1.57	—	1.90	
CH_4	16.041	55.54	50.06	3.99	13.28	17.27	2.74	2.25	13.28	
C_2H6	30.067	51.92	47.52	3.73	12.39	16.12	2.93	1.80	12.39	
C_3H8	44.092	50.38	46.39	3.63	12.07	15.70	2.99	1.63	12.07	
C_4H_{10}	58.118	49.56	45.78	3.58	11.91	15.49	3.03	1.55	11.91	
C_5H_{12}	72.144	49.06	45.40	3.55	11.81	15.35	3.05	1.50	11.81	
C_2H_4	28.051	50.34	47.21	3.42	11.39	14.81	3.14	1.29	11.39	
C_3H_6	42.077	48.94	45.80	3.42	11.39	14.81	3.14	1.29	11.39	
C_4H_8	56.102	48.47	45.35	3.42	11.39	14.81	3.14	1.29	11.39	
C_5H_{10}	70.128	48.18	45.04	3.42	11.39	14.81	3.14	1.29	11.39	
C_6H_6	78.107	42.36	40.66	3.07	10.22	13.30	3.38	0.69	10.22	
C_7H_8	92.132	42.89	40.98	3.13	10.40	13.53	3.34	0.78	10.40	
C_8H_{10}	106.158	43.38	41.31	3.17	10.53	13.70	3.32	0.85	10.53	
C_2H_2	26.036	50.01	48.32	3.07	10.22	13.30	3.38	0.69	10.22	
$C_{10}H_8$	128.162	40.24	38.86	3.00	9.97	12.96	3.43	0.56	9.97	
CH_3OH	32.041	23.86	21.12	1.50	4.98	6.48	1.37	1.13	4.98	
C_2H_5OH	46.067	30.61	27.75	2.08	6.93	9.02	1.92	1.17	6.93	
NH_3	17.031	22.49	18.61	1.41	4.69	6.10	—	1.59	5.51	
S	32.06	9.264	9.264	1.00	3.29	4.29	—	—	3.29	2.00
H_2S	34.076	16.51	15.22	1.41	4.69	6.20	—	0.53	4.69	1.88

Adapted from *Fuel Flue Gases*, American Gas Association, New York (1940).

Table A.26 Force constants for the Lennard-Jones potential model

Species	σ Å	ϵ/k K	Species	σ Å	ϵ/k K	Species	σ Å	ϵ/k K
Al	2.655	2750	CH_3CCH	4.761	252	Li_2O	3.561	1827
AlO	3.204	542	C_3H_8	5.118	237	Mg	2.926	1614
Al_2	2.940	2750	$n\text{-}C_3H_7OH$	4.549	577	N	3.298	71
Air	3.711	79	$n\text{-}C_4H_{10}$	4.687	531	NH_3	2.900	558
Ar	3.542	93	$iso\text{-}C_4H_{10}$	5.278	330	NO	3.492	117
C	3.385	32	$n\text{-}C_5H_{12}$	5.784	341	N_2	3.798	71
CCl_2	4.692	213	C_6H_{12}	6.182	297	N_2O	3.828	232
CCl_2F_2	5.25	253	$n\text{-}C_6H_{14}$	5.949	399	Na	3.567	1375
CCl_4	5.947	323	Cl	3.613	131	Na Cl	4.186	1989
CH	3.370	69	Cl_2	4.217	316	NaOH	3.804	1962
$CHCl_3$	5.389	340	H	2.708	37	Na_2	4.156	1375
CH_3OH	3.626	482	HCN	3.630	569	Ne	2.820	33
CH_4	3.758	149	HCl	3.339	345	O	3.050	107
CN	3.856	75	H_2	2.827	60	OH	3.147	80
CO	3.690	92	H_2O^a	2.641	809.1	O_2	3.467	107
CO_2	3.941	195	H_2O^b	3.737	32	S	3.839	847
CS_2	4.483	467	H_2O_2	4.196	289	SO	3.993	301
C_2	3.913	79	H_2S	3.623	301	SO_2	4.112	335
C_2H_2	4.033	232	He	2.551	10	Si	2.910	3036
C_2H_4	4.163	225	Hg	2.969	750	SiO	3.374	569
C_2H_6	4.443	216	I_2	5.160	474	SiO_2	3.706	2954
C_2H_5OH	4.530	363	Li	2.850	1899	UF_6	5.967	237
C_2N_2	4.361	349	LiO	3.334	450	Xe	4.047	231
$C_2H_2CHCH_3$	4.678	299	Li_2	3.200	1899	Zn	2.284	1393

[a] For μ and k.
[b] For $\mathcal{D}$.
Taken largely from Svehla, R. A., *Estimated Viscosities and Thermal Conductivities of Gases at High Temperatures*, NASA TR R–132 (1962).

Table A.27   Molar volume at the normal boiling point

Species	$\tilde{V}_b$ m^3/kmol × 10^3	T_b K	Species	$\tilde{V}_b$ m^3/kmol × 10^3	T_b K
Hydrogen	14.3	21	Carbon tetrachloride	102	350
Oxygen	25.6	90	Methyl chloride	50.6	249
Nitrogen	31.2	77	Ethyl acetate	106	350
Air	29.9	79	Acetic acid	64.1	337
Carbon monoxide	30.7	82	Acetone	77.5	329
Carbon dioxide	34.0	195	Ethylpropyl ether	129	335
Sulfur dioxide	44.8	263	Dimethyl ether	63.8	250
Nitric oxide	23.6	121	n-Propyl alcohol	81.8	370
Nitrous oxide	36.4	185	Methanol	42.5	338
Ammonia	25.8	240	Methane	37.7	112
Water	18.9	373	Propane	74.5	229
Hydrogen sulfide	32.9	212	Heptane	162	372
Chlorine	48.4	239	Ethylene	49.4	169
Bromine	53.2	332	Acetylene	42.0	190
Iodine	715	458	Benzene	96.5	353
Hydrochloric acid	30.6	188	Fluorobenzene	102	358
Methyl formate	62.8	305	Chlorobenzene	115	405
Bromobenzene	120	429	Iodobenzene	130	462
Dichlorodifluoromethane	80.7	245			

Table A.28 Collision integrals for the Lennard-Jones potential model

kT/ε	$\Omega_\mu = \Omega_k$	$\Omega_\mathcal{D}$	kT/ε	$\Omega_\mu = \Omega_k$	$\Omega_\mathcal{D}$	kT/ε	$\Omega_\mu = \Omega_k$	$\Omega_\mathcal{D}$
0.30	2.785	2.662	1.60	1.279	1.167	3.80	0.9811	0.8942
0.35	2.628	2.476	1.65	1.264	1.153	3.90	0.9755	0.8888
0.40	2.492	2.318	1.70	1.248	1.140	4.00	0.9700	0.8836
0.45	2.368	2.184	1.75	1.234	1.128	4.10	0.9649	0.8788
0.50	2.257	2.066	1.80	1.221	1.116	4.20	0.9600	0.8740
0.55	2.156	1.966	1.85	1.209	1.105	4.30	0.9553	0.8694
0.60	2.065	1.877	1.90	1.197	1.094	4.40	0.9507	0.8652
0.65	1.982	1.798	1.95	1.186	1.084	4.50	0.9464	0.8610
0.70	1.908	1.729	2.00	1.175	1.075	4.60	0.9422	0.8568
0.75	1.841	1.667	2.10	1.156	1.057	4.70	0.9382	0.8530
0.80	1.780	1.612	2.20	1.138	1.041	4.80	0.9343	0.8492
0.85	1.725	1.562	2.30	1.122	1.026	4.90	0.9305	0.8456
0.90	1.675	1.517	2.40	1.107	1.012	5.0	0.9269	0.8422
0.95	1.629	1.476	2.50	1.093	0.9996	6.0	0.8963	0.8124
1.00	1.587	1.439	2.60	1.081	0.9878	7.0	0.8727	0.7896
1.05	1.549	1.406	2.70	1.069	0.9770	8.0	0.8538	0.7712
1.10	1.514	1.375	2.80	1.058	0.9672	9.0	0.8379	0.7556
1.15	1.482	1.346	2.90	1.048	0.9576	10.0	0.8242	0.7424
1.20	1.452	1.320	3.00	1.039	0.9490	20.0	0.7432	0.6640
1.25	1.424	1.296	3.10	1.030	0.9406	30.0	0.7005	0.6232
1.30	1.399	1.273	3.20	1.022	0.9328	40.0	0.6718	0.5960
1.35	1.375	1.253	3.30	1.014	0.9256	50.0	0.6504	0.5756
1.40	1.353	1.233	3.40	1.007	0.9186	60.0	0.6335	0.5596
1.45	1.333	1.215	3.50	0.9999	0.9120	70.0	0.6194	0.5464
1.50	1.314	1.198	3.60	0.9932	0.9058	80.0	0.6076	0.5352
1.55	1.296	1.182	3.70	0.9870	0.8998	90.0	0.5973	0.5256
						100.0	0.5882	0.5170

Taken from Hirschfelder, J. O., Curtiss, C. P., and Bird, R. B., *Molecular Theory of Gases and Liquids*, John Wiley & Sons, New York (1954).

Table A.29   Constant-pressure specific heat capacity c_p [J/kg K]: Database in GASMIX

Species	Formula	Temperature [K]								
		200	300	400	600	800	1000	1500	2000	2500
Air		1009	1005	1009	1038	1089	1130	1202	1244	1285
Aluminum	Al	820	793	783	776	773	772	771	771	771
Aluminum	Al2	646	714	764	795	785	770	750	748	756
Aluminum oxide	AlO	686	719	756	807	838	866	961	1051	1095
Argon	Ar	520	520	520	520	520	520	520	520	520
Carbon	C	1741	1735	1733	1732	1731	1731	1733	1745	1769
Refrigerant–12	CCl_2F_2	485	601	681	774	820	844	871	881	885
Carbon tetrachloride	CCl_4	455	543	596	628	670	681	693	697	699
Methane	CH_4	2087	2226	2525	3256	3923	4475	5396	5884	6157
Ethanol	C_2H_5OH	1810	1972	2126	2408	2652	2857	3262	3540	
n-Pentane	n-C_5H_{12}	1983	2216	2438	2843	3188	3467	4011	4380	
Cyclohexane	C_6H_{12}		1272	1781	2676	3319	3769	4405		
n-Hexane	n-C_6H_{14}	1983	2215	2433	2833	3170	3444	3981	4341	
Carbon monoxide	CO	1039	1040	1048	1087	1139	1183	1257	1294	1315
Carbon dioxide	CO_2	735	844	937	1074	1168	1232	1329	1371	1397
Chlorine	Cl_2	447	479	498	515	523	528	535	542	516
Hydrogen	H_2	15910	14780	14400	14400	14530	14760	16000	17050	17779
Water	H_2O	1851	1870	1900	2003	2130	2267	2594	2832	2992
Hydrogen peroxide	H_2O_2	1084	1271	1424	1637	1759	1848	2009		
Hydrogen sulfide	H_2S	980	1004	1044	1143	1248	1344	1511	1604	1660
Hydrogen chloride	H Cl	799	799	800	811	836	868	934	976	1003
Helium	He	5200	5200	5200	5200	5200	5200	5200	5200	5200
Mercury	Hg	104	104	104	104	104	104	104	104	104
Iodine	I_2	142	145	147	148	149	150	156	168	178
Nitrogen	N2	1045	1043	1047	1075	1123	1167	1244	1287	1307
Nitrous oxide	N_2O	764	877	970	1099	1187	1247	1327	1363	1381
Ammonia	NH_3	1982	2096	2273	2659	3008	3317	3908	4277	4499
Neon	Ne	1030	1030	1030	1030	1030	1030	1030	1030	1030
Nitric oxide	NO	1014	995	998	1041	1092	1133	1192	1221	1238
Oxygen	O	1421	1390	1343	1320	1312	1370	1303	1302	1303
Hydroxyl	OH	1810	1763	1743	1736	1759	1804	1936	2039	2110
Oxygen	O2	900	920	945	1000	1050	1085	1140	1180	1214
Sulfur dioxide	SO_2	568	624	679	766	819	851	890	909	920
Uranium hexafluoride	UF6		369	399	422	432	439			
Xenon	Xe	158	158	158	158	158	158	158	158	158

Various sources, primarily Chase, M. W., Jr., et al., *JANAF Thermochemical Tables,* 3rd ed., American Chemical Society (1985), and Rossini, F. D., et al., *Selected Values of Properties of Hydrocarbons,* National Bureau of Standards Circular C461 (1947).

Table A.30 Slip factor C, Brownian diffusion coefficient $\mathcal{D}_p$, and particle Schmidt number Sc_p for spherical particles in air at 1 atm and 300 K

d_p μm	C	$\mathcal{D}_p$ m²/s	Sc_p
0.001	222.3	5.30×10^{-6}	2.96
0.002	111.4	1.33×10^{-6}	1.18×10^1
0.005	44.9	2.14×10^{-7}	7.31×10^1
0.01	22.7	5.42×10^{-8}	2.89×10^2
0.02	11.7	1.39×10^{-8}	1.12×10^3
0.05	5.07	2.42×10^{-9}	6.48×10^3
0.1	2.92	6.95×10^{-10}	2.25×10^4
0.2	1.89	2.26×10^{-10}	6.94×10^4
0.5	1.34	6.38×10^{-11}	2.45×10^5
1	1.17	2.78×10^{-11}	5.62×10^5
2	1.08	1.29×10^{-11}	1.21×10^6
5	1.03	4.93×10^{-12}	3.18×10^6
10	1.02	2.43×10^{-12}	6.46×10^6

Table A.31 Thermodynamic properties of water vapor–air mixtures at 1 atm

Temp °C	Saturation Mass Fraction	Specific Volume m³/kg		Enthalpy[a,b] kJ/kg		
		Dry Air	Saturated Air	Liquid Water	Dry Air	Saturated Air
10	0.007608	0.8018	0.8054	42.13	10.059	29.145
11	0.008136	0.8046	0.8086	46.32	11.065	31.481
12	0.008696	0.8075	0.8117	50.52	12.071	33.898
13	0.009289	0.8103	0.8148	54.71	13.077	36.401
14	0.009918	0.8131	0.8180	58.90	14.083	38.995
15	0.01058	0.8160	0.8212	63.08	15.089	41.684
16	0.01129	0.8188	0.8244	67.27	16.095	44.473
17	0.01204	0.8217	0.8276	71.45	17.101	47.367
18	0.01283	0.8245	0.8309	75.64	18.107	50.372
19	0.01366	0.8273	0.8341	79.82	19.113	53.493
20	0.01455	0.8302	0.8374	83.99	20.120	56.736
21	0.01548	0.8330	0.8408	88.17	21.128	60.107
22	0.01647	0.8359	0.8441	92.35	22.134	63.612
23	0.01751	0.8387	0.8475	96.53	23.140	67.259
24	0.01861	0.8415	0.8510	100.71	24.147	71.054
25	0.01978	0.8444	0.8544	104.89	25.153	75.004
26	0.02100	0.8472	0.8579	109.07	26.159	79.116
27	0.02229	0.8500	0.8615	113.25	27.166	83.400
28	0.02366	0.8529	0.8650	117.43	28.172	87.862
29	0.02509	0.8557	0.8686	121.61	29.178	92.511
30	0.02660	0.8586	0.8723	125.79	30.185	97.357
31	0.02820	0.8614	0.8760	129.97	31.191	102.408
32	0.02987	0.8642	0.8798	134.15	32.198	107.674
33	0.03164	0.8671	0.8836	138.32	33.204	113.166
34	0.03350	0.8699	0.8874	142.50	34.211	118.893
35	0.03545	0.8728	0.8914	146.68	35.218	124.868
36	0.03751	0.8756	0.8953	150.86	36.224	131.100
37	0.03967	0.8784	0.8994	155.04	37.231	137.604
38	0.04194	0.8813	0.9035	159.22	38.238	144.389
39	0.04432	0.8841	0.9077	163.40	39.245	151.471
40	0.04683	0.8870	0.9119	167.58	40.252	158.862
41	0.04946	0.8898	0.9162	171.76	41.259	166.577
42	0.05222	0.8926	0.9206	175.94	42.266	174.630
43	0.05512	0.8955	0.9251	180.12	43.273	183.037
44	0.05817	0.8983	0.9297	184.29	44.280	191.815
45	0.06137	0.9012	0.9343	188.47	45.287	200.980
46	0.06472	0.9040	0.9391	192.65	46.294	210.550
47	0.06842	0.9068	0.9439	196.83	47.301	220.543
48	0.07193	0.9097	0.9489	201.01	48.308	230.980
49	0.07580	0.9125	0.9539	205.19	49.316	241.881

[a] The enthalpies of dry air and liquid water are set equal to zero at a datum temperature of 0°C.

[b] The enthalpy of an unsaturated water vapor–air mixture can be calculated as $h = h_{\text{dry air}} + (m_1/m_{1,\text{sat}})(h_{\text{sat}} - h_{\text{dry air}})$.

Bibliography

CHAPTER 1

Most modern treatments of diffusive mass transfer derive from:

Bird, R. B., "Theory of Diffusion," in *Advances in Chemical Engineering*, vol. 1, Academic Press, New York (1956) (see pp. 170 et seq.).

Essentially the same material, together with worked examples, is given by:

Bird, R. B., Stewart, W. E., and Lightfoot, E. N., *Transport Phenomena*, John Wiley & Sons, New York (1960).

A text having a treatment of mass transfer similar to that in this text is:

Lienhard, J. H., *A Heat Transfer Text Book*, 2d ed., Prentice Hall, Englewood Cliffs, N.J. (1987).

Texts written for chemical engineers include:

Cussler, E. L., *Diffusion: Mass Transfer in Fluid Systems*, Cambridge University Press, London (1997).

Geankoplis, C. J., *Transport Processes and Unit Operations*, 3d ed., Prentice Hall, Englewood Cliffs, N.J. (1993).

Hines, A. L., and Maddox, R. N., *Mass Transfer*, Prentice Hall, Englewood Cliffs, N.J. (1985).

Sherwood, T. K., Pigford, R. L., and Wilke, C. R., *Mass Transfer*, McGraw-Hill, New York (1975).

Skelland, A. P., *Diffusional Mass Transfer*, R. E. Krieger, Melbourne, Fla. (1985).

The text by Geankoplis contains a nice introduction to the drying of porous solids.

More advanced texts containing analytical solutions for many steady and transient mass diffusion problems include:

Crank, J., *The Mathematics of Diffusion*, 2d ed., Clarendon Press, Oxford (1975).

Gebhart, B. J., *Heat Conduction and Mass Diffusion*, McGraw-Hill, New York (1993).

CHAPTER 2

The Spalding formulation of steady convective heat and mass transfer is presented in detail in:

> Kays, W. M., and Crawford, M. E., *Convective Heat and Mass Transfer*, 3d ed., McGraw-Hill, New York (1993)

A chemical engineering perspective is given in:

> Deen, W. M., *Analysis of Transport Phenomena*, Oxford University Press, New York (1998).

A wide range of engineering problems involving convective mass transfer are treated using the simple Reynolds flux model in:

> Spalding, D. B., *Convective Mass Transfer*, McGraw-Hill, New York (1963).

A number of interesting advanced mass transfer problems are analyzed in:

> Rosner, D. E., *Transport Processes in Chemical Reacting Flow Systems*, Butterworth-Heinemann, Stoneham, Mass. (1986).

CHAPTER 3

Texts written for chemical engineers dealing with mass exchangers include:

> Geankoplis, C. J., *Transport Processes and Unit Operations*, 3d ed., Prentice Hall, Englewood Cliffs, N.J. (1993).
> Treybal, R. E., *Mass Transfer Operations*, 3d ed., McGraw-Hill, New York (1980).
> Walas, S. M., *Chemical Process Equipment: Selection and Design*, Butterworth-Heinemann, Stoneham, Mass. (1988).

A great variety of polution control equipment is nicely described in:

> Liptak, B. G., *Municipal Waste Disposal in the 1990's*, Chilton, Radnor, Pa. (1991).

Additional references for electrostatic precipitators include:

> Robinson, M., "Electrostatic Precipitation," in Strauss, W., ed., *Air Pollution Control: Part I*, Wiley-Interscience, New York (1971).
> Strauss, W., *Industrial Gas Cleaning*, Pergamon Press, New York (1975).

Design methods for electrostatic precipitators are described by:

> Turner, J. H., Lawless, P. A., Yamamoto, T., Coy, D. W., Greiner, G. P., McKenna, J. D., and Vatavuk, W. M., "Sizing and Costing of Electrostatic Precipitators. Part I: Sizing Considerations," *J. Air Pollut. Control Assoc.*, 38, 458–471 (1988); "Part II: Costing Considerations," *J. Air Pollut. Control Assoc.*, 38, 715–726 (1988).

Design methods for gas absorption are given by:

> Zenz, F. A., "Design of Gas Absorption Towers," Sec. 3.2 in Schweitzer, P. A., ed., *Handbook of Separation Techniques for Chemical Engineers*, McGraw-Hill, New York, (1979).

A complete reference on filters is:

> Matteson, M. J., and Orr, C., eds., *Filtration: Principles and Practice*, Marcel Dekker, New York (1987).

Reference works for humidifiers and cooling towers include:

American Society of Heating, Refrigerating and Air Conditioning Engineers, *ASHRAE 1992 Systems and Equipment Handbook (SI Version)*, ASHRAE, Atlanta (1992).

Baker, D., *Cooling Tower Performance*, Chemical Publishing Co., New York (1984).

Hill, G. B., Pring, E. S., and Osborn, P. D., *Cooling Towers: Principles and Practice*, 3d ed., Butterworth-Heinemann, London (1990).

Johnson, B. M., ed., *Cooling Tower Performance Prediction and Improvement*, vols. 1 and 2, EPRI GS–6370, Electric Power Research Institute, Palo Alto, Calif. (1990).

Kelly, N. W., *Kelly's Handbook of Cross-Flow Cooling Tower Performance*, Neil W. Kelly and Associates, Kansas City, Mo. (1976).

Singham, J. R., "Natural Draft Towers" and "Mechanical Draft Towers," Secs. 3.12.2 and 3.12.4 in Hewitt, G. F., Coord. ed., *Hemisphere Handbook of Heat Exchanger Design*, Hemisphere, New York (1990).

Stoeker, W. F., and Jones, J. W., *Refrigeration and Air Conditioning*, 2d ed., Chapters 3 and 9, McGraw-Hill, New York (1982).

Drying theory and practice are described in:

Keey R. B., *Drying Principles and Practice*, Pergamon Press, Elmsford, N.Y. (1972).

Keey R. B., *Drying of Loose and Particulate Materials*, Hemisphere, New York (1992).

Mujamdar, A. S., ed., *Handbook of Industrial Drying*, Marcel Dekker, New York (1987).

Strumillo, C., and Kudra, T., *Drying: Principles, Applications and Design*, Gordon & Breach, New York (1986).

APPENDIX A

The most comprehensive compilation of thermophysical property data available is:

Touloukian, Y. S., and Ho, C. Y., *Thermophysical Properties of Matter*, 13 vols., IFI/Plenum Press, New York (1970–1977).

Other useful sources include:

American Society of Heating, Refrigerating and Air Conditioning Engineers, *ASHRAE Handbook of Fundamentals*, ASHRAE, New York (1981).

American Society of Metals, *Metals Handbook, Vol. 1: Properties and Selection of Metals*, 8th ed., ASM, Metals Park, Ohio (1961).

Eckert, E. R. G., and Drake, R. M., Jr., *Analysis of Heat and Mass Transfer*, McGraw-Hill New York (1972).

Hewitt, G. F., coord. ed., *Hemisphere Handbook of Heat Exchanger Design*, Hemisphere, New York (1990).

Irvine, T. F., Jr., and Lilley, P. E., *Steam and Gas Tables with Computer Equations*, Academic Press, Orlando, Fla. (1984).

McAdams, W. H., *Heat Transmission*, 3d ed., McGraw-Hill, New York (1954).

Norris, R. H., et. al., eds., *Heat Transfer and Fluid Flow Data Books*, General Electric Co., Schenectady, N.Y. (1943, with supplements to current date).

Vargaftik, N. B., *Handbook of Physical Properties of Liquids and Gases*, 2d ed., Hemisphere, Washington, D.C. (1983).

MASS TRANSFER JOURNALS

The heat transfer journals listed in *Heat Transfer* also publish papers on mass transfer and simultaneous heat and mass transfer. Other relevant journals include:

> *AIChJE Journal*
> *ASHRAE Journal*
> *Aerosol Science and Technology*
> *Chemical Engineering Progress*
> *Chemical Engineering Science*
> *Combustion Science and Technology*
> *Drying Technology*
> *I&EC Fundamentals*
> *Journal of Aerosol Science*
> *Progress in Energy and Combustion Science*

Nomenclature

A	area, m^2
A_c	cross-sectional area; area for flow, m^2
A_{eff}	effective area, m^2
A_f	fin surface area, m^2
A_{fr}	partial surface area, m^2
a	surface area per unit volume, m^{-1}; speed of sound, m/s
$\mathcal{A}$	Avogadro's number, molecules/kmol
B	blowing parameter
BP	boiling point
Bi	Biot number, Eq. (1.50)
$\mathcal{B}$	mass transfer driving force
b	height of air inlet for a natural-draft cooling tower, m
C	slip correction factor, Eq. (1.128)
C_D	drag coefficient, Eq. (4.68) in *Heat Transfer*
C_{He}	Henry constant, bar or atm
c	specific heat, J/kg K; average molecular speed, m/s; molar concentration, $kmol/m^3$
c_p	specific heat at constant pressure, J/kg K
c_v	specific heat at constant volume, J/kg K
D	diameter, m
D_B	shell base diameter for a natural-draft cooling tower, m
D_{ij}	multicomponent diffusion coefficient, m^2/s
D_i^T	thermal diffusion coefficient, kg/m s
D_p	nominal size of a random packing, m
D_h	hydraulic diameter, m
DF	decontamination factor
$\mathcal{D}_{12}$	binary diffusion coefficient, m^2/s
$\mathcal{D}_{\text{im}}$	effective binary diffusion coefficient, m^2/s
$\mathcal{D}_p$	particle Brownian diffusion coefficient, m^2/s
d	molecule diameter, Å
d_f	fiber diameter, μm
$\mathbf{d}_j$	driving force, Eq. (2.215)
d_p	particle diameter, m or μm
E	energy, J; emissive power, W/m^2; voltage, V; electric field, V/m
E_a	activation energy, kcal/mol or kJ/kmol

e	elementary electric charge, A s
F	force, N; function; Faraday's constant
Fr	Froude number, Eq. (3.38)
f	dimensionless stream function; friction factor, Eq. (4.15) or (4.119) in *Heat Transfer*; friction coefficient, kg/s
$\mathbf{f}_i$	external force per unit mass, N/kg
G	irradiation, W/m^2; mass velocity, kg/m^2 s; gas stream superficial mass velocity, kg/m^2 s; correction factor
G_m	mole transfer conductance, kmol/m^2 s
Gr	Grashof number, Eq. (4.27) in *Heat Transfer*
g	gravitational acceleration, m/s^2
$\mathbf{g}$	gravity vector, m/s^2
g_m	mass transfer conductance, kg/m^2 s
g_h	heat transfer conductance, kg/m^2 s
H	elevation or height, m; total enthalpy, J/kg; molar enthalpy J/kmol; hydrodynamic factor for Kuwabara flow, Table 3.2
He	Henry number, Eq. (1.13)
ΔH_c	heat of combusion per kmol, J/kmol
h	heat transfer coefficient, W/m^2 K; enthalpy, J/kg
h_c	convective heat transfer coefficient, W/m^2 K
h_r	radiative heat transfer coefficient, W/m^2 K
h_{fs}	enthalpy of solidification, J/kg
h_{sg}	enthalpy of sublimation, J/kg
Δh_c	heat of combustion per unit mass, J/kg
Δh_d	heat of dissociation per unit mass, J/kg
I	electrical current, A
$\mathbf{i}$	unit vector in the x direction
J	diffusion molar flux, kmol/m^2 s
J	molecular flux, molecules/m^2 s; particle flux, particles/m^2 s
Ja	Jakob number, Eq. (3.90) in *Heat Transfer*
j	diffusion mass flux, kg/m^2 s
$\mathbf{j}$	unit vector in the y direction
K	dimensionless flow rate in flooding correlation, Eq. (3.65)
K_{ij}	friction coefficient, J s/m^5
Ka	Kapitza number, Eq. (3.45)
Kn	Knudsen number, Eq. (1.129)
k	thermal conductivity, W/m K
k_T	thermal diffusion ratio
k''	rate constant for a heterogeneous reaction, m/s if first-order
$\hat{k}$	Boltzmann constant, J/K
$\mathbf{k}$	unit vector in the z direction
L	length, m; liquid stream superficial mass velocity, kg/m^2 s
Le	Lewis number, Eq. (1.57)
$\mathcal{L}$	characteristic length
ℓ	collision mean free path, m
ℓ_t	transport mean free path, m

M	molecular weight, kg/kmol
$\dot{M}$	molar flow rate, kmol/s
$\dot{M}_L^{\dagger}$	molar flow rate of liquid stream divided by Henry number, kmol/s
m	mass fraction
$\dot{m}$	mass flow rate, kg/s
$\dot{m}''$	mass flow rate per unit area across a phase interface (mass transfer rate), kg/m^2 s
m	mass of a molecule, kg
MP	melting point
N	absolute molar flux, kmol/m^2 s; number of velocity heads
N_{tu}	number of transfer units, Eqs. (2.62), and (2.95)
Nu	Nusselt number, Eq. (4.19) in *Heat Transfer*
$\mathcal{N}$	molecule number density, molecules/m^3; particle number density, particles/m^3
n	absolute mass flux, kg/m^2 s
n	number fraction
N	absolute molar flux vector in a mixture, kmol/m^2 s
n	absolute mass flux vector in a mixture, kg/m^2 s
P	pressure, Pa
Pe	Peclet number, Section 3.3.4
Pr	Prandtl number, Eq. (4.18) in *Heat Transfer*
p	pitch, m
$\mathcal{P}$	perimeter, m; permeability, m^3 (STP)/m^2 s (atm/m); conserved property
Q	thermal energy, J; particle charge, A s
$\dot{Q}$	rate of heat transfer into a system, rate of heat flow, W
q	heat flux vector, W/m^2
q	heat flux, W/m^2
R	radius, m; gas constant J/kg K; interception number, Section 3.3.4
$\dot{R}'''$	molar rate of species production in a homogeneous reaction, kmol/m^3 s
Ra	Rayleigh number, Eq. (4.29) in *Heat Transfer*
Re	Reynolds number, Eq. (4.13) in *Heat Transfer*
$\mathcal{R}$	universal gas constant, J/kmol K
r	radial coordinate, m
r_e	effective pore radius of a catalyst, μm
$\dot{r}'''$	mass rate of species production in a homogeneous reaction, kg/m^3 s
S	surface area, m^2; contact area
$\overline{S}_i$	partial molal entropy, J/kmol K
S_p	pellet surface area, m^2
S'	surface area per unit width, m
Sc	Schmidt number, Section 1.2.3
Sh	Sherwood number, Eq. (1.41)
St	Stanton number, Eqs. (4.21) in *Heat Transfer* and (1.41)
Stk	Stokes number, Section 3.3.4
S	solubility, m^3(STP)/m^3 atm
S'	solubility coefficient

T	temperature, K or °C
T_{HW}^+	hot water correction factor for cooling tower packings
t	time, s; thickness, m; temperature, °C
t_c	time constant, s
U	overall heat transfer coefficient, W/m^2 K
u	specific internal energy, J/kg; velocity component in x direction, m/s
u_b	bulk velocity, m/s
u_i	liquid ion mobility, m^{2}kmol/J s
V	velocity, m/s; volume, m^3
V^E	electrical migration velocity, m/s
v	specific volume, m^3/kg; velocity component in y direction, m/s
$\bar{v}$	average molecular speed, m/s
$\mathcal{V}$	characteristic velocity, m/s
$\mathbf{V}$	velocity vector, m/s
$\mathbf{v}$	mass-average velocity vector in a mixture
$\mathbf{v}^*$	mole-average velocity vector in a mixture
W	width of a surface, m; mass, kg; molar content of a system, kmol
We	Weber number, Eqs. (7.136) in *Heat Transfer* and (3.38)
w	mass content of a system, kg; velocity component in z direction, m/s
$\mathbf{X}$	external force per molecule, N
x	rectangular coordinate, m; mole fraction; liquid-phase mole fraction
y	rectangular coordinate, m; gas-phase mole fraction
Z_p	height of packing above basin in a natural-draft cooling tower
z	rectangular coordinate, m; elevation, m

Greek Symbols

α	thermal diffusivity, m^2/s
α_{12}	thermal diffusion factor
β	thermal coefficient of volume expansion, K^{-1}; burning rate constant, m^2/s
Γ	flow rate per unit width, kg/m s
γ	single-fiber collection efficiency
Δ	finite increment
Δ_{m2}	convection thickness for mass transfer, Eq. (2.138)
δ	film thickness, m; hydrodynamic boundary layer thickness, m
δ_f	equivalent stagnant film thickness, m
ε	mass exchanger effectiveness; maximum energy of attraction for molecules, J/molecule; ionization fraction
ε_v	void fraction
η	dimensionless spatial coordinate; similarity variable
η_p	pump efficiency; catalyst pellet effectiveness
θ	angle, rad; dimensionless temperature
Λ	Thiele modulus, Eq. (1.81)
μ	dynamic viscosity, kg/m s; gaseous ion mobility, m^2/V s; chemical potential, J/kmol

ν	kinematic viscosity, m^2/s
ξ	unheated starting length, m; dimensionless spatial coordinate; general spatial variable, m
ρ	density, kg/m^3
σ	surface tension, N/m; Stefan–Boltzmann constant, W/m^2 K^4; collision diameter, Å
σ_c	critical surface tension, N/m; collision cross section, m^2
σ_t	transport cross section, m^2
τ	shear stress, N/m^2; time period, s; tortuosity factor
Φ	dimensionless enthalpy; drag factor for random packing; electrostatic potential, V
Φ_{ij}	factor in Wilke's mixture rule, Eq. (1.118)
ϕ	angle, rad; dimensionless mass fraction; association parameter for a solvent
χ	dynamic shape factor, Eq. (1.132)
ψ	stream function, m^2/s
Ω_μ	collision integral for viscosity and thermal conductivity
$\Omega_{\mathcal{D}}$	collision integral for diffusion
ω	humidity ratio

Subscripts

b	bulk or mixed mean value for a stream; boiling point
c	convection; centerline; critical; coolant
conv	convection
e	external; free-stream
es	free-stream composition at the surface temperature
f	forced; friction
fr	frontal
fu	fuel
G	gas stream
g	gas
h	based on enthalpy
i	inside; internal; interfacial; intitial; species i
j	species j
L	liquid stream
l	liquid
lm	logarithmic mean value
MP	melting point
m	mean; mass transfer
N	normal
n	natural
o	outside; reservoir
ox	oxidant
p	pumping; catalyst pellet

prod	products of a reaction
r	radiation; reservoir; reference
rad	radiation
s	s-surface (in a fluid, adjacent to an interface or wall)
sat	saturated
T	thermal
t	transferred state
tr	transition
u	u-surface (in a condensed phase adjacent to an interface)
v	vapor phase
w	at a solid wall; wall material
0	initial
12	species 1 and 2 in a binary mixture
∞	far away; far upstream
(α)	chemical element

Superscripts

E	forced diffusion in an electrostatic precipitator
F	forced diffusion
oa	overall
T	reference temperature
0	reference state
$*$	reduced value; dimensionless; relative to mole-average velocity; limit of zero mass transfer
$'$	differentiation with respect to η; per unit length
$''$	per unit area
$'''$	per unit volume

Overscores

$^-$	average
$^\cdot$	per unit time
$^\wedge$	relative

Index